Sociality Revisited?
The Use of the Internet and Mobile Phones in Urban Cameroon

Bettina Anja Frei

Langaa Research & Publishing CIG
Mankon, Bamenda

Publisher
Langaa RPCIG
Langaa Research & Publishing Common Initiative Group
P.O. Box 902 Mankon
Bamenda
North West Region
Cameroon
Langaagrp@gmail.com
www.langaa-rpcig.net

Distributed in and outside N. America by African Books Collective
orders@africanbookscollective.com
www.africanbookcollective.com

ISBN: *9956-728-41-1*

DISCLAIMER
All views expressed in this publication are those of the author and do not
necessarily reflect the views of Langaa RPCIG.

Table of Contents

Acknowledgements

A range of people have contributed in different ways to the writing of this book. I would like to thank following people: My advisors Prof. Dr. Till Förster, and Prof. Dr. Judith Schlehe. My family for their patience and support: my spouse Tarek Moussalli, my mother Marianne Frei, my sister Maja Frei, my friends Bettina Vogt, Tabith Ruepp, Cigdem Scarpatetti-Böke, Rebecca Szedivy, Maria Ingletti, Daniela Häberli, Tobias Jäggli, Guido Happle, and others, who shared in each step concerning my research and the writing of this book, my friends from Rigpa and Calcutta Project Basel for their patience regarding my absences. Thanks to Gareth Jones for proof-reading the book, my colleagues in the VW Foundation research group "Passages of Culture" for fruitful collaboration, above all my "tandem-partner" Primus Tazanu. Furthermore Prof. Adamu Abdalla, Prof. Bole Butake, Prof. Elizabeth Gunner, Jennifer Musangi, Jendele Hungbo, Nura Ibrahim, Mainasara Kurfi, Kenneth Tume, Pani Fomiyen and Paul Tafor. Thanks to colleagues and staff at the Institute for Social Anthropology, Fiona Siegenthaler, Kathrin Heitz, Michelle Engeler, Andrea Kaufmann, Rita Kesselring, Barbara Heeb, Jana Gerold, Piet Van Euwijk, Brigitte Obrist, Lucy Koechlin, Kerstin Bauer, Peter Lindenmann, Sabine Schultz, and others. Special thanks to Sandra Burri-Furler for coordinating the "Passages of Culture" Research Project. Special thanks to Eric Chefor for supporting me in my fieldwork. Thanks to other people who have inspired and encouraged me in fruitful discussions, Divine Fuh, René Egloff, Ephrem Temesa, Veit Arlt, Balz Andrea Alter, or regarding their works, Francis Nyamnjoh, Mirjam De Bruijin, H.-P. Hahn, Michaela Pelican, and other authors whose work I have cited.

Special thanks go to so many people in Cameroon and beyond, who have contributed to my research. Celestine Ebie, David Sama, Doreen Bieri-Ngafor, Frank Mbock, Elvis Awa, Jacob Ache, Clotaire Yanze, Vivian Nchang and family, Ezichiel Atonehe and family, Miranda Nche, Alex Che, Delphine Maaya and family, Ivo Ngade, Eunice Tita, Bertrand Taboh, Cletus Tamoh, Cletus Anye, Isidore Mbianda and family, Stanley Lema, Pascal Chefor and family, Patience Sirri and family, Valerie Awubung, Ernest Che, Clement Chia, Lambert Monji,

Divine and Louisa Kintashe, Aboubakar Jaja and family, Ernest Nkwenti, Richard Tambe, René Yang, Emile Ngah, Derick, Elvis Yengi, Killy Roland, Uginia Urwick, Elvis Khan, Simon Muluh Ngwi, Charles Mobit, Louis Verdzekov, Anabel Bih, Peter Takwi, Anita Ngum, Nicoline Suh and family, David Banye, Clovis Nforsi, Patience Acho, Mathew Suh, Raymond Tita, James Tambang, Norbert, Yannick Ndatoua et famille, Precillia Gahke and family, Anne-Cécile Nonga et famille, Clauvis Ndumbi, Serge Kika, Chris Ndifor, John Neba, Briant Ngu, Pride Akuma, Michel Sugnin, Schubert Forhnjah, Elaine Tamnjong, Misba and family, my landlord Mr. Kimbeng and family, and all I might have forgotten… Thanks to the VW Foundation in Hannover, Germany, for sponsoring this research.

Abstract

In my thesis I have examined the meaning of New Media for transnational migration, and how the use of New Media is interrelated with negotiations of sociality. Bringing these two views together, I have examined how the mediality of New Media effects on transnational sociality. I have thereby addressed the topic from the perspective of non-migrants, of urban youth in Bamenda, in the Northwest Province of Cameroon. As a counter perspective, I have integrated the views of Cameroonian migrants in Switzerland. New Media of communication and information offer a broad range for users in order to pursue an ideal of connectedness, to social others and life chances, which are constitutive regarding their adopting of New Media technologies.

Social liveness emphasizes the possibility of a connected presence, or a quality of interaction in New Media of communication. Liveness, therefore, has different dimensions, as a potential, as a sensory experience, and as work or effort invested in mediated social ties. Likewise, liveness can be reversed or avoided, and liveness is a matter of degree and intentionality of New Media users. Negotiations of liveness in mediated transnational social ties have a decisive effect on notions of sociality and solidarity.

Transnational relationships are framed by conditions, the most important examples of which are physical dislocation and differing lifeworlds. As well they are superimposed by strong imaginaries of great potentials abroad. Notions of an ideal sociality are re-evaluated – and super elevated - vis-à-vis the perception of slippages, which are likely to be experienced in mediated social interaction. These slippages in mediated communication derive from limited social and emotional cues, as well as an often only partial understanding of migrant's life conditions abroad. From the perspectives of the migrants, exaggerated expectations and claims towards them lead to their adopting of strategies of New Media use, which seem to oppose those of non-migrants. Uses of New Media are then strongly related to dealing with these tensions in transnational social ties between migrants and non-migrants.

Such negotiations are likely to come to the fore in practices of New Media use, in which notions of sociality are revised according to the mediated conditions of social interaction. These are then evaluated in modes of conduct in New Media social interaction. This conduct is related to the media's specific mediality and conditions of New Media use in the Cameroonian context, such as financial constraints, limited computer literacy and habits of media use, as well as imagined potentials for liveness.

List of Acronyms

ADSL   Asymmetric Digital Subscriber Line
AES   Sonel American Electricity Services
BFM   Bundesamt für Migration
CAB   Central African Backbone Programme
CASA net   Cameroonian Skills Abroad Network
CFA-Franc   CFA Franc BEAC (Coopération Financière en Afrique
Centrale, Banque Centrale des États d'Afrique Centrale)
CIA   Central Intelligence Agency (CIA World Fact book)
CISCO   Computer Information System Company
CPDM   The Cameroon People's Democratic Movement
(French RDPC, Rassemblement démocratique du Peuple Camerounais
CRTV   Cameroon Radio and Television
DSRP   Document de Stratégie de Réduction de la Pauvreté
(English PRSD Poverty Reduction Strategy Declaration
DV Lottery Diversity Visa Lottery (US Greencard)
ECOWAS   Economic Community of the West African States
ENS   École Normale Supérieure (Teacher's Training College)
GNP   Gross National Product
ICT   International Communication Technologies
ID   Identifier, e.g. user ID, chat ID, messenger ID
ILO   International Labour Organization
IMF   International Monetary Fund
IOM   International Organization for Migration
IP   Internet Protocol (Address)
ITU   International Telecommunication Union
LAN   Local Area Network
LOL   Laughing out loud ("chat language")
MINEDUC   Ministry of National Education Cameroon
NGO   Non Government Organization
NICI   Cameroon National Information and Communication
Infrastructure
OECD   Organization for Economic Co-operation and Devel-
opment
SCNC   Southern Cameroons National Council

SDF	Social Democratic Front
SIM	Subscriber Identity Module (SIM card)
UNDP	United Nations Development Programme
UNECA	United Nations Economic Commission for Africa
UNESCO	United Nations Educational, Scientific and Cultural Organization

List of Figures

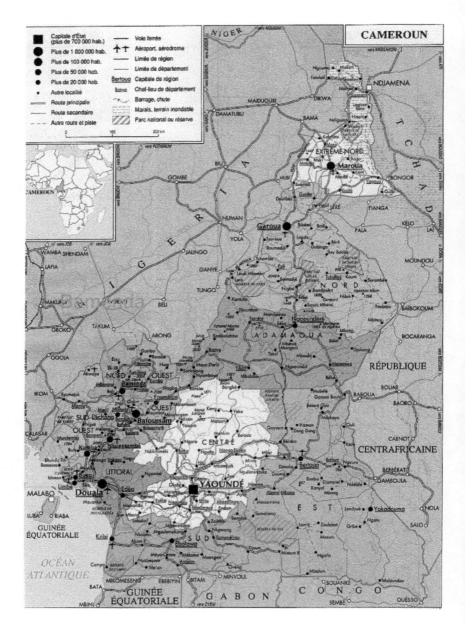

Figure 1: Map Cameroon, indicating regions (provinces)

Sociality revisited? Liveness, and the use of the internet and mobile phones in urban Cameroon

Introduction, ethnographic background and theoretical framework

In this introduction chapter I will first relate to the relevance of research and give a short insight about my fieldwork. Then I will introduce the field site and look into the ethnographic background, and the history and situation of mobility and of New Media[1] in Bamenda in order to give the reader an idea of the developments and recent transformations in this field. I will introduce the concept of "liveness"[2], which will serve as a guideline to examine sociality[3] in relation to New Media use and mobility throughout this book. Furthermore, as theoretical background, I will revise the state of the art concepts of mobility, transnationalism, and media studies, and I will discuss basic ideas about sociality, communities, and social norms and roles - notions, which serve as an introduction to the guiding questions of this book.

General introduction

The setting of my research offers interesting reference points for the aforementioned fields – migration, New Media use, and transformations of sociality, providing a specific empirical example for these intersections in a particular locality. I suppose that the site of my research – the city of Bamenda in the North West Province of

---

[1] I will use the notion of "New Media" interchangeably with the notion "ICT" (International Communication Technologies). New Media encompass electronic technologies, based on digitalization, and thus also include media such as DVD, MP3 or MP4, and others. For this book I will focus on the internet and mobile phones.

[2] I will introduce the concept of „liveness" later in this chapter.

[3] By the notion of sociality I understand a tendency and disposition of being sociable and relating to social others, and the situatedness of actors in multiple temporally evolving relational contexts (Emirbayer & Mische 1998, Mead 1932).

Cameroon – stands out and is particularly interesting in regard to New Media technologies and mobility[4]. Migration and the use of New Media are very current topics, which impact on many fields of public discourse. Also New Media technologies continue to develop at a high pace, and migration movements are transforming in their characteristics and impacts. This research is thus a momentary inventory of the situation[5]. However, since my research also addresses the historical set up of migration and mediated social interaction, the findings might provide a perspective from which to understand related transformations in this context, even though a complementing of the findings will become necessary in the near future.

Relevance of the research

I hope this book renders insights into how New Media technologies in today's "globalized world" impact on mobility, communication and imagination, which often do not occur in ways they are commonly expected to do so. The so-called "digital divide", and also a "mobility divide", often do not function along the lines of simple presumptions of the impact New Media may or may not have. Here I suppose I can fill some gaps in empirical research with my study.

Despite being favourable means of staying in contact over distance, New Media of communication also contribute to a negotiation of social ties, which can lead to tensions and frictions, on the basis of potentially being connected. Likewise, knowledge conveyed by New Media of information does not necessarily lead to a re-evaluation of opportunities, but often enhances the reproduction of strong societal imaginaries. The opportunities to stay in close contact with those abroad, as well as an apparent access to life opportunities through such connections, can also imply a strong feeling of immobility and disconnectedness. So far, what has been neglected in the discourses about liveness regarding practices of connecting, is the impact of the differentiating locations – and likewise social status - of the communication partners. These include imaginations and valuations, which are negotiated in mediated

---

[4] In view of different modes of mobility, we could speak of mobility in plural form, of mobilities (Adey 2010:7-9,18). Close interconnections exist between physical, social and virtual mobility. See later in this chapter and chapter 2.

[5] Although I might use an ethnographic present tense, the reader should bear in mind that I am describing the situation as it was between 2008 and 2011.

communication and have an impact on the perceived qualities of mediated social relations. In this sense, practices of New Media users – in the framework of sociality and mobility – could be examined in view of pursued habitual practices, orientations to life projects, and the adopting of New Media in the course of unfolding possibilities.

Fieldwork in Bamenda - different field stays and activities

The reason why I have decided not to relate in detail to fieldwork methods at the beginning of this work, is the fact that I have dedicated a whole chapter about issues of research methods and self-reflection at the end of this book so that they may be more easily understood after having read through my work. However, I would like to briefly give an overview over different field-stays and activities here, as background orientation for the reader. During fieldwork and the writing of my book, I have collaborated and exchanged with my research partner Primus Tazanu, who was part of the research group "Passages of Culture", and has worked on a similar topic for his PhD. Throughout this book I will make reference to his work (Tazanu 2012).

For my work for the topic of the PhD, I have completed four field-stays in Bamenda, and five field-stays altogether. The first 6 weeks field-stay in Bamenda in 2003 was an internship for fieldwork, in the framework of my studies of Social Anthropology at Basel University (Frei 2003). Thereby, I focussed on social practices of communication by internet and contributions of this media towards the negotiation of identities. The following master's thesis (Frei 2005) was also based on the topic and focussed on New Media's impact on transformations of social relationships and identity, taking concepts of space and time as a background. These hitherto existing findings and experiences in the field have also served as a background for my later research for my PhD, when returning to Bamenda ending 2008.

My stay in the field from October to December 2008 was mainly dedicated to gaining an overview about the transformations of New Media developments as a kind of exploratory phase of the research. The changes that had occurred in the field since 2003 were astonishing. It turned out that the internet had spread significantly, and the role of mobile phones in interpersonal communication had become crucial, whereas it was only at an initial stage in 2003. I was mapping cyber cafés and other New Media related sites, and collected background information. Furthermore, I started to observe the operation of cyber cafés, interviewed key persons in different fields, and also talked to media users and observed interactions in cyber cafés and other media sites[6]. I encountered new or shifted modes of using New Media. In particular internet use seemed to be strongly influenced by the desire of reaching out and linking up to opportunities, which were often related to migration and a general notion of "a good life"[7]. This impression has influenced my subsequent approach[8]. After an analysis of my data, the next field-stay followed six months later, from June to November 2009. In this second stay, I concentrated on interviews, sorting out dimensions of people's internet and mobile phone use, by studying different practices and imaginaries. Also, I included a group of scammers[9]. I furthermore included people's personal life histories and stories of their relationships with relatives and friends abroad. Thereby, I tried to sort out guiding principles in mediated social interaction and communication, such as feelings of closeness, sociality and solidarity. I covered these issues by conducting 52 systematic and rather formal interviews[10]. Moreover, I started to spend time in cyber cafés more systematically as I had done before, and mainly concentrated on a sample of ten different cyber cafés. The next field-stay arose after almost one year, from October 2010 to January 2011. Meanwhile, I had been proceeding in the writing process. In this fieldwork stay I intended to "round up", and to clarify some notions and concepts of sociality and solidarity, by

---

[6] See figure F1, a mobile phone repairer in his workshop, Commercial Avenue, Bamenda.

[7] See figure F2, opportunities for mobility.

[8] Compare to Tazanu's (2012) thesis, which places greater emphasis on migrant's perspectives.

[9] Young New Media users involved in fraudulent business practices in the internet. See chapter 3.

[10] I ordered these interviews into three categories of internet users. I will indicate these interviews accordingly in the book, by: Name, Interviews 2009, and the category O: for occasional, R: for regular, and F: for frequent internet user. In the category of frequent internet users I have integrated a group of "scammers", indicated by the addition: FS. I have included an appendix at the end of this book, where the reader can find the most crucial information about the interviews: about the person interviewed, his or her familial situation, education, occupation, as well as some information regarding the interviewee's relationship with the interviewer and the situation in which the interview was conducted.

4

addressing young people's integration into networks of social ties on a local and translocal level, by specifically exploring their notions of responsibilities and duties, and levels of social and emotional closeness. These notions proved to be central, but I had the feeling I had not addressed them systematically enough. Also I probed deeper into mobile phone use as part of socializing. In addition to participatory observation in youth social spaces I conducted 26 formal interviews[11]. This was coupled with many informal conversations[12], and I integrated interviews done by a fieldwork assistant[13]. When coming back from this field-stay of about three and half months, I integrated the findings into my written work, reworked my book and concluded the writing process. I only returned to the field site a fourth time in September 2011, when I had almost concluded my writing, in order to clear a few questions, among others things relating to concerns over copyright on quotes and pictures.

I have as well included a perspective "from the other side" of Cameroonians in diaspora, in Switzerland. I built up a social network in my home-town and beyond, and lead 17 interviews on internet use, also integrating mobile phones, migrant's relationships to their relatives and friends back home in Cameroon and living conditions in Switzerland.

F1. Mobile Phone repairer in Bamenda
F2. Signboard for DV lottery as an opportunity for mobility, Buea

---

[11] These interviews were conducted partly with new interview partners and partly with some of the persons I had interviewed in 2009. I have indicated these interviews by: Name, Interviews 2010/11.

[12] For informal conversations, I have indicated my field notes as source, including the date.

[13] His interviews I have indicated in the book by the abbreviation EC (Name, Interviews 2010/11/EC), see appendix.

The setting – The site of fieldwork, and youth as a social category

Between my first stay in Bamenda in 2003 and coming back in 2008, the developments in the field of telecommunication had been amazing. New Media technologies seemed to have an outstanding importance for many people, especially for youth. Thereby the ways, how these media were used regarding opportunities to "reach out" beyond the local context attracted my attention.

Bamenda - the site of my fieldwork

I will give a first impression about Bamenda, the site of my fieldwork, in order to explain the importance of "practices of connecting" in a framework of particular socio-cultural conditions. Bamenda is located in the Western highlands[14]. Regarding political administration, Bamenda is the capital of the North West Province, together with the South West Province composing the English speaking part of the country, within Francophone dominated Cameroon (compare Jua & Konings 2004)[15]. Indications regarding the number of inhabitants figure around 270,000[16]. The city is a stronghold of the opposition party SDF[17], and the birthplace of the SCNC, struggling for the restoration and independence of what was known as the British Southern Cameroons before Cameroons' independence in 1960, opting

---

[14] In the so-called „Grassfields", from the German "Grasland". This notion is designating parts of the Western Provinces. Mbaku (2005) relates to it by „Western highlands", cutting through English and French speaking regions, relating to a comprehensive area, which is shaped by centralized chiefdoms (Mbaku 2005:10).

[15] See figure F1, a political map including the provinces of Cameroon, page xi.

[16] Such indications vary depending on the source. It is also not clear, to what extent suburbs are included or not, and some internet sites do not state their original source, (see e.g. http://www.bamendauniversity.com and http://en.wikipedia.org/wiki/Bamenda (source dated 20.08.2011). According to Awambeng, in 1954 Bamenda's population was estimated to be about 18'000, in 1970 of around 41,000. Strong population growth started from the 1970's after the consolidation of Bamenda as the headquarters of the North West Province in 1972, which is evident from the 1976 census (Awambeng 1991:22). According to the census of 2005, Bamenda's number of inhabitants is estimated to be approximately 270,000 (see: http://en.wikipedia.org/wiki/Bamenda), and for the Mezam department - with Bamenda as capital - approximately 525,000 (see: http://www.geohive.com/cntry/cameroon.aspx).

[17] Opposing the ruling party CPDM.

to belong to Cameroon – and not to Nigeria – in 1961. By the replacement of the federal by a unitary state in 1972, Southern Cameroons was denominated as North West and South West Provinces, which had definitively lost their autonomous status. Likewise, due to its past of having belonged to different colonial systems, and politically exceptional position, Bamenda was often involved in tug-of-war and subject to repressive measures

F3.Commercial Avenue: The main business street in Bamenda
F4. Cyber cafés and mobile phone services

Despite the characteristics of an urban centre – also compared to cities of similar size in Cameroon – regarding infrastructure, job opportunities within the industrial or service sector, and scale and turnover of business, Bamenda has a rather poor economic performance. The sector of – at times less formally organized, but usually registered and taxed – small-scale businesses by far outweighs the sector of government administration and private formal entrepreneurship on a greater scale (compare Awambeng 1991, Fokwang 2008:95)[18]. More recently, petty traders and small-scale

---

[18] In Cameroon, it is estimated, that approximately 90,4 per cent of the population live from economic activities in the informal sector in diverse fields, which only contribute 29 per cent to the GNP (Le quotidian, Le Messager, 9 Octobre 2007; INS, 2005, Evina 2009:33). However, reliable statistics in this field are hard to obtain. In general, as for many African countries, the formal employment sector has been declining in relative and absolute terms (compare Fluitman & Momo, ILO, 2001, Kinge 2004) since the 1990's. It is questionable to make the distinction formal/informal here, since however small, these private business initiatives are all taxed by tax collectors, but

commercial manufacturing and services have also integrated new goods and services[19], of which the category of electronics could be considered as vital for the city's economy.

Youth as a social category and dimensions of agency

In my research, in particular related to the use of New Media and related narratives of mobility – migration, imaginations, and social achievement – I relate mainly to young interviewees, who form the main part of New Media users in the context of urban Bamenda. My work is, in particular, addressing dreams, narratives and social practices of youth, thereby relating to habitual and projective dimensions of agency, as well as to their handling of daily life situations and lines of sight (Emirbayer & Mische 1998, Förster 2010, Simone 2005, Simone & Abouhani 2007). Emirbayer & Mische (1998:962) term these agentic dimensions "iterational", "projective" and "practical-evaluative". In this sense, any social action is embedded within the flow of time and thus oriented to the past, future and present. "Since social actors are embedded within many such temporalities at once, they can be said to be oriented toward the past, the future, and the present at any given moment, although they may be primarily oriented toward one or another of these within any one emergent situation" (Emirbayer & Mische 1998:963,964)[20]. Furthermore, social actors may switch and recompose their temporal or relational orientations, and thus they are capable to change their relation to structure. Through the interplay of habit, imagination and judgement, these temporal-relational contexts are both reproduced and transformed (Emirbayer & Mische 1998:970, Förster 2010). In regard to a capacity to act in an unfolding temporal-relational context of action, Emirbayer & Mische (1998:974) emphasize the importance of intersubjectivity, social interaction and communication; the authors see agency as an inherently social phenomena or process, "by and through which actors immersed in temporal passage engage with others within collectively organized

---

also partly circumventing their incorporation into official registration and structures, as also large scale businesses do in Cameroon. Compare to figure F3, where on the right side of the storey building, a range of typical small-scale businesses are visible.

[19] See figure F4

[20] In practice, these dimensions are intertwined and not easy to decompose. In order to analytically examining them, we need to look at them as separate agentic dimensions, which I will discuss in further depth in the conclusion.

contexts of action". In this book, I thus relate to social practices and their transformation, which are more or less oriented toward one or several of these dimensions. In this sense, the dimensions of agency can serve as a kind of grid, which relate to different dimensions of mobility – as practices to deal with space – and liveness – as a modus of sociality. Also, these agentic dimensions transform from local to transnational mobilities, and from face-to-face to mediated sociality.

The discourses in scholarly work about youth range around youth's exclusion from power and accumulation and their integration in gerontocratic power structures, where social position is strongly tied to age – and gender – and their adopting of alternative social spaces. Youth's taking over an important part in society is increasingly postponed to a higher age, related to a lack of access to economic opportunities, and therefore to marriage[21], starting a family and taking over a position of responsibility, transforming youth into adulthood. Argenti (2007:7) relates to youth's being at the bottom of social hierarchies. Warnier introduces the category of "cadets", unmarried men, who are "void of transmissible life essence, as symbolically impotent" (Warnier 1993a:305). Cadets are seen as classificatory children, independent of their biological age (Bayart 1985 (1979)). Various social scientists emphasize the markers of a category of youth (Diouf 2003, Cruise O'Brien 2003, Fokwang 2008), as differing in age, gender, ethnic group, level of education, occupation, and material interests (Fokwang 2008). However, even though not a coherent social category, the socio-historical generation goes along with a feeling of commonality of a group in a specific region and time, sharing common influences and concerns, or in Mannheim's (1936) terms, sharing a common destiny of a historical and social unit. The age of my interviewees ranges between twenty up to mid-thirty - they fit into the category of Warnier's "cadets" regarding their social position. We could relate to them regarding some shared characteristics: being urbanites, internet and mobile phone users, rather educated, living in - more of less

---

[21] For young men, this is related to the issue of bride wealth, which a man needs to hand over to the family of his future wife, in earlier times paid in kind, later on monetized. Since it is difficult for young men to acquire sufficient wealth, some of them might remain bachelors up to their thirties or even forties (Argenti 2007:184, Warnier 1993a). Of course, there are also alternative ways how to deal with bride price, concerning agreements between the respective families.

- unstable economic conditions, apt to being open towards global influences of alternative realms of success, and their "will to consume" not matching with "the opportunity to earn" (Comaroff & Comaroff 1999:293). They could also be seen as a technological generation, having been influenced by telecommunication and information media in the last years and often picking up these skills relatively quickly, which has influenced their life experience in specific ways, in comparison, to the generation of their parents. Especially related to the New Media technologies, for example, users of internet in cyber cafés are a dominantly young clientele, which is also a reflection of the demographic situation[22] in Cameroon in general, and moreover of an urban area: young people come to cities in search for jobs, opportunities for business, education and "urban lifestyle". Political and economic marginalization is clearly felt by many people in Cameroon, but it seems to culminate in the expression of dissatisfaction by in particular young people. When we want to understand youth's practices of media use and the importance of issues surrounding mobility in Bamenda, we need to consider such specific preconditions.

A local history of migration

Even though migration is such an important topic in Bamenda and Cameroon, and a central issue in everyday narratives and imaginations of opportunities, it seems that the importance given to migration in the local context does not apply to difficult conditions and the respective small numbers of those who "make it". I will postulate a prevalent "normality" of migration in Bamenda, also deriving from the past, which impacts on the "attitude towards migration" up to today.

Migration history in Cameroon and the Western highlands

I want to relate to the background of the history of migration in Cameroon, specifically concerning the region of my fieldwork. Even though absolute numbers of African immigrants to Europe have – moderately – increased in the last two decades, compared with the

---

[22] According to CIA World Factbook, 61 per cent of Cameroonians were below 14 years, and the estimated median age of the population was 19.2 years (compare to Switzerland: 15.6 per cent below 14 years, median age 41 years) in 2008.

overall percentage of immigration to Europe in general the numbers are decreasing. The number of internal migrants has to be considered seven times higher than migration from Africa to the rest of the world: Migration in Africa has often been inner-continental in nature, and to a large extent occurring within national boundaries, especially to urban areas (Bakewell & De Haas 2007:111, Aina & Baker 1995).

The history of migration in Cameroon is complex, and often lacking definitive figures and historical evidence. We could differentiate types of migration movements, regarding their temporal and conditional frame – also partly overlapping. One such example of migration was Islamic warfare and conquest – such as several Fulani Jihads – and related claims for political control and domination, reaching back for centuries, and influencing migration movements in the region up to the 19th century. Another issue was the pre-colonial Trans-Saharan Slave Trade, followed by – and coincided with – the emerging Trans-Atlantic Slave Trade. The beginning of greater influence of colonial powers clashed with the slave trade[23], and the start of permanent settlement by Europeans in the area was also related to the consequent abolishment of slavery by the British in 1804 (compare Mbaku 2007). In seminal works authors have always pointed to the extraordinary situation and intense migration movements originating from the Western highlands, compared to other Cameroonian regions. The Grassfields continue to be the most densely populated area in Cameroon[24]. However, these historical migration movements lack evidence and documentation.

Migration movements were intensified from the time of the German annexation from the late 19th century onwards. Some documentation of migration movements in Cameroon was effected at the beginning of the German colonization period (Chilver 1967, Ardener & Warmington 1960, Den Ouden 1987). In 1902 a German military station was established in Bamenda: the Germans were interested in the Grassfields in general as a source of labour for coastal plantations (Goheen

---

[23] Warnier (1995:255) has estimated that about 0.5per cent of the population left the Grassfields annually as slaves, taken out by slave traders, or by labour recruitment for workers in the plantations or carriers, from 17th until early 20th century.

[24] With a population density of 99 per square kilometre, the North West Province is by far one of the most densely populated areas in Cameroon, compared with the national average of 22.6 people per square kilometre. See: http://www.populationlabs.com/Cameroon_Population.asp (compare Mbaku 2007).

1996:65)[25]. Colonial times were influenced by enforced labour schemes – a kind of continuity of slave trade – under the Germans, and from the 1920's onwards, after their defeat in the First World War, by wage labour migration to the plantations. In 1954, according to a survey of Ardener and Warmington (1960), there were about 42,000 wage earners in Southern Cameroons – today's South West Province – 31,400 of them were employed in the plantations of Victoria (Limbe) and Kumba divisions. The intensification of migrations from the Grassfields during the colonial period went along with an increasing permeation of a cash economy. It was then possible to evaluate many goods and transactions in monetary terms, which in turn enforced migration movements in search of cash income, which could not be obtained in the villages and local economies of goods exchange (Ardener & Warmington 1960:221, Warnier 1985:268,269, Mbaku 2007). Ardener & Warmington (1960) state that in 1955, 60 per cent of the employees in the plantations in Southern Cameroon were of British Cameroon origin. Out of these, the three Bamenda plateau divisions alone – Bamenda, Wum and Nkambe – supplied 32.8 per cent of all workers (Ardener & Warmington 1960:27). In 1955 about 80 per cent of the workers were below the age of 35 years, and a large percentage of them belonged to the social category of youth – most importantly they were not yet married (Ardener & Warmington 1960:70,71). As these figures show, for many youth it was worth taking the risk of venturing on new pathways, in view of lacking opportunities in their home societies. According to Warnier (in Argenti 2007, 1975:375), apart from the attraction of cash earnings and economic independence, specific local conditions in the Grassfields have contributed to migration. Such were systems of inheritance, which go together with centralized systems of control over land, enforced by a growing importance of the use of land for cash crops and not for subsistence. These systems had the potential to become problematic along with a growing population and pressure on land resources. Also political power was highly centralized, and thus the opportunities for accumulation. Further reasons could be a flight from witchcraft

---

[25] Enforced plantation labour had reached its peak during colonial times under the Germans, involving about 18'000 workers. The German colonization period ended with the end of the First World War.

accusations, family tensions, and a high awareness for status[26]. The emergence of a new force, the colonial authorities, and the partial weakening of the regional powers, created a vacuum of power (Bayart 1993:258, Warnier 1985). At least for a small minority of youth, colonial times offered a chance to free themselves of the subordination by elders, seeking Western education, status and wealth in the array of skilled labour in colonial bureaucracy, or work in coastal plantations (Argenti 2007:8, Warnier 1993:318).

In literature sources about migration in the Grassfields, most information concerns male migration movements. There are only a few sources, which explicitly concern female migration in colonial times before independence. Goheen (1996) writes about women's integration in rural economies, as subsistence producers, but with limited possibilities to gain access to cash income. Women seemed to hardly migrate independently in these times, which was also due to a lack of female work for cash income, and migrating women were often suspected of engaging in immoral practices, such as prostitution. In this sense, Ardener & Warmington (1960:240-242) speak about women who migrated - often temporarily - to the coast for marriage, following their husbands[27].

The rural exodus towards towns and cities continued, in particular during the period of the emergence of a new independent Nation State from the 1960's, and with economic upturn from the 1970's onwards, following "the chimera of economic salvation presented by emerging neoliberal free-market economies (…)" (Argenti 2007:10,11), which soon lead to problems. A concentration of unemployed youth in the cities, whose labour market could not absorb educated school and university leavers, and masses of unskilled workers, was leading to social unrest, petty crime and a staggering informal sector (Närmann 1995:167)[28]. According to Warnier (1993a:52,61) about half a million

---

[26] I will refer in more detail to the motivations for migration in the next chapter 2.

[27] Compare to the role of women in contemporary migration ventures, in chapter 6. Furthermore compare to Tazanu's revision of female migration (2012:21).

[28] Närmann (1995) relates to Africa's experience of an expansion of educational facilities since the independence era, going along with a „promotion of a basic mass education". Cameroon has one of the highest schooling rates in Africa. For adults, the literacy rate is 59,8per cent for women and 77per cent for men. The school enrolment

migrants from the North- and South West Provinces were found in urban centres such as Douala and Yaoundé in the mid 1970's. Regarding transnational migration, only a limited number of African students and workers migrated to industrialized countries in the labour migrant movements of the 1960's and 1970's, mainly following the French-English colonial divide. From the 1980's, increasing numbers of African migrants were comprised, mainly from Nigeria, Ghana, Senegal, but also from Cameroon, of high skill labourers and students on the one hand, and on the other hand low skilled and also irregular migration (Bakewell & De Haas 2007, Mbaku 2007).

A history of social connections over distance

Migration movements in the Grassfields, and likewise the experience of transformations in social relationships due to dislocation, were intensified from the colonial period onwards. Labour migration was highly encouraged by elders and parents (Ardener & Warmington 1960:241), due to the high evaluation and expectations such opportunities of work could offer, especially in monetary terms. The position of being an absentee had gained valuation, regarding the attributed status of those who lived and worked in the "realm of the white man" (Ardener & Warmington 1960:239). Returning was usually the aim: "Generally speaking, working away from home is regarded as a feature of young manhood, a period in which a man establishes himself economically and socially to a greater or lesser extent. He generally hopes to return home permanently in middle age to take his part in village and tribal life" (Ardener & Warmington 1960:69).

During a migrant's absence, a strong emphasis was laid upon their duty to support their families back home. In these times, the first monetary remittances were sent from the coast to the villages, with visiting kinsmen who were travelling back. Ardener & Warmington (1960:176-181) and Den Ouden (1987:15,16) relate to how money was saved by migrants in saving clubs[29], based on ethnic ties, or they engaged in small businesses or farms, to back up their income. Tazanu explains

---

rates are 76,2per cent for girls (6 to 14 years) and 81,3per cent for boys in 2003 (Rép. du Cameroun-DSRP 2003:18, Evina 2009:29).

[29] These associations are known as "Njangi", in Anglophone, and as "Tontines", in Francophone parts of Cameroon.

14

how remittances contributed to tensions and frictions in social ties between migrants and non-migrants, which were at these times mainly to be comprised in rural-urban linkages (2012:16-20). Undoubtedly, migrants and non-migrants also communicated during these times. They sent letters, messengers, or relatives visited each other. Furthermore, they would usually stay away only for a limited time and intensified their ties, by getting married to women from their community, or encouraging migration of other family members (Chilver 1967, Ardener & Warmington 1960).

However, it was a challenge to maintain the contact with migrants and secure their loyalty, as it was deplored by Bali people vis-à-vis the German colonial authorities[30], that migrants would neither send back remittances, nor return home (Chilver 1967:505). The experience of becoming out of reach was common. In extreme cases, some going to the plantations to work never gave any sign of life, and if they returned years later, they had become old, hardly recognizable to their fellow villagers, as if coming from a realm outside of this world (Nyamnjoh 2005:243). From the perspective of non-migrants, the migration destinations at the coast evoked strong connotations of ambiguous realms of great potential but also dangers, alluding to witchcraft narratives[31]. Many imaginations circulated about these realms, similar to today's narratives about "white man's kontri". Thus, the notion that – from the perspective of people in the village – wealth originates outside of the local community, and attempts to control those moving out in the sense of making claims for redistribution – which could turn out as a difficult task – are rooted in historical developments (Nyamnjoh 2005:244,251), which influence contemporary migration movements and imaginaries.

---

[30] This had been during the period of the German's enforced labour schemes. Bali Nyonga is located close to Bamenda.

[31] This was the case since the experience of slave trade and violence of pre-colonial times, when slaves were traded to the coast, or regarding the enforced labour schemes under the German administration. In coastal societies' occult associations – of which some still exist - the shore and the watery realm of the sea are associated with mythical realms of the white man, where Western goods are obtained in exchange for "zombies", bewitched kin, whose labour is exploited (Nyamnjoh 2005:242, Comaroff & Comaroff 1999:289). In other witchcraft beliefs, such as "Nyongo" of the Bakweri, the Bamiléké "Famla'" and others, wealth is obtained by occult forces in exchange for selling one's kin for labour to imaginative plantations (Geschiere & Fisiy 1994:323).

For example Cameroon: figures of contemporary emigration

Cameroon held censuses in 1976 and 1987 – although the results were never published - and in 2005[32]. OIM states that 170'363 emigrants, or 0,009 per cent of Cameroonians were living abroad in 2007[33] (Evina 2009:58), however, statistics and numbers differ. Thereby, France is the main destination for Cameroonian migrants. In 2008, 38,530 Cameroonians resided in France, 30'216 in Gabon, 16'890 in Nigeria, and 12,835 in the US (Evina 2009:29). Main destinations within Africa are Chad, Congo Brazzaville, the Central African Republic and the neighbouring countries in general. In Europe it is the United Kingdom, Germany, and Italy (Sources UNDP). The net migration rate[34] is relatively equal, which means that emigrants and immigrants do more or less balance.

In absolute figures, the emigration of Cameroonians towards Western countries has decreased slightly between 1997 and 2007, this is probably due to the more strict policies of immigration in particular to European countries, for reasons such as conditional barriers regarding qualification of immigrants, for family reunion and the fight against illegal immigration[35] (Evina 2009:62,65). Increasingly popular is studying abroad: according to UNESCO (2008), in 2006 about 16,000 Cameroonians were studying abroad, above all in France, Germany, Italy and the US (Evina 2009:53,54). For the year 2000, about 17 per cent of Cameroonians with superior education had left the country (Evina

---

[32] General results of the 2005 census were released in April 2010 (see: www.nationsencyclopedia.com/Africa /Cameroon-MIGRATION.html). It sparked a discussion in Cameroon, as to whether the results were manipulated by the government in order to gain an advantage for the 2011 presidential elections. In the 2005 census, for the first time, figures of emigration as well as main incentives and countries of destination were asked for. In addition the share of female emigration was considered. However, the detailed results of the census were still lacking at the end of 2011.

[33] See also http://www.iomdakar.org/profiles/content/migration-profiles-cameroon

[34] The net migration rate is the difference of immigrants and emigrants per 1000 inhabitants over a period of time. A negative value indicates that more people leaving the country than entering it, in the case of Cameroon, immigrants and emigrants almost balance, due to refugees from Chad, Sudan, Congo Kinshasa, and so on (UNDP).

[35] Efforts for a cooperation with immigration countries is only at an initial stage: Some political efforts have been made between Cameroon and certain states, such as France and Switzerland. These agreements are related to facilitated repatriation to the country of origin, and DNA tests in the case of family reunion (Evina 2009:79).

16

2009:16). In 2005 there were, according to IOM, 57,050 migrants of Cameroonian origin[36] counted in OECD countries, of whom 42.3 per cent were highly educated with university degree: Cameroon is experiencing a considerable „brain drain". But there are also positive influences regarding emigration: According to IOM (2009:24) „Cameroon considerably relies on diaspora and its human and financial investment capacity for the country's development by creating incentives, even though Cameroon is at the initial stage of programmes aimed at mobilizing diaspora." According to the World Bank, there were $167 million for private investments transferred to Cameroon from abroad in 2008, which represents 0.8 per cent of the GNP in the same year (Evina 2009:17). From 1985 Cameroon entered economic recession, due in part to a decrease in world market prices for primary raw materials, and a devaluation of the Franc CFA in 1994, which was part of structural adjustment programmes of the IMF and World Bank. The IOM report principally cites the decline of living conditions as a reason for emigration.

Apart from the view of the ones who emigrated, I will focus very much on the perspectives of the non-migrants in this book, as historically as well as today the great majority of people, who were and are part of translocal and transnational social networks of mutual relations and flows, and whose perception of "the world" is accordingly shaped by and through these interconnections. Thus I argue, that in particular their view and perspectives on sociality and life opportunities have transformed with the advent of New Media of communication and information in the setting.

---

[36] One difficulty that makes the significance of these figures questionable is that there is a significant number of Cameroonians of origin, who have naturalized and are citizens of other countries, and thus not officially registered as Cameroonians (compare to Fleischer regarding Germany, 2007:414). According to the Swiss Migration Authorities BFM, there were 4070 Cameroonian nationals living in Switzerland by the end of 2010. Fleischer estimates in regard to Germany, that Cameroonians might be approximately 50per cent more than in the indicated numbers. See http://www.237 online.com/ 2008062193/ Actualites/Economie /cameroonian-diaspora-exhorts-govt-to-encourage-local-investment.html.

The situation of New Media technology in Bamenda – history and background

Although the presence of mobile phones, cyber cafés and the importance of New Media communication seems to be exceptional in Bamenda, we cannot lose sight of certain facts of what is often called the "digital divide". Despite all the enthusiasm for the high rates of growth of telecommunication markets and new subscribers to mobile phone and internet: In the same way as people's opportunities "to move", are restricted by various conditions, we have to keep in mind that in so-called developing countries, opportunities to connect through New Media are limited.

In the mid-nineties internet technology first came to Bamenda, consisting of offers of email sending and receiving services, which still exist today, in "documentations", combined with fax, scanning, laminating, binding and stationary[37]. These are also services, which are today included in many cyber cafés. In the year 1997 the first cyber café – Paul's Computer Institute - was opened by a US national. By 2003 there were already more than 20 cyber cafés located in the core of the city, and by 2008 there were 60-70 altogether, including those on the outskirts of town. The first mobile phones attracted attention in 2003, whereas by 2008 most young people in the urban area possessed a mobile phone. Here I benefitted from the fact that my work was on New Media, and specifically the internet, when I was in the field in the year 2003, when this development was about to emerge, in the sense that I could observe the development over time, coming back in 2008 for my PhD research. In order to explain the spread of New Media in Bamenda, I need to relate to the background information, which could serve as an explanation for such transformations.

A history of media connectivity and mediated sociality

In former times, before the advent of new communication technologies, staying in contact with those out of presence availability was more difficult. However, hardly any scientific work exists on how mediated connectivity has effected and been organized (compare Ardener & Warmington 1960, Nkwi 2009). Postal services and landline

---

[37] See figure F5.

phones were the only possibilities to maintain contact between migrants and non-migrants[38]. The first post office was established by the Germans in the year 1887, it however mainly served to handle mails between the colony and Germany (Rudin 1938:216). The landline phone was introduced in the year 1893 by the German colonial authorities in Buea, the capital under German rule, to connect them to Germany (Rudin 1938:217, Nkwi 2008:54, Tazanu 2012:26). Further telephone lines were installed later on, but these were reserved for the elite, for the colonial administration and business people. Despite the long history of the landline phone, we could say that communication technology in Cameroon has leaped directly to wireless technology – access to phones for a wide range of people has only become possible with the advent of the mobile phone in the beginning of the new millennium.

Today, postal services and landline phones still play a role in interpersonal communication in Cameroon. However, deficiencies concerning these communication media are prevailing. Postal services have always had the reputation of being unreliable and slow. Better established than the State's postal service are the private postal services, linked to travel agencies. In order to send parcels and letters through travel agencies, they are collected, the names of the sender and recipient registered, and the fee is paid in advance by the sender. In this sense, coordination of postal deliveries is necessary, since the receiver needs to know that a delivery is about to arrive[39], a coordination which was more difficult in earlier times. As an interviewee said, remembering times without mobile phone: "I wrote letters on a weekly basis, and I always went there (to the agency) to check for mails" (Barbara. Interviews 2009/R)[40]. Regarding sending deliveries from or to abroad, the postal

---

[38] Radio announcements could also serve to spread news, mainly about births, deaths, marriages, and other updates, however, these are no media related to interpersonal exchange. Compare Tazanu (2012:21).

[39] Additionally the mobile phone number of the recipient is written on the parcels or letters.

[40] Even in urban areas people do not usually have a personal post box. Often aspiring migrants, who expect a delivery of for example visa or application forms for universities, which needs to be filled out on paper, indicate a business post box of an institution in their location. Also stealing documents – for instance when concerning visa procedures – from post boxes is common, and interrelated with it blackmailing in the sense that stolen documents are only handed out to the addressee against a sum of money.

19

services are seen as unreliable, and in this regarding private but expensive postal services are favoured, such as for example DHL. Another way to transfer parcels, letters and oral messages, is through people moving to and fro between places, as messengers, in the country and to and from abroad. On the level of personal messages, before the era of mobile phones, posting messages at public message boards was also common, for example concerning messages for students studying at universities in distance from their hometowns.

Landline phones have always been scarce, usually only owned privately by economically viable individuals or businesses, and also before the era of the mobile phone, only a few were publicly available (Nkwi 2009). Additionally, fixed telephone lines do not offer personalized contact numbers – a coordination of contacting was thus difficult. For example, as some interviewees told me, one had to arrange a call beforehand, or had to rely on persons who had a phone on their disposal[41]. The contacting thus involved several steps, was not spontaneous and it was expensive. However, this is also the case today, when somebody does not dispose over a personal phone, or when the network connection in the village is limited. In such cases, additional coordination of space and time is necessary in order to be able to speak to somebody on phone, whereas with a personal mobile phone, coordination has become easy: "Now it is just a call, asking, are you in the house? Yes I am there, then you go and see the person. Before you could spend for transport for nothing" (Elene. Interviews 2009/O)[42].

New communication media was embraced quickly from the beginning, even though – in the earlier stages of this development – only few had the know-how needed, the new technologies were integrated into the communication practices of different parts of the population. In order to make this possible, various services were established, for example, in the case of email an employee would type a message from a written document or dictate, and send it through the email address of

---

[41] In this sense, these means of communication do not provide immediate and personal reachability, but people need to move: for example to check their post box, to the agency, or to the next accessible landline phone.

[42] It is an interesting discussion whether New Media partly replaces mobility and transport in a short and middle – national – distance range, or if they – due to maintaining a greater number of contacts – rather enhance mobility in this range. I will not look at this issue in detail, but briefly relate to it in the conclusion.

the cyber café or the documentation service for a client, similar to former letter writing services. In this sense, in urban areas and in inter-urban communication, email has been embraced replacing the former postal services, being far less expensive, more reliable and faster. Likewise, mobile phones – and creative ways how to connect to them[43] – circumvented the lack of adequately distributed and available landline phones (Nkwi 2009). Mobile phones have spread widely among large parts of the population today, also because they circumvent abilities to read and write to a large extent, and are also profited from in rural areas. They have facilitated inter-linkages and communication between distant places, up to remote areas, which have earlier been "out of reach".

A digital divide and its differentiation – New Media connectivity in Cameroon in figures

Despite rapid growth, we should not lose sight of the fact that Africa's New Media penetration lags far behind the rest of the world. In 2008, Africa as a whole had around 12 Gbps of international bandwidth, less than one third of India's total international connectivity (ITU 2009:10). Nevertheless, the mobile sector in Cameroon had reached a penetration rate of about 31 per cent in 2008, which is the average rate in Africa's mobile market, of course with differences regarding urban and rural areas. Regarding internet, Africa had 2008 nearly eight times as many internet users than in 2000, headed by Nigeria (ITU 2009:4). Statistics show that in Cameroon, only about 3 per cent of the population were internet users in September 2008 (ITU). In other sources they counted 725,000 internet users in September 2009, which is 3.8 per cent of the population (Internet World Stats), which is clearly below the African average of 6.8 per cent (Internet World Stats) not to mention the world's average of 23 per cent internet users in 2008 (ITU 2009:5). However, concerning such figures, one should always note their limited validity. In the ITU ICT Report and other statistical sources it is acknowledged, that "(...) public internet access centres and other points of access account for a significant part of internet usage in Africa". (ITU

---

[43] Through practices such as phone sharing, or when somebody - in the village for example – does not have a phone, one needs to call somebody possessing a phone who is close to the person who is the intended recipient. Then this person needs to inform when he/she is close to the intended recipient of the call, prearranged, or indicated by beeping.

2009:7) [44]. Moreover, in the case of mobile phones such statistics are not always significant, and do not predicate about different mobile phone uses, uses of double SIM, switching SIM cards and different practices of sharing mobile phones (compare to De Bruijn et al. 2009). As Nyamnjoh states: "Although connectivity in Africa is lower than anywhere else in the world, local cultural values of solidarity, interconnectedness and interdependence make it possible for people to access the internet and its opportunities without necessarily being directly connected. In many situations, it suffices for a single individual to be connected in order for whole groups and communities to benefit" (Nyamnjoh 2005a:16). A significant range of work has been done, not least by Social Anthropologists induced by mobile phone companies in order to examine the market to sound out their clients' potential and needs - for example by Jonathan Donner and Jan Chipchase. Costs for mobile telephony and internet airtime are, compared to other African countries, relatively high in Cameroon, even though they have dropped significantly in the few last years. The "digital divide" thus continues to exist however, and it is not only dividing developed from less developed countries, but it is also to be found within countries, dividing regions, urban and rural areas and groups of people.

In urban areas, such as Bamenda, two factors contribute most importantly to a "hype of ICT", such as computer education and falling prices for technological devices and costs for communication, developments, which favours the access to ICT for a wider range of

---

[44] Furthermore, most statistics do not provide gendered figures; however there are a few: According to the survey of Gilwald, Milek & Stork (2010) there are no considerable differences between the internet use of females and males in Cameroon. Nevertheless, in other statistics for gendered use of the internet for other African countries, females have generally a lower share (see also: http://www.itu.int/ITU-D/ict/statistics/Gender/index.html, Sciadas 2005, Hafkin & Huyer 2008). From my own experiences in Cameroon, and overall in Bamenda, males and females in cybercafés are by far not equated - see later in this chapter. Regarding mobile phones, some research points out to drawbacks for females regarding access and use of phones, due to lacking economic means, access to cash, and social power (see e.g. Gillwald, Milek & Stork 2010, or http://www.dirsi.net/english/files/backgroundper cent20papers/070216--dunn.pdf). Other studies, such as Horst & Miller 2006, Caron & Caronia 2007, or Mbarika & Mbarika 2006, however rather point out to equalized uses of the mobile phone between the genders. From my own experiences, in urban areas of Cameroon, the access and use of mobile phones among men and women seems to be more or less balanced.

people. Nevertheless, I would argue, that even though internet users might be privileged in the sense of having access to media and a certain level of education, the majority of the users cannot be called "an elite" (compare Burrell 2009), but instead they use these media because they feel disadvantaged - young, jobless and without prospects. In this sense, the situation in Cameroon and other countries contradicts the assumption that New Media development goes inherently along with developments in economy or in general betterments in people's everyday lives.

Computer education in schools and private ICT training centres

One reason for the increase in the numbers of young people in cyber cafés is the continually better computer education in schools at secondary level, and Cameroonian government policy follows in that direction. ICT was officially introduced in 2001 by the presidential statement for secondary and tertiary education (Tetang 2007), in harmony with the World Bank's programme for integration of New Media into schooling curriculums. For this task, a delegate for each province was appointed by the Ministry for Education MINEDUC, in order to organize and manage introduction and quality control for general computer training in schools, at ordinary levels (O-levels) and advanced levels (A-levels). In a key document in 2004, regarding strategies of using ICT in education, the Cameroon National Information and Communication Infrastructure taskforce (NICI) prepared a plan and policy, with support from UNDP and UNECA. In this document, the Cameroonian government recognised ICT as a national priority. A curriculum was provided, which was standardized on different levels of computer education. In 2010, computer sciences as a curriculum was integrated into the formation at ENS, the higher teacher's training colleges[45]. Nevertheless, the organization was characterized by unsatisfactory coordination and preparation, and a general underestimation of the situation and difficulties of the stakeholders, above all a lack of teachers trained in ICT, infrastructure

---

[45] My contact person, the delegate of the National Pedagogy Support Unit for computer studies in the Northwest Province, was part of a team, who taught a series of compulsory courses for computer teachers, in summer 2009. These courses were supposed to guarantee a certain quality level, and to harmonize the curriculums.

and financial drawbacks. By 2006 a basic computer education for all students should have already been implemented in secondary schools. By 2008 media labs should have been a standard for every school at that level. In 2009, many schools were still only in the implementation stage, if at all – in rural areas, many schools, including government schools, do not even have electricity[46]. In the case of private schools, even though they are also required to introduce computer studies, they often do not meet these criteria, creating a divided school system (compare Tetang 2007). Even in the urban areas, where the situation has to be considered better compared to rural areas, there are huge differences in standards between schools: it seems that the governments' aims had been maybe too ambitious.[47].

F5 : Documentation services, Meta Quarter, Bamenda
F6 : Small private ICT training centre, Bali Nyonga, suburb

Furthermore, private ICT training centres have enforced their operation[48]. Compared to 2003, when most ICT courses were offered by cyber cafés, there were many specialized schools offering computer formation in 2008: Three major ICT training centres offered a wide

---

[46] Financial means are not provided by the government, it is up to the schools to finance the equipment and staff, in most cases, these costs are then passed on to the student fees.

[47] Compare www.bc.edu/bc_rog/avp/soe/cihe/inhea/profiles/Cameroon.htm. In general, the schooling system in Cameroon is facing difficulties and decreasing standards of education, at least since the 1990's. More institutions of private education have arisen on all levels, in particular in the last decade. However, very often they do not offer a better education, but instead contribute to the inconsistencies and further differentiation of educational institutions (Tetang 2007).

[48] See figure F6.

range of courses to be completed in depth over periods of 9 to 12 months, or 3 month's crash courses, their student's numbers in a teaching period ranging from 80 to 200[49]. For people interested in ICT training, it is generally not a matter of offers, but of financing their formation. In 2010, prices varied according to duration, formation and school, for example a crash course of 3 months cost 50,000 FCFA, training as an accounting secretary at Paul's computer Institute cost 217,000 FCFA[50]. By then, apart from a few medium sized computer schools – with less than 50 students – there were various small institutions with few students and an irregular curriculum, most often in backrooms of cyber cafés. In this sense, the quality of the teacher and learning schedule varied. In general, the private ICT training centres, according to some owners and teachers, are beginning to feel the impact of computer education in schools, but they have specialized in offering a range of formations in different fields[51].

Sinking prices for communication – better access
I have already highlighted the decreasing prices for New Media use, be it for airtime for internet use and credit for phone calls alike. In regard to the internet, this refers to rates for private subscription as well as for professional service solutions. Providers are the telecommunication companies MTN, Orange, and Camtel, but also direct satellite connections to servers located in the UK, the US, or China[52]. Although it was still a high expenditure for cyber cafés, in 2008, due to the latest inexpensive offers for bandwidth, most cyber cafés had taken the opportunity to buy bandwidth directly from the telecommunication companies, sharing accesses was not as common anymore as it was in 2003. Prices varied from 150,000 FCFA to 900,000

---

[49] See also http://flexcominstitute.com, http://www.paulscomputerinstitute.com, http://lauratebusinesscollege.com.

[50] I am relating to price indications of price lists obtained in 2010. 100,000 FCFA is about 200 Euro. A young woman in a supermarket, as well as an employee - even with certification - in a cyber café, can make 30-40,000 FCFA per month.

[51] These are for example networking, hard- and software specializations, accounting, marketing and financing, secretarial duties, and as well graphics and web design courses.

[52] In other Provinces – and cities such as Douala or Yaoundé - Ringo, a new telecommunication company offering internet-, not yet mobile phone services, is currently entering the market.

FCFA per month, depending on what service solution a customer wanted, for how many computer and the connection speed. New devices such as routers and Wi-Fi technology allowing access to the internet were launched by Orange in 2009, called "livebox", and similar offers by MTN, as business subscription solutions for cyber cafés. Prices for airtime for clients have also dropped accordingly. Whereas in 2003 prices varied between 350-500 FCFA, in 2008 they varied between 150-300 FCFA - 35-70 cents - for one hour airtime, depending on the speed of the connection and also the quality of the computers, keyboards and the convenience of the working space in general. In this sense, different segments of cyber cafés have developed[53]. Due to the fact that prices for internet subscription for private persons and businesses have also dropped, business people did not frequent public cyber cafés anymore in 2008, whereas they had been a group of regular customers in 2003. In many private businesses' offices, internet access had become standard. Private subscription to the internet are offered by all three providers Camtel, MTN and Orange in various categories, from ADSL up to a subscription of a certain number of hours per month, offered in different systems such as wireless, for telephone fixed-lines with modem, or, most popular, a starter kit for mobile phone users, as subscription or in the form of a prepaid internet card. In 2009 devices serving as modems, which have an appearance similar to that of a flash drive, were launched. The price ranged around FCFA 25,000 - or 50 Euro - and airtime costs per month depending on the service, access hours and speed of lines, ranged between FCFA 10,000 up to 75,000 for private persons. However, in the lower price brackets, many users are disappointed with the quality of the access and the lines are considered to be too slow in the sense that certain activities are difficult.

---

[53] However in Bamenda the range is not that significant. In 2009 in Yaoundé I had come across high a standard, and even luxurious cyber cafés, where airtime cost FCFA 1000 up to 1500 per hour. It was mainly business people and government employees who frequented them. In Bamenda, there was a cyber café syndicate operating in 2003, responsible for price agreements. The syndicate was said to be still existing in 2008, but not exerting any tasks – it probably never did: Prices differed in 2003, as well as in 2008 and later years. Many business branches have syndicates, but not many are actively representing interests. However, it was a must for cyber café owners to belong to those syndicates.

The lowering of prices also concerns the mobile phone users. Not only have the prices of the mobile phones themselves fallen, but also the cost of phone calls and text messaging. However, financial means are the most important issue here: People seem to be aware of the various special offers and advantages. In 2009, offers included advantages according to the type of calls people preferred or needed to make: For example, they could prefer to call at the call box under certain conditions, rather than using their personal phone[54]. For calls from their personal phones it was, for example, the opportunity to share the costs, or the taking over of the costs of the call by communication partners[55], free call-me-back text messages when credit had finished, calling a favourite number for free, lowered prices at night, or calls of a specific duration[56],

At the level of national policies regarding telecommunication, Cameroon is only hesitatingly opening up the telecommunication market. There are many rules and conditions to foreign investors, and maintaining advantages for the State's telecommunication company Camtel, so far hinders true competition in the field[57]. However, prices have already been drastically lowered[58] since the liberalization of the telecommunication market, under pressure from the World Bank's privatization policies, in 1998. Still, in Cameroon, people pay considerably more for telecommunication than in other African countries adopting more liberalized policies[59]. Thus, whilst on the one hand ICT formation as well as lowering prices provides access to these

---

[54] Depending if they are subscribed to billing per second or per minute, on the length of the call, or if they have to call from MTN to Orange or to a number in the same provider network.

[55] The offers called 50/50 (MTN) or „pay for me" (Orange): the person being called has to agree to take the costs, first.

[56] Such as the MTN 8 minute calls in the night – adopted by people by dropping after 8 minutes and calling again.

[57] Still holding the monopoly on the fixed line telephony, which is not a lucrative market, Camtel is re-entering the mobile phone telephony market, but with not as strong position as the foreign private providers MTN and Orange.

[58] For example a domestic phone call was 500 FCFA per minute in 2002, and 50 FCFA per minute in 2010, an hour internet airtime could cost up to 800 FCFA in 2002, or up to 500 FCFA in 2003, and 250 FCFA in 2010.

[59] The World Bank and African Development Bank have introduced a Central African Backbone programme in October 2009, to bring lower costs and higher internet speed to the region, with Cameroon, Chad and the Central African Republic participating so far.

technologies for wider groups of people, on the other hand, there are some difficulties and deficiencies when accessing and using ICT. However, these technologies are granted great attention and importance in the local context.

New Media – a promising business?

With ICT technologies, new and promising business opportunities and professional fields have arisen. Many young people feel the need to have at least basic computer skills and some take their training to an advanced level. International degrees are thereby increasingly seen as desirable, also with a view to migration, but specialized ICT training also enhances the chances to obtain a visa to go abroad for specialized training. One such example would be CISCO, which is an advanced computer networking specialization, which could be trained in India. This "hype" regarding ICT is enforced through school education and government policies, also in the announcement of the introduction of ICT at a government's administration level[60].

However, the main perception is that "ICT knowledge is indispensable nowadays", especially with a view to securing one of the scarce jobs in the private sector, or to start a business. In this field, regarding opportunities for employment, probably the biggest employers in Bamenda are the telecommunication provider companies (compare Mbarika & Mbarika 2006). Between 2008 and 2011 they often searched for new staff for office work, client's services and publicity campaigns. In this sense, some computer school graduates are able to find a job or an apprenticeship in a computer or mobile phone repair workshop, a cyber café or in another private business. In the field of hardware and software repair business, many intend to open their own business, after they have complemented their theoretical knowledge with practical experience in a workshop. Opening a cyber café is also still seen as a good business, even though the risks sometimes seem to be underestimated. Given the high fixed costs, such as rent, electricity, and the maintenance of the technical equipment, expenditures for the

---

[60] In the view of a country where the main employer is the government, this addresses the hopes of young educated people, regarding the chance that "a new computer literate generation" could replace the old bureaucratic generation.

connection and bandwidth, salaries for employees, and payment of taxes, the business needs to be well run in order to make break even. In many cases, migrants who have returned or operate the cyber cafés from afar – are involved in the cyber café businesses. They seem to be potentially viable in order to raise the capital, which is needed for such a venture. Other opportunities are businesses concerning dealing with electronics – for example, second hand computers, mobile phones, and respective spare parts. The fact that electronics are top value items makes them attractive to deal in. Contributing to the popularity of these businesses is the developing market, the concentration of these businesses in urban areas, and the absence of other opportunities. Competition is high, and especially in the computer field, investment is considerable. To buy and ship computers needs start-up capital as well as either a reliable and financially viable business partner abroad – for example a migrant – or the persons involved in business have to travel themselves.

The field of electronics in general has changed significantly in the most recent times, shifting from importing Western second hand electronics to new electronics from China, Korea, and Dubai. Prices for electronics have dropped considerably in the last few years, due to Chinese products entering the market – especially this is true and remarkable in regard to mobile phones. However, importers and consequently the consumers in Cameroon pay considerably and comparatively much more for electronic devices. One of the major reasons is the high custom duty added to imported goods. In 2008, second hand computers could cost 50-80,000 FCFA, and for a second hand laptop price ranges started from 90-130,000 FCFA. Although new computers are gaining ground, the imported second hand computers still formed the major part of the available computers in 2008. However, shops dealing specifically with new goods, computers and related electronics, have opened up along the main commercial street in Bamenda during the last years.

Along with the importance and spread of computer equipment, there is a growing need for computer repair specialists. Apart from computer repair workshops, most of the private computer training centres and also schools and bigger cyber cafés, have their own technicians, who maintain the equipment accordingly. Furthermore, many young people who are versed in computing or have a qualification in this sector are complementing their income by doing repairs, above all in the domain of software, since not much infrastructure is needed. Businesses related to computers are mainly – but not only – a male domain. However, young women are increasingly doing hardware courses and I met a few female apprentices in computer repair workshops. In software courses, women already make up a considerable number of the participants. As employees in cyber cafés women are common.

Mobile phone technology has spread widely, and it is still spreading. The greatest changes concerning the mobile phone market are the increasingly dominant Chinese mobile phone brands. "Original" mobile phones – European or US brands – have become scarce, but due to their better quality however, they are still desired. Whereas the "original phones" are second hand phones, the Chinese phones are new. Prices for mobile phones have fallen considerably in the last few years and even the most recent time. Whereas one had to pay at least 50,000 FCFA for a second hand phone at the beginnings of mobile technology at the turn of the millennium, in 2010 Orange or MTN were selling their new simple Chinese phones for 10,000 FCFA[61]. For a "reasonable price"[62] – from 30-40,000 FCFA upwards - a Chinese mobile phone can be bought today with a variety of technological features: they have become multipurpose technological gadgets being used as MP3 and MP4 players, camera, and storage device for music and pictures. The demand for these types of phones is high. Their disadvantage, and of Chinese products in general - is their lesser quality.

The repair sector has become of vital importance. Mobile phone repairers often collaborate, as many do not have all the equipment needed - some do not have a computer and are therefore not able to solve software problems. Thus there are many small – mostly one-man - enterprises, collaborating with each other, although the competition for clients is high. The mobile phone repair business has had to adapt swiftly to the new Chinese phones, to a confusing number and range of brands and models, and the fact that these mobile phones all need

---

[61] 10,000 FCFA corresponds approximately 20 Euro.

[62] However, this is an average month's revenue for a young employee in the private sector.

different spare parts, which are not widely available on the market, and even software that is different from those of the original mobile phones[63]. Until recently mobile phone repairers have only learnt from each other, but in the last few years some computer training schools have begun to offer courses in phone repair. The mobile phone sector is a promising business field, also initial investments are not that high. The same is the case regarding other mobile phone related businesses, such as, prominently, call boxes – selling credit on the street or in shops – and the downloading of music[64]. Mobile phone repair is a male domain, but it is also slowly changing. In 2011 there was only one woman active as a mobile phone repairer in the hardware field in Bamenda. In the software domain few women are active in mobile phone repairs. Women are mainly vendors at call boxes (Mbarika & Mbarika 2006). Some women are dealing with mobile phones, but those I met only did so on a small-scale level.

The situation of New Media in Bamenda – a stand out example?

The strong orientation towards New Media business branches stands out in Bamenda. Part of the explanation for this could be the medium size of the city, and the lack of other opportunities, as for example compared to Yaoundé, which is the capital and administrational hub, and Douala, which is the business and industrial centre. Apparently, considering their wide distribution and accumulation, these New Media sites and businesses have become an important factor in the city's economy[65]. The differences to other cities are most striking concerning the cyber cafés. In Bamenda, new cyber cafés are constantly being set up and opened. In 2011, there was a high fluctuation of cyber cafés in town, but overall, their numbers were still increasing slightly[66]. Explanations

---

[63] From around 2010, the situation of the spare parts has been improving: Stocks of spare parts have been developed, which are taken from discarded phones. Another business has therefore arisen, which is dealing with discarded Chinese phones.

[64] As for example downloading sound on mobile phones mainly needs a computer and a space to offer the service. Mobile phones could also be bought and sold on a minimal scale and range of profit, and also operating call boxes does not need much of initial investment. However, in this range these businesses are also only generating small income.

[65] See map of the distribution of businesses in chapter 3.

[66] Regarding the cyber café's numbers and diffusion, it was not easy to obtain reliable information about these developments over time. However it was possible to

for the density of cyber cafés could be the urban area, with a relatively high attraction vis-à-vis the surroundings[67], a high proportion of youth, urban education facilities - also in the field of computer education - and therefore a high potential for more or less internet literate youth. It is the search for opportunities in general, which motivates – especially young people – to go to the cyber cafés.

Partly interrelated with this, or at least related in regards to "looking out for greener pastures", are the infamous "scamming practices"[68]. The high density of the cyber cafés in Bamenda must to a large extent be explained by the hype regarding these fraudulent internet related business practices. At the period of my fieldwork, a great percentage of the young, mainly male internet users in the public cyber cafés in Bamenda seemed to be involved in activities of "scamming", on different levels of intensity and "professionalism"[69]. I relate here specifically to Bamenda, and the practices that were observable in public cyber cafés, and in this sense to a "small scale scamming" on a rather informal level[70]. As cyber café owners and employees stated: "If one goes into this business one must be aware of that the main customers will be the scammers" (Jude. Field notes 2009. 18.07.09). The question, why the scamming issue is that important in Bamenda in these years, and apparently to a larger extent than in other cities, could only be

---

find out regarding single cyber cafés, but apart from that, it was more the subjective remembrance of those who witnessed the transitions. The mapping I did in 2003 focussed on the core of the town and is therefore only comparable to a certain extent. At that time, I had counted about 20 cyber cafés. Nevertheless, it seemed that since 2003 the number of cyber cafés had greatly increased, including phases of stagnation.

[67] In 2008 I had investigated in some bigger villages in the surroundings of Bamenda, as Bali-Nyonga and Bafut, where cyber cafés did not have a good stance. People preferred to go to Bamenda to use the internet. In Bafut two cyber cafés had closed down in 2009, and in Bali there was one left where there were two previously. In other villages, there were no cyber cafés at the time.

[68] "Scamming" can be related to a set of specific business practices, the notion derives from "scam" - fraud, cheat, deceit or swindle. More encompassing terms are "feymania" (Ndjio 2008, Malaquias 2001) or "419"-practices, "(…) relating to the respective section of the Nigerian criminal code that deals with business malpractices at large and in specific fraud" (Ndjio 2008:4).

[69] In some cyber cafés I estimated that about 70-80 per cent of internet users were involved in scamming practices, especially in the years 2008 and 2009.

[70] Regarding the internet related fraud practices in Bamenda, most activities are related to the field of the so called "puppy-fraud", which included advertising non-existent puppies - or other animals or goods - on ads sites.

answered by making speculations. One factor is certainly the proximity to Nigeria, where scamming practices seem to have come from. Language is certainly also a factor, English being the "global language" of the internet, which makes it easier for users to move in these virtual spaces, accessing a higher number of potential scam victims. Another issue is the relatively high number of well-educated and jobless youth. It seems that there are differences regarding the situation of cyber cafés and their clienteles, in other cities such as Yaoundé or Douala. In the capital, which is attracting people for business or jobs in public administration, as well as university students, the composition of people in cyber cafés is different, and the share of women is higher. However, especially in student's quarters, there is a culmination of small cyber cafés and also the scamming phenomenon is prevalent. Also in Douala, cyber cafés are not that spread out, they are concentrated around the university and suburbs. Douala is a hub for business, also in regard to computers and mobile phones. Limbe has more similarities with Bamenda however – a coastal town in the Anglophone Southwest Province. Although it is a very small town, it has a high density of cyber cafés, and scamming practices are prevalent. Buea is also a small town, and it is special because its population is to a large extent dominated by students during the teaching period. During the university term, scamming practices are widely spread over town, however, the number of cyber cafés is however limited. This could be one explanation why Tazanu (2012) has paid less attention to internet use than I have done so in my book, since he was conducting fieldwork in Buea.

Probably due to the scamming phenomenon, which is an almost exclusively male domain, the proportion of male internet users in cyber cafés is high in Bamenda, also compared to Yaoundé, where there were higher numbers of young women to be seen in the public cyber cafés[71]. In 2008 it seems that fewer women were to be found in cyber cafés than in 2003, at least in proportion to male users. Roughly estimated - and of course with differences regarding different cyber cafés – females make up not more than one quarter of the customers. "I think that there are

---

[71] However, regarding age and gender, I cannot provide figures, and the comparisons with the situation in other cities, especially Yaoundé, where I spent some weeks checking on cyber cafés, can only be evaluated as estimations and impressions, which, apart from observations, I obtained in conversation with cyber café owners, employees and users.

not many girls and women in the public cyber cafés because of this scamming issue. I do not know whether there exists any female scammer… I do not think so, only boys involve with it. But this scamming makes cyber cafés become irresponsible places, also it is noisy and insecure especially for girls" (Linda. Interviews 2009/F). Apart from this explanation, other reasons could be that from some people's perspective a negative image of girls in cyber cafés still exist, as they are suspected of engaging in immoral practices, browsing on dating sites or prostituting themselves (compare Tufte 2002:251). In such narratives people relate to discourses on controlling female sexuality and morality, also in the sense of parents exerting greater control upon girls. Other explanations are related to women's' lower general level of education and disposition to technology, or, as uttered by a (female) interviewee: "Girls have a mentality that does not really expose them to the upgraded world. That's called gender inequality (…). They like to depend on men. They rarely like to learn more, they do not give importance to that" (Barbara. Interviews 2009/R). Another interviewee also mentioned a high competition among young men, which motivated them to "exploit the internet", and "to use the internet as a tool to struggle" (Clothilde. Interviews 2009/F)[72].

The average age of the internet users in public cyber cafés has dropped compared with 2003, when cyber cafés and internet knowledge were not yet wide spread. In 2008, young people – and especially males – in their twenties and early thirties, but also many teenagers, were by far the majority. This could be explained by the increasing numbers of young internet literates and the enhanced computer awareness among youth, the scamming issue and the fact that business people and elderly, more economically well off persons, who had still formed a part of internet users in 2003, hardly work in public cyber cafés anymore, but in their offices or homes – also to delineate themselves from young people in the public cyber cafés. The clientele in public cyber cafés thus seems to have transformed in the sense of reflecting the economic disadvantages of youth: the falling age and also the activities of a

---

[72] Contrarily to the internet, regarding mobile phone technology there appears to be no outstanding differences to mobile phone uses in other parts in Cameroon – I am speaking here of urban areas, and mobile phone technology seems to be adopted more or less equally among male and females. Compare earlier in this subchapter.

majority of the users, proving that the internet is viewed as a strategy to escape poverty and dissatisfactory life circumstances (compare Burrell 2009).

Framing the research topic: liveness as a lens

Having introduced the ethnographical background and an appraisal of migratory movements and New Media technologies in Bamenda, at this point, I will address the concept of "liveness"[73], which I will adopt as a lens throughout this book. It is a concept, which can capture a specific type of social experience within a given framework of media use and mobility, in relation to specific time-space configurations. It could be seen as a specific mode of sociality. As such, it can serve to approach empirically observable "social situations" in order to enlighten the view of the specifics of mediated sociality in the site of my fieldwork[74].

The concept of "liveness"

Liveness as I am referring to here, presumes mediated sociality under conditions of not being co-present. The concept of liveness is fluid and has transformed in the course of time and development of various new technologies. It has been embraced by researchers from various disciplines[75] - from studies of performance and the arts, to media studies, and law – and thus ranging from examining the "live" in media as theatre, life performance as an art form, dance, music, sports, trials, theatre, cinema, television, radio to informatics, and eventually New Media such as the internet and mobile phones. Thus it ranges from performances in real time with simultaneous co-presence of performers

---

[73] I could however translate "liveness" into "liveliness", in order to use a better English term, but for my work I will adopt the notion of "liveness", relating to the already existing discussions and conceptualizations of the term.

[74] I refer here to my guiding questions and related sub questions.

[75] I will mainly relate to definitions and notions of liveness related to by Philip Auslander and Nick Couldry. But compare as well to, for example, Feuer 1983, Hoskins 2001, McCarty 2001, Burt 2002, Klein 2000, Sarbaugh-Thompson & Feldman 1998, Turner 2011, Zhao 2003, Leader 2010, Ytreberg 2009, Moores 2004, and White 2004.

and audiences, up to diverse modes of mediation (compare Auslander 1999:4).

Both, "co-locality" and "co-presence" point out to physical encounters in face-to-face situations[76]. These concepts have been developed and reworked in particular – among others – by representatives of Sociology, Social Anthropology, and Philosophy concerned with Social Phenomenology, in what Alfred Schütz relates to as "life-worlds", the inter-subjectivity of social experience and everyday life conduct. Some famous representatives of Social Phenomenology are Edmund Husserl, Alfred Schütz, Max Scheler, Peter L. Berger, Thomas Luckmann, and Erving Goffmann. Encounters, as Goffmann calls it, fulfil the conditions of an overall experience of the "other", including all senses, in a specific context and situation. Thus, in the words of Berger and Luckmann (1980:35), the encounter under conditions of co-locality is the basic form of social interaction[77], providing the least mediated impression of the other person by including behaviour, mimic, conduct, and reactions. In order to argue, why I intend to operationalize the concept of liveness rather than directly building on co-locality and co-presence, I want to state a few points here.

The notions of co-locality and co-presence imply being – together - in a certain place at the same time, which points out to objective coordinates of space and time, rather than to a subjective "sense" of being close. Liveness is a perception, a feeling, and as I will try to show, liveness is not necessarily bound to co-presence, but can be mediated through media. A sense of being close can in this way – ideally – occur while being physically separated in space. Liveness can in this sense serve to conceptualize a normative ideal of a mode of sociality. Co-presence does not necessarily mean to experience an encounter, human to human, considering that people can be together in a room, perceiving each other, but having no interest to get into contact. Liveness in media is – most often – related to a deliberately sought out social encounter.

---

[76] I will use the notions of co-locality and co-presence interchangeably, even though, co-locality emphasizes a spatially confined sharing of a local space, whereas co-presence could also be used to indicate a temporally shared moment, both relate to an association of a point of intersection of spatial and temporal coordinates.

[77] Nevertheless, there have been attempts to widen the concept in the sense that co-presence is not only seen as a mode, but also as a feeling or sense of being with others, in order to include mediated communication (compare Zhao 2003).

36

Liveness contains both, spatial and temporal aspects and qualities, which can be employed under conditions of physical absence and separation. However, place plays an important role, for example in the sense that communication partners are able to situate each other in space. Liveness has a strong relation to temporal simultaneity, of synchronism, through the simulation of "togetherness" in simultaneous interaction. Liveness is also an ideal state, it encompasses actuality as well as the potentiality of the liveness of situations. Liveness is – usually – not an aim in itself, but constituting itself and emerging in interactions, as a qualitative characteristic of the interaction, and of the social relation itself, when mediated. However, the quality of liveness could consciously be induced, enhanced or manipulated by the help of technological features. It includes the motivations of actors by an impulsion of wanting to create a kind of closeness, to contact, to be connected, but it can also mean that one tries to avoid too much closeness, or liveness, in certain instances. It is thus important to see liveness as a continuum of the perception of an "intensity of liveness", up to the extreme of a negation or cutting off the experience. Liveness is not a "solution" to overcome - spatial, temporal and social - distance, but solely an approximation, and can even generate tensions and disharmony. Media liveness thus offers the opportunity to negotiate closeness and distance in several ways.

To sum up, liveness is an indicator of the quality of a social encounter – mediated through media or in co-presence. It is an expression of a lively being connected to people, situations and options, and, as Zhao (2003) expresses, we could see liveness as a sense of being with others. Altogether, the subjective relational sense of "feeling closeness", the profitableness of the concept related to technological mediation, the integration of the concept as relating to both, potentiality and actuality, and the ability to integrate a relatively wide scope of social interactions, are reasons for adopting the concept of liveness for my work. However, the understanding of the specifics of co-presence, of face-to-face situations and social encounters, is helpful in order to understand a basis of sociality, to which all forms of sociality seem to refer.

Mediated liveness

The live concept has switched its reference, as media technologies have transformed, and have thus transformed individuals' perception and notion of "live" and sense of liveness. Auslander (2008) distinguishes six forms of liveness[78] (Auslander 2008:61): first, "classic liveness" as liveness in an original sense denotes a presence of an audience and a physical co-presence of an audience. It is also the feeling of a kind of communality among the members of the audience, or including the performers, and the characteristics of an event as such, as something outstanding[79]. Peggy Phelan (1993) states, that a live performance is only possible by the presence of living bodies, experienced by the audience in the very moment of its realization. Critique was among others coming from cognitive science, claiming that all experience is mediated through our senses. Auslander (1999, 2002, 2008) does not look at the "live" and the "mediatized" as oppositions, but as mutually dependent on each other. Differences between witnessing a live event and watching it on TV need to be examined here. Provided that it is a live broadcast, it is a simultaneous experience of consumption and production, as it is for example the case when watching a football match on TV. This form relates to Auslander's second type of liveness, "broadcasted liveness". Nevertheless, with other live events, it might be of less importance, whether it is real time or broadcasted with a temporal delay, when recorded, filmed and put on DVD, or shown on TV. This is what Auslander relates to as "live recording", as his third type of liveness. It is certainly related to the intensity and variety of media productions mediated by all kinds of different media, that definitions of liveness have become less rigid (compare Auslander 1999:2).

In fact, the feeling of liveness seems most importantly to be attributed to the desire of the audience for proximity and intimacy transferred by media contents, or likewise communication acts. Seminal earlier work about TV generally related to TV's essential properties as a medium of immediacy and intimacy, due to the proximity of the event

---

[78] I will discuss these forms in the current and the following subchapters.

[79] As such, live performances have been looked at from a perspective of ritual theories, the performance as a ritual, the time of the event happening as a liminal phase, and the audience's communality has been seen as an example of communitas (Auslander 1999:3,4, compare Victor Turner, Arnold Van Gennep, Bourdon 2000).

to the viewer, and the domestic realm of viewing (1999:16). White (2004:81) writes in her examination of TV liveness: "(...) liveness is the master term or key word that subsumes a host of other qualities and characteristics. Liveness serves as the conceptual anchor for the properties considered essentially televisual – immediacy, presence, reality effects, intimacy, and so on." Scannell's (1996:84) "presencing", or Tomlinson's (2007) "immediacy", could produce a feeling of "being there" and being involved. The audience, or the media user, as well as the event, the media production – or mutual communicative acts – come together by mediation. Their meeting needs a certain temporal and spatial configuration.

The first three of Auslander's six types of liveness rather relate to the interrelation between media productions and media users, and less to interpersonal communication and to interactivity. I will examine the further three types of liveness in the following sections.

Online or internet liveness

Auslander (1999:7, 2008) and also Couldry (2003, 2004) speak of a redefinition of liveness in the digital age. Couldry (2004) relates to what he calls "online liveness", and states, that online liveness often overlaps with liveness in an original understanding: He mentions thereby live impressions of weather conditions, animals in a zoo, or live-cams watching events in the Big Brother container (Lawson 2001, Vickers 2001). Both, Auslander and Couldry also extend the notion and state, that "internet liveness" – how Auslander calls his fourth type of liveness – is more than consuming media productions in real time, but it is also about social interaction and a feeling of co-presence among users. Hereby, included is thus also interpersonal communication with known or unknown – mere online relations – social others in internet based communication media. Also related to the internet, Auslander (2008) mentions – his sixth type of liveness – what he calls "website goes alive" (compare Couldry 2004). He relates to media users' reaching out to news on news sites, which are regularly updated, or to interactive sites, as an interaction between user and technology. The internet makes a specific difference compared to other media here: its news feeding and information generation is decentralized, its contents sheer endless, it is

39

not bound to broadcast features (compare Turner 2011), but information can be retrieved at any time.

However, it seems that both, Auslander (2008) as well as Couldry (2004) include in their respective notions of online liveness or internet liveness, online contents as well as interpersonal communication, ranging from small groups who interact in a blog, up to a global audience related to news sites, integrating private personal communication acts with news feeds for the publics. For my work, I will mainly relate to interpersonal communication, and related perceptions of liveness, and I therefore prefer to relate to sociality online under the notion of social liveness – Auslander's fifth type of liveness – as I will describe in the following subchapter, and thus separate media contents from mediated interpersonal communication.

Social liveness

What Couldry (2004:356,357) relates to as "group liveness", and Auslander (2008:61) as "social liveness" as his fifth type of liveness, points to the potential of connectedness, and a feeling of liveness in the sense of being interconnected. Couldry introduces group liveness as relating to mobile subjects who are continuously in contact with other – mobile – subjects, in the sense that each member of a group is continuously accessible to every other member – via mobile communication technology. This sense of liveness has become possible due to the mobile phone. Why I hesitate to follow Couldry's notion of group liveness, is the sense of social networks that are implied here, which have the connotation of a local social network within which everybody is connected to each other. I prefer Auslander's social liveness, which he describes simply as a sense of direct connection and simultaneity in calls and text messaging. I would prefer to speak of social liveness, instead of group liveness, in order to emphasize the individuality of social networks. Cross-connections are not always given, and social networks ought rather to be seen as continuums transcending potential presence availability, to solely mediated social ties.

Even though it makes also sense to distinguish forms of liveness in view of their specific mediality, I do not follow Auslander's (2008) separation line of internet liveness and social liveness – the types of liveness which play the most important role in my work. The main difference of social liveness – based on mobile telephony - to internet based communication media is that the mobile phone is a mobile technology and thus connects people wherever they are. However, I tend not to separate internet based and mobile phone communication, for several reasons. First, I suppose that social ties are maintained by a variety of media, and the experience of liveness is generated through a general communicative performance. Secondly, I suppose that both media are used on a level of presence availability up to a translocal level. And thirdly, I think that internet based and mobile phone based technological features blur – the internet can increasingly be accessed through mobile phones and calls can be effected through the internet. Furthermore, an aspect, which is not mentioned by Couldry or Auslander, but around which central claims of my work are built up, is how specific places and their conditions have an impact on communication partners, their use of technology and communication habits, which has an impact on the quality and not least the sense of liveness in mediated social relationships, as I will show in the course of this book.

Online and social liveness and three dimensions of descriptive qualities

Overall, after an examination of Auslander's and Couldry's versions of online or internet and social liveness, I have argued how I would remodel these two categories. I will relate "social liveness" to mediated communication with known or unknown social others, be it by mobile phone or the internet. I will take over "online liveness" as a sum of media content and information retrieval activities in general, which is above all related to internet based technology. Then, I would like to add further "dimensions of liveness" to both of the categories – the two categories of "social liveness" and "online liveness" being overlaid with the three dimensions as a kind of grid: by applying them, I will try to obtain a conceptual framework in order to have a tool to examine empirically observable situations of New Media use and perceptions of liveness.

Grid dimensions of liveness

	Sensory and Emotional Liveness	Liveness as Imagined Potential	Liveness as Work and Virtual Presence
Social Liveness	-Simultaneous communication, interaction -Simulation of co-presence -Feelings of emotional closeness -Negotiating "levels" of liveness regarding different types of social relationships	-Potential to "link up" with known and unknown social others -Purposeful activities of connecting -Pursuing social capital	-Maintaining mediated social relationships -Creating a virtual presence by various mediated interactions -Maintaining social capital
Online Liveness	-Immediacy of responses -Opportunity to generate effects	-Potential to "reach out" to media contents -Imagined opportunities -Purposeful activities of connecting -Pursuing human capital	-Pursuing projects -Regularity of habits -Staying updated -Maintaining human capital

The first dimension is what I want to name "sensory and emotional liveness". It relates to the feeling of co-presence in a virtual sense when simultaneously communicating with physically absent others by mobile phone or on the internet. Thus, this dimension is very much related to the senses involved in communication acts, and references to space and time. It is hearing each other's voice, or responding to each other in real time chat communication, seeing each other by a web-cam while communicating and so on. It also involves the complementary use of media in order to enhance this sense of liveness, which is enhanced by the multimediality of, for example, the internet. We could speak of a simulated co-presence or face-to-face interaction, which is given by simultaneity and integration of different sensory media to enhance the "live" sensation, of actually being close to a person in an emotional sense while being physically distant. This dimension of liveness has a habitual component, in regard to following up sociality, however framed differently. Concerning "online liveness", the dimension relates to the feeling of liveness in the sense of immediate response to an action on the internet, news sites are continuously updated, a request on the internet is answered instantly, using search engines serve to orient oneself, and so on.

The second dimension I will call the dimension of "liveness as potential" (compare to Couldry 2004), which contains a strong projective element, expressing a perception of a potential to connect to "the world", as to media productions and information – referring to "online liveness" – and to physically absent others - referring to "social liveness". It gives an internet user the perception of enhanced opportunities "to reach out", and that such potential, consisting of social, human, and symbolic capital, can be realized when needed. Media users have the opportunity to reach out to what is happening in the world, or to whatever they are interested in. They do it selectively, since the scope of media productions is infinitely vast. Potentially, everything is in their reach. This feeling might be deceptive at times, as the conditions of media user's environments include constraints, which cannot be neglected. Also, the perception of potential and related imaginations, can have very tangible effects on practices, for example of media use in the lives of media users[80].

The third dimension then relates to what I want to call: "liveness as work and virtual presence", represented by continuous efforts and work as social practices, in order to maintain liveness. This dimension relates to a practical and evaluative adopting of media in specific situations and contexts (compare Emirbayer & Mische 1998). Thus, this dimension relates to a potential of being able to extend connectedness by maintaining it over time. Regarding "social liveness" I relate to „work" in the sense of ensuring a regular communication in appropriate temporal intervals, by diverse media of communication, cultivating mediated social relationships and thus fostering social capital. Concerning "online liveness", "work" relates to continuous accessing of media contents, in the sense of "staying up to date" regarding what is happening in the world or trying to keep up with projects that one has resolved to do, and a continuous coming back to online habits. This

---

[80] It influences, for example, how internet users perceive their chances to migrate, what courses of actions they follow in this field, and it for example creates new practices of business activities, such as scamming, and it influences imaginaries of the Western world by the purposeful consumption of information about Western lifestyles, and so on.

dimension includes as well as local implications related to the work of "staying connected", which could be related to as "negotiating virtual presence". It is related to various strategies of people to maintain their influence, be it on certain persons, or as active involvement on a local level. It thus represents the presence of a physically absent person in the perception of the ones present in a certain location. These activities manifest on a local level in various ways, also materially. It can describe how migrants may engage at home, being involved not only by maintaining contacts with various people, but also an engagement in organizing work, transfers, business, and travelling.

I have related to liveness as a quality of social experiences in specific time-space configurations. Liveness then represents a shared reality at a particular moment due to mediation. It guarantees a potential connection to shared social realities as they are happening and it is at the same time a means to negotiate these experiences. New Media, such as the internet and mobile phones, are said to be characterized by enhancing interconnections: In temporal terms, by frequency, immediacy, and simultaneity of interaction, and in spatial terms, by enhancing intimacy, a sense of closeness, as well as the integration of different senses, which enforce or simulate a the sense of co-presence in a situation of physical separation. Stretched that far the concept of liveness is closely related to media's role in the temporal and spatial organization of the social world[81].

Theoretical framework – mobility, transnationalism, New Media, and social transformation

In this subchapter I will relate to other background theoretical concepts. In particular will I examine concepts of mobility and transnationalism, the state of the art in African media studies and specifically in relation to the internet and the mobile phone, as well as to central notions concerning sociality and social transformation. Hereby, I will direct the reader's attention to the guiding research questions, which derive from the discussion of these theoretical concepts and their employment for empirical research.

---

[81] I will relate to concepts of space and time in the course of this book in different chapters and in the conclusion.

Concepts of mobility and transnationalism – state of the art

Transnationalism as well as mobility studies have striven to prevent shortcomings regarding the complexities of spatial practices and movements of actors. Both of these concepts try to capture more adequately, configurations of space and time, and their interrelations with movement, flows of people, goods, and communication. In the area of intersection of mobility and New Media use, on the one hand, a broad approach to "mobility", as "mobility practices" serves the purpose in order to integrate movements and exchanges in a virtual and imaginational sense (compare Sheller & Urry 2006, De Bruijn, Van Dijk, and Foeken 2001). In a similar sense, the notion of "transnational activities" or practices can be useful in order to capture the transnational aspect of diverse border crossing activities (Kennedy & Roudometof 2002, Pries 1997). On the other hand, transnationalism studies focus on social networks and transnational social activities, and thus tend to include a multiple perspective on relations between those who migrate and those who stay behind. In this sense, both concepts prove to be useful for my work.

The concept of "mobility" can be used in order to replace the notion of migration. It relates to different scopes of movements, from a local, to a national, and an international or global scope, and furthermore, there are different categories of movement, as related to in the definition of De Bruijn, Van Dijk, and Foeken (2001:1): "Mobility as an umbrella term encompasses all types of movement including travel, exploration, migration, tourism, refugeeism, pastoralism, nomadism, pilgrimage and trade". Approaches to mobility have become popular in order to transcend older approaches emphasizing deficiency and the shortcomings of the view of mobility as an aberration of normality, whereas being sedentary was considered as habitual and the prevailing state of human's being in the world (compare Hahn & Klute 2007). It is also shifting away from the notion that mobility means to be perceived as a rupture, "as a problem", but rather mobility is seen as a part of everyday life and making a livelihood. Contrary to the deficiency model, Hahn (2007) speaks about migration as a strategy to complement space, as an enlarged experience by living in more than one locality (Hahn 2007:150, compare De Bruijn 2007, Cresswell 2002), as Pfaff (2007:63) writes about mobility, in the case of Zanzibar traders, as a "way of being in the world", as a necessity and freedom in the traders' lives. Furthermore, the term "mobility" is unconstrained regarding the embedding of movement in spatial and temporal terms: the focus shifts away from a simple movement from A to B in a spatial, and A replaced by B once and for all, in a temporal sense, towards a more encompassing idea of continuous, cyclical or multidimensional movements of people. Furthermore, it encompasses the mental "images" and ideas intrinsic with these practices. "A cultural perception of mobility implies a close reading of people's own understandings of the spaces and places in which they move and the experiences these movements entail." (Bruijn, Van Dijk, Foeken 2001:2). The concept of mobility thus also relates to the agency of subjects, and various "motivations to move" (compare to De Brujin, Van Dijk & Foeken 2001, De Brujin, Van Dijk & Gewald 2007), which consequently also include feelings of immobility (compare Grätz 2010). Different "dimensions of mobility", how I will use them for my work, do influence and permeate each other in different ways: Physical mobility can be aspired to in a sense of a social mobility, to raise in status, which might in turn be enhanced by virtual mobility – media use – and media of communication and information can be a tool to inform, to network, to connect over distance, and to trigger imageries of "the world", which influences in return physical mobility[82].

In transnationalism studies, according to Levitt and Glick-Schiller (2007), there are, on the one hand, individuals embedded in a transnational social field, with existing connections across borders, and on the other hand, transnational activities, maintaining such connections in practice. Transnational activities relate to several exchanges within social networks, as regular communication and interaction between people related to communication through media and remittances, travels, visits, as well as return migration and movements in different directions, for different purposes (Glick-Schiller, Basch & Blanc-Szanton 1992, 1997, Pries 1997, Kennedy & Roudometof 2002). Therefore, cultural belonging and relations to "here and there", to "home" and "diaspora" have gained increased importance accordingly (Mohr de Collado 2005, Ong 1999). They include individual and collective networks and activities, informal or institutionalized migrants

---

[82] Compare to a detailed examination of these different forms of mobility, in the introduction to chapter 2.

organizations related to both, the community of origin and of the "receiving country" (Larsen, Urry, Axhausen 2006:11). Ortiz also relates to the importance of processes of perception and imagination, in the sense that transnational processes and linkages build the basis for everyday life experiences of transnational migrants[83]. For example Smith (1997:198) relates to this by the notion of "transnational lives", often analysed within transnational social spaces or transnational social fields or -spheres (Glick-Schiller et al 1992, Massey et al 1987, Goldring 1997, 1998, Rouse 2004, Guarnizo & Smith 1997:198, Förster 2004, Smart & Smart 1998:105). Notions of the social field, stemming from Bourdieu's concept, have influenced studies of transnationalism. Transnational fields or spheres seem to be useful concepts to explain the global interdependencies of local phenomena, transnational relations and social networks, as well as ideologies of community and solidarity "reaching out in space". Transnational social fields integrate people who themselves never went across the boundaries of their country or region, but are connected to different people at different places in various ways. "Transnational ways of being" relates to social practices of creating and maintaining transnational networks, including the exchange of ideas, information, business, political activities, and consumption. "Transnational ways of belonging" relates to processes of representation, ideology and identity, which people pass through and within they often gear towards "home" and "abroad" as an ideal or concept (Glick-Schiller, Basch & Szanton 1997).

However useful for my work, the concepts of mobility and transnationalism also need to be critically examined. I will relate to mobility as a complex whole of individual and collective decisions, and in this sense to the agency of actors. Youth – as most potential migrants and users of New Media technologies belong to this category in my research context – are attempting to mobility practices as creative strategies (Hahn & Klute 2007:10). However, I intend to emphasize, that mobility or being mobile is dependent on individual and collective circumstances and various underlying restrictions[84].

---

[83] Also Ortiz (2005:28) emphasizes the interrelation of the dynamics of a social network with the capacity for the movement of people belonging or relating to it.

[84] Thus, we cannot equate this with mobility in a postmodern sense, that has a connotation of an independent, dynamic, and flexible way of life, in the sense of being relatively independent of constraints, because one can access all kinds of mobility

A popular strain of critique of the concepts transnationalism and mobility goes into a similar direction: A critique of mobility approaches is that they often treat individuals as free floating and overlook the "relational commitments and obligations to family members, partners and friends that connect people and their networks" (Larsen, Urry, Axhausen 2006:11), and as well, political and economic macro constraints that influence the mobility of individuals and groups of people. Similarly, with the concept of "transnationalism", the "trans-" is often related to "a space in between", deterritorialization, spatial unboundedness, related to notions of "flux and flow". Concentration on "trans-" might be sometimes overstated, since processes are directional and people are relating themselves to spaces and places. The dangers of such notions slipping into a merely analytical and intangible level is given and should be antagonized through a differentiated look upon structures and processes on the empirical grounds, regarding agency and an emphasis on people's practices. I want to emphasize here that "place matters": differentiating life-worlds of communication partners have an impact on social relationships, which continue to be negotiated in frameworks of social networks and related moralities. Furthermore, life-worlds are determined by specific conditions.

Further critique of transnationalism relates among others, to the discussion of the role of the Nation-State. Jensen (2004:37,38) and others criticise the concept of transnationality for its strong relation to configurations of Nation-States, which always relate to a physical, territorialized space. In order to deal with such critiques, Appadurai introduces the notion of "translocality", thereby stressing the spatial references of migration, which can be other than the Nation-State (Appadurai 1996:12, compare Hahn 2007:165). The notion of mobility seems to be more open, as it does not make an allusion to "borders" in a physical or political sense. When intending to emphasize long distance or border crossing migration movements, for example, it is often spoken of as "transnational mobility". I intend to show in my book that, on the one hand, that practices and activities, which transcend space occur in a continuum from a local, national, up to a transnational level. Also New

opportunities as transport and communication, and free movement across borders (compare Urry 2002).

Media are used in social relations in presence availability (Giddens 1992) as well as in situations of durable physical absence. On the other hand, transnational relations gain their specific meaning from the perspective of New Media users in Cameroon in an imaginative sense (compare Smith & Guarnizo 1998), due to certain attributions made to faraway foreign countries and dislocations which are implied by political borders.

"Mobility" does not imply a confined space where practices take place, but rather stresses the situational and imaginary construction of space within the minds of mobile subjects[85]. Rather, moving is seen here as a potential and a project, related to a social mobility. In this sense, the notion of mobility can also be related to division, a kind of "mobility divide", or involuntary immobility, whereas "transnationalism" here fails to capture such inversions (Larsen, Urry, Axhausen 2006:53, 54). This is an important point for my research: yet, even though people are involved in "transnational practices", immobility might be felt as a strong all-pervading feeling.

The question arises then how mobility and transnational activities are influenced by New Media of information and communication. I have paid attention to the intersections of New Media use and practices of mobility in the first guiding question and related sub-questions:

1. → What role does New Media (internet and mobile phone) and their opportunities for "liveness" play in relation to practices and imaginaries of mobility?

■ 1.1. How are New Media used intentionally for mobility – and migration – purposes?

■   1.2. What local narratives and imaginaries relate to mobility in the Cameroonian context and how do New Media possibly contribute to them?

Media studies, internet and mobile phone – state of the art

There are several seminal studies regarding media in Africa (compare Hyden, Leslie, Ogundimu 2003, Nyamnjoh 2005a, Beck & Wittmann 2004, Mytton 1983, Armbrust 1996) mostly covering Mass Media,

---

[85] Through the concepts of transnational ways of being and belonging, transnationalism studies give however a compilation of transnational practices, which interrelate with imagination of mobility, but confined to a specific social field.

especially radio, video and TV (e.g. Abu-Lughod 2002, Fardon 2000, Spitulnik 2002a, Larkin 1997). These studies usually discuss Mass Media in regard to cultural hegemony, their role within processes of democratization, appropriation and identity formation (Spitulnik 2002, Larkin 1997, Mankekar 1993). Within media studies, the production side of the media was taken into consideration at first, followed by the analysis of media text, the examination of the audience or recipients and the reception of media contents (Watson 1998, Murphy & Kraidy 2003, Lull 2000). Furthermore, the internet as information media could be treated as part of Mass Media[86] regarding hegemonic tendencies[87]. However, according to Castells (2001) New Media are less mass oriented as their consumers are split up into different segments. Furthermore, there are additional distinct, direct and two-sided communication opportunities and –flows - one-to-one, not only one-to-many and many-to-one: the internet also proves to be a very individualistic media, serving specifically interpersonal communication interests. However, the great agency of New Media users does not oppose to the fact that the internet seems unexpectedly, to be a media which contributes greatly to the reproduction of societal imaginaries. Referring to interrelations of New Media use and mobility, scientific work on the one hand considers New Media to provoke a substitution of mobility (Plaut 2004), and on the other hand, New Media of information and communication is seen as the base for an enhancement of mobility practices (compare Lyons & Urry 2005, Castles 2007). In this book I will not specifically relate to this topic[88], but nevertheless it shows that New Media use enhances mobility and its imagination to a great extent. Regarding interrelations of social relations, networks, mobility and media, works range from assumptions related to the corruption of social relations in a mediated setting and undermining solidarity (compare Putman 2000, Albrow 1997, Cresswell 2002) up to the view of the potential of media practices to strengthen and resurrect social ties (compare McLuhan 1996 (1967)). Here I intend to point out to both tendencies. New Media is the base for an intense being in contact despite spatial distance. However, my findings similarly

---

[86] The phone or letters as mere instruments of communication are not to be considered as Mass Media.

[87] Compare to the conclusion chapter regarding hegemonic tendencies of New Media versus creativity and agency.

[88] Compare to a short reflection on this topic in the conclusion.

raise aspects of tensions and frictions in mediated social relationships, in view of physical absence, differentiating life-worlds of communication partners, imaginations and expectations.

The internet

The issue of ICT-media – International Communication Technologies – and digital media has only slowly been taken up by ethnographers, as it is a relatively new phenomenon whose relevance was continually underestimated in regard to so called "peripheral" societies in which ethnographers traditionally worked in. Especially during the initial stages of research, together with its characteristic as a communication tool, the internet was considered as a virtual space, in which people and machines are networked (Castells 1996). The internet was hereby understood as new social space or context. "Cyberspace" was considered as a more or less "parallel world" in contrast to the "real world" (compare Hakken 1999, Rheingold 1993, Haraway 1991, Bell & Kennedy 2000). "Cyberculture" is the corresponding notion for a new field of research, pointing out a more or less distinct "culture" and formation of communities of internet users within cyberspace (Ardevol 2005). Here, numerous studies of online-identities and online-communities as new forms of identity construction and interaction have been effected (compare Markham 1998, Massey 1997, Thompson 2001, Turkle 1995, Uimonen 2001, Smith & Kollock 1999, Jurriëns 2004). "Hypertext" has been highlighted as describing an associative, cross-linked and non-linear thinking, influenced and facilitated by interactions and technological features (compare Beisswenger 2002, Crystal 2001, Calhoun 1992, Ong 1999). In development discourses the internet has been regarded as a threat as well as a chance. Similarly to media in general, and especially TV, the internet was described and criticized as a hegemonic instrument (compare Silver 2000, Schröder & Voell 2002, Miller & Slater 2001, a.o.).

The internet has certainly been described as a mode of information technology *and* as a communication tool. Empirical ethnographic studies of social practices on internet use, often examine the appropriation of technological features in different societal contexts and the transformation of forms of communication and interaction, as well as in social relations over distance (e.g. Miller & Slater 2001, Schröder & Voell

51

2002, Jules-Rosette 1990, Hawisher & Selfe 2000, Tufte 2000, Wellman & Haythornthwaite 2002, Burrell & Anderson 2008, Tsatsou 2009). Studies regarding processes of cultural negotiation on a societal and individual level seem to be dominant. In particular there have been studies within different societal spheres – for example within ethnic groups in diaspora, partly influenced by studies of migration, and in regard to the constitution of identity and transnationality (Glick Schiller, Basch, Blanc & Szanton 1992, Jackson, Crang & Dwyer 2003, Smith 1998, Thompson 2001), identity and religion (compare Goldlust 2001), "nativeness" or ethnic minorities (compare Landzelius 2006), conflicts (compare Bräuchler 2003), and the use of the internet within economy and crime (compare Rivers 2005, Ndjio 2008, Glickman 2005, Malaquais 2001, Smith 2007). There is a tendency within studies on the internet to give greater emphasis to the internet's use in diaspora than its use among non-migrants[89]. In my research I will relate more heavily to the latter perspective.

The mobile phone

Regarding studies on the phone and mobile phone, historical outlines regarding the landline phone as well as its impact on transformation processes of social communication and sociality have been done (compare Fischer 1992, Baumann & Gold 2000), along with publications which provide an overview regarding the technology of the mobile phone (compare Agar 2003). One special issue in a range of literature is multimedial features which include different opportunities for communication and the transformation of communication by phone as a consequence of these new technological features – for example text messages/SMS and MMS – which have gained significant importance beneath oral communication (Ling 2004 for Norway, Ito et al. 2005 for Japan, both highly industrialized countries, Brown, Green, Harper 2001, Castells, Fernandez-Ardevol, Qui & Sey 2007). There have been discourses on the influence of the mobile phone on social interaction, and corresponding transformation of sociality patterns (compare Horst & Miller 2005, 2006, Nafus & Karina 2002, Brown, Green, Harper 2001, Ling 2004, Ito et al. 2005, Caron & Caronia 2007, Katz & Aakhus 2002,

---

[89] With some exceptions, for example Miller & Slater (2001), Horst & Miller (2006).

Brinkman, De Bruijn & Bilal 2009, Fortunati 2002, Kriem 2009, Humphreys 2005, Sey 2011). Furthermore, studies have been conducted regarding the mobile phone's role within societal spheres, such as in health, education, economy, and crime (Galambos, Abrahamson 2002, Horst & Miller 2006, Navas-Sabater, Dymond & Juntunen 2002), as well as on the impact of mobile phone technology on development processes.

From various sides it has been stressed, that in low-income countries, and more specifically in an African context, the mobile phone is to be considered as *the* future technology[90], rather than the internet (compare to Horst, Miller 2006, Navas-Sabater, Dymond & Juntunen 2002). Its impact is more powerful: to some extent, it is easier to access and affordable, its handling does not need as much know-how. Furthermore, it is a mobile object and, in comparison to the computer, the mobile phone is a personal good, which may have different implications regarding processes of its appropriation, its corresponding use and impact considering the processes of the constitution of identity (compare Hahn 2004, Spittler 2002). Mobile phone technology develops quickly and increasingly transforms mobile phones into multimedia tools, portable personal computers with access to the internet[91]. In most studies on the use of the mobile phones, its use within local communities is emphasized. In my research, I will relate strongly to its use in transnational social relationships and communication.

In order to penetrate, understand, and interpret New Media uses, I need to examine them in a specific locality, thereby integrating the agency of users as well as structural conditions and context of their New Media practices. This leads me to the second guiding question and sub-questions:

2. → How do practices of New Media use impact on sociality and social space(s) at a local level?
■ 2.1. How does the materiality of New Media technologies influence their use, and vice versa?

---

[90] See figures F7, and F8
[91] Yet, concerning my site of research, for certain social groups, the internet is also widely accessible. See figure F9.

■ 2.2. How do mediated and face-to-face social interaction intersect on a local level?

■ 2.3. What narratives of social change are provoked by the materializations of imaginaries?

Characteristics of New Media and impact as social media

New Media set up a veritable new field of media studies: technologically New Media are computerized and access digitalized data. Howells describes New Media as new delivery systems (Howells 2003:221, Sturken & Cartwright 2004, Döring 1999). As such, they can be used to access familiar media forms such as photographs, radio programmes, music, videos and live action. Having said that, the internet is also more than a new gateway to existing forms. The fact that different medial forms have been brought together in a "bundle" of different media, is presumably a main condition for their specific characteristics as "New Media": they are characterized by a high level of multimediality. The switching between different media – for example on the internet – is called intramediality; combinations of different media is related to as intermediality. One of the central characteristics of digitalizing is the reproducibility of content[92], which renders contents therefore more easily accessible. Furthermore, reproducibility moves away the production or use of media contents from professionals towards non-professional media users (Howells 2003:230-232, Spitulnik 2002), providing opportunities for manipulation, transformation, juxtaposition and reworking of content. Overall, this leads to a large increase of media productions of all types, and an acceleration of media news. A central feature of new technologies – the internet and the mobile phone – is their interactivity (Rafaeli 2009, Licoppe 2009, Katz & Rice 2002, Pertierra 2005). Every communication act is an interactive act, is an interaction with communication partners, each adding or transforming of a text or picture, could be seen as an interaction (Hjarvard 2002:81, Miller & Slater 2001).

The technological characteristics of New Media seem to be perfectly adapted for sociality and diverse modes of social interaction and exchanges, and media technologies are purposefully adopted according

---

[92] What was true by the invention of book printing and photography, has been boosted by the internet (Darley 2000:125).

to the user's compulsion for interactivity (Licoppe 2009, Horst & Miller 2006, Krotz 2001). The mobile phone has become, most importantly, a part of people's daily practices of socializing. It gives the impression that, on the one hand, these media fit the needs of people, and are meaningfully integrated in social practices of networking, that have existed before the advent of these media (Horst & Miller 2006, Miller & Slater 2001)[93]. On the other hand, these technologies of communication and information undoubtedly will have had, and have, a decisive impact on sociality, although social transformations might not easily be traced back to singular technologies, but are intertwined with other factors. To these we could generally relate to as globalization processes – without taking the argument any further at this point – which integrate these transformations in a wider scope of discourses, to which I will refer in the conclusion of this book. Here, I will relate to notions of sociality, whose discussion serves as a background for the examination of social transformations through New Media use and mobility in this book.

F7: Phone booth and services
F8: Call box, Bamenda
F9: Imported second hand computers

---

[93] Compare to the conclusion chapter, regarding discourses about the novelty of New Media and its impacts on sociality.

Communities, social networks, and mediated social ties

We can approach the notion of community in three ways (compare e.g. to Crow & Allen 1995, Bell and Newby 1976). First of all, communities, which share a place or territory in a geographical sense, based upon geographical propinquity and co-presence. Secondly, communities, which share a common interest, in the sense of a local social system, with its own set of relationships, groups and institutions. Thirdly, communities and "communion", which could be most simply explained as the sense of a community belonging to a place, group, and ideology or "culture" (Smith 2001), and thus (compare Crow & Allen 1995) a community established through close personal ties, belonging and a sense of duty between its members. Yet "community" is related to several normative attributions as it has been used for example within dichotomies of "community" and "society", "emotional affiliation" vis-à-vis a "rationally motivated group of interest", "mechanical" versus "organic solidarity" for example, and so on (compare Ferdinand Tönnies 1988, Max Weber 1968, Talcott Parsons 1969, Emile Durkheim 1953,1964).

The fact that communities – and families – are not necessarily connected to one place, and can also be dispersed in space, is an idea with a long history – it was already mentioned by Tönnies – but has gained more importance with the emergence of global communication and information media and migration practices. Therefore, community has to be understood in a broad sense, in order to include the conditions of new forms of social and spatial organization (compare to Goldlust 2001, Rheingold 1993, Barnes 1972): more recent understandings of community are not so much related to a nexus of place and social groups, but rather to social practices of members of a community, which again constitute membership of groups. Thus, communities have been empirically approached most commonly by exploring the social relationships and networks between people. Wellmann (1999) distinguishes so-called "weak ties" – casual friends and groups of interest – and "strong ties" – long-term relationships and kin (compare Döring 1999). Furthermore, communities have been looked at regarding a quality of connectedness, which contributes to people's experiences of community (Castells 1996, Hogett 1997, Bukow 2000, Baros 2001, Bian 1999, Lee & Campbell 1999). Qualities that shape communities regarding intrinsic norms and habits could be captured in notions of tolerance, reciprocity, solidarity and trust, conditional for the social cohesion of a community (Smith 2001). However, I want to emphasize in my book, that these qualities can be under considerable pressure when social ties are mediated.

Discourses on communities have often been normative, for example related to the influence of transnational interaction, media use and migration, which have often been seen to cause indirect and mediated social relationships to increase in importance, to the expense of local face-to-face social interaction – something which my findings do not confirm. People have worried about community before the advent of the internet[94], on the one hand, but on the other hand, discourses evolved that the internet would enhance community building (Wilding 2006). Prominently Marshall McLuhan (1999, 1996) has discussed New Media's simulation of face-to-face interaction, bringing along a new sense of community. We have to admit that in fact new forms of social interaction have arisen: the internet offers a wide range of opportunities to connect to different groups of interest, and serves as tool in order to create and maintain structures for community organization, while the mobile phone escalates social networking practices (compare Wellman 1999, Wellman, Carrington & Hall 1988, Horst & Miller 2005, Slater 2005, Levitt & Glick-Schiller 2007, Larsen, Urry & Axhausen 2004, 2006, Wellman & Haythornthwait 2002). I make reference to Slater (2005), Horst & Miller (2005), and Miller & Slater (2001): in particular to their notion of "link-up" and addressing of the interrelation between New Media use and migration, and taking social networking as a continuum from a local to a translocal level. For my work, I concentrate on mediated social ties with those previously known face-to-face,

---

[94] Toennies (1988) relates to transformations from communities (Gemeinschaft) to societies (Gesellschaft) as a decline of "natural bonds of relatedness" to self-interest. This dualism is similar to Durkheim's "mechanical solidarity" based on personal social relationships between individuals, versus "organic solidarity" of modern societies. Weber makes a similar typology, but acknowledges conflict and friction in communal relationships also. According to Weber, motivations are mixed, containing self-interest as well as interests based on a mutual consent of obligations, in the sense that individuals have an interest in the continuity of this social agreement (compare to Weber 1968, Miztal 1996:57).

because it seems that such ties are of greatest importance for New Media users in the context of my research[95].

Direct and indirect social interaction, primary and secondary social ties

Social relations are dependent on and transform with physical presence or absence in a context of social interaction. Underlying here are differentiations of direct and indirect – or mediated – interaction. Direct interaction supposes a shared life-world and co-presence. Direct interaction with social others remains, as Calhoun expresses "emotionally central to people in most advanced modern societies and at the heart of most people's evaluative frameworks" (Calhoun 1991:100). Direct interaction intersects with primary and secondary social relationships. Primary relationships are "emotionally close" social relationships with family and friends, secondary relationships are "emotionally distant" neighbours, work and school mates, and even people with whom one only interacts briefly, in the sense of coincidental encounters, or interactions defined by roles of individuals representing institutions and organizations in our everyday lives[96]. Calhoun states (1991:101) that secondary relationships in direct social interaction are marked by the "potential for expansion direct interaction provides", and he even goes further in the sense that sharing a life-world always provides direct social interaction as a potentiality: "The world of mere contemporaries, those with whom we are not at the present time in contact, is nonetheless itself defined by the face-to-face situation that remains a possibility". This is also reflected in Giddens' (1984) "presence availability", as a potential to reach out to others, which depends on spatial proximity (compare Endress & Srubar 2003, Raab 2008, Berger & Luckmann 1980, Calhoun 1991). Schütz relates to a "vividness" of face-to-face situations, as crucial for the directness of experience, regardless of whether the apprehension of the other is central or peripheral (Schütz 1967:219). In this sense, within direct social

---

[95] For this book, I will differentiate the following terms: face-to-face social ties in the sense of direct social interaction, mediated social ties in the sense of prior face-to-face social relationships, and online social ties in the sense of online social interaction without face-to-face knowledge of each other. The latter will only play a minor role in this book.

[96] Compare to forms of trust, personal and social trust, in chapter 5.

58

interaction, primary and secondary relationships mark a "continuum of decreasing closeness, multiplexy, and completeness of grasp of the other" (Calhoun 1991:101).

Indirect or mediated social relationships are also contained in such a continuum of decreasing closeness from primary to secondary relationships: they integrate the wide range of primary social relationships with those emotionally close, and more distant social others or representatives of functional roles, in mediated interaction. The realms of indirect and direct social interaction are intertwined: indirect relationships in the realm of primary social relations are still built in physical co-presence at a moment in time. Furthermore, the potentiality of social interaction can even be transcended to indirect social relations, in the sense that people can build intimate indirect relationships by means of communication technology, as for example in chat rooms: a sense of online presence availability can thus also be achieved through media to a certain extent.

Undoubtedly, the part of mediated social relationships in "complex" societies has increased considerably, maintained by means of communication and transport, enhancing far flung networks of social relations in space – also necessary in order to integrate into economic and political systems and institutions which are increasingly transnational (Calhoun 1991:102). Through electronic media, spatial proximity has become less important, connectivity depends instead on accessibility and mastering of media technologies than on space[97]. However, social interaction is situated in space and time, be it face-to-face or technologically mediated (Giddens 1984:110).

Social norms and roles – situational and reproduced in social interaction

For Heinrich Popitz (2006), social norms are the basis for the social existence of human beings; he has worked out the central principles of construction, by which social norms are built. These are based on a triple typology of actions, situations and persons as well as a third authority, which is intervening and sanctioning in the case of a violation

---

[97] For example, one might more easily communicate with an uncle in the US who is constantly online, than with a friend in Yaoundé who hardly uses the internet, and one cannot always call him, because it might be too expensive.

of norms. Popitz points to inner disparities, and he hereby stresses the diversity and variability of social norms, the creativity and productivity of people. Berger and Luckmann (1969), proponents of a phenomenological view, relating to a cognitive order of processes of norm building, highlight processes, which constitute morality and societal consent. Processes of typification, as a coherent perception of social reality beyond individual phenomena in the sense of a cognitive template, occur related to human action in a specific society. Meaningful social action could be viewed as an anticipation of the future in reflexive orientation to social others' actions and constructions of meaning. Thereby, for phenomenologists, in particular contained in Schütz's life-world, the anticipation of the future is based on an ontological security of the duration of life-world including its organization (compare Giddens 1991), values and codes. Processes of habituation are then based on actions, which consolidate into models of action due to repetition, which can be followed by an institutionalization of the same – likewise institutions or normative regulations direct the actions of people. Coming along with the institutionalization is the need for legitimization, in the form of cognitive and normative interpretations legitimizing a specific morality. Morality and social order are part of social knowledge, which is acquired in processes of socialization within a society, and an internalization in which an individual takes over a social and moral world, rendering the base for the acquisition of a meaningful societal reality. As meaning is continuously constructed and re-evaluated, socialization is in this sense never total and complete (Stegmaier 2008:269). Norms can thus not just be seen as institutionalized entities, but norms and their institutions are continuously produced and reproduced by agents in inter-subjective exchange, and they cannot provide normative direction in all situations of life, but there is a realm of insecurity and precision lacking, which is left to be meaningfully interpreted, even more so under conditions of physical absence and mediated social interaction (Stegmaier 2008:265).

The notion of „role" (compare to Goffmann 1973) is derived from the notion of „norm", and is a condensed and specified form of what is signified by "norm". Each individual in a society is a member of different social units and therefore an agent of different social roles, which can also result in a conflicting negotiation of norms. According to

60

Zifonum (2008:314,315), life-worlds are fragmented in sub-life-worlds related to different roles, which are split into various sequences: the performance of different roles is ordered in time. Roles are performed situationally, or roles are situated (Goffmann1973:109), and they attain their legitimacy in a certain time-space and social context. Social roles are encompassed in activities, in the sense of normative expectations and duties. Roles are related to positions in society, or to profession, and status, being a mother, a chief, a blacksmith, and so on. In this sense, individuals can assume various roles (Goffman 1973:95). Roles often overlap in certain circumstances, which depends on the individual's prioritization of roles, or societal expectations and valuations: for example a male adult is a father in his compound, but an elder in local royal associations. Moreover, being a father is a base in order to take over the position of an elder in the palace. Furthermore, roles are attributed to symbols of status as expressions of these roles in terms of conduct and style of clothes, and so on, and roles are furthermore hierarchically positioned within society and valuated respectively. The fragmentation of individuals into different roles relating to place and time sequences, leads Goffman to relate it to as a performance of roles with a role-playing game - Rollenspiel. Social roles also relate to social norms, which are interdependent of these roles and social positions (Goffman 1973:96).

Within mediated social interaction, roles have an important stance, including over distance and under conditions of lacking co-presence. From this point of view, the valuation of roles and norms tend to be maintained in social interaction with those, who are not physically present. However, situated roles become more prone to negotiation and they can gain new meaning and importance. Roles are linked up with positions and notions of solidarity and reciprocity, as well as different forms of capital (Bourdieu 1986, 1998). I mainly relate to relations between migrants and non-migrants here (compare Bukow 2000, Baros 2001, Wilding 2006, Fleischer 2006, Miztal 1996, Faist 1997, Coleman 1990, Nyamnjoh 2005, Diefenbach & Nauck 1997, Massey et al. 1998, Espinosa & Massey 1997). For example, a young man abroad is still a younger brother as a socially and hierarchically subordinated member of a wider social network of kin. Nevertheless, his position as a migrant abroad can induce an ambivalent relationship towards his kin by, on the one hand, being constrained to fulfil moral notions of solidarity, but on the other hand, his situation places him in new frameworks of status, as he slips into the role of a provider, despite his young age and deviant social position. He could use this opportunity to on his part claim a more important role and position within his family depending on him. Such issues would be negotiated in mediated communication and social interaction, which could accordingly lead to frictions and tensions in transnational social ties (compare Massey & Espinosa 1997, Faist 1997, Baros 2001, Warnier 1993a, Geschiere 1997, Geschiere & Fisiy 1994, Nyamnjoh 2005, Bayart 1993). Here, Gidden's notion of "conduct and tact" (1984) can be adopted usefully in order to describe such negotiations of social interaction and transformations in mediated communication (compare to Miztal 1996, Rasmussen 2002, Hjavard 2002, Auhagen 2002, Wilding 2006, Licoppe 2009, Hantel-Quitmann & Kastner 2002, Emirbayer & Mische 1998). This leads to the third guiding question of my research:

3. → How do users of the internet and mobile phones negotiate "liveness" in mediated social relationships?
■ 3.1. How do practices of sociality and solidarity transform from a local to a translocal level?
■ 3.2. How are qualities attributed to social ties transforming and how are they evaluated when mediated?
■ 3.3. What kind of frictions and tensions can occur in social relationships due to mediation?

Outlook

In the following chapters of this book I intend to develop my central arguments and answer my research questions in relation to my empirical findings. The following second chapter deals with mobility practices and how different forms of New Media are being adopted by youth, by orienting their use to these mobility aims, and related imaginaries and discourses in regard mobility. The third chapter will then examine the local impacts of these practices, how these are reflected materially in the local sphere, including a detailed description of the materiality of New Media, furthermore relating to local performances and discourses. In the fourth chapter I will relate to social networking practices on a local and translocal level through New Media, and notions of solidarity. The fifth chapter will explore communication practices and the negotiation of qualities of social ties in mediated social interaction, and in the sixth chapter I will look at occurring frictions and tensions in communication, due to mediated interaction and physical dislocation. The seventh chapter will be dedicated to fieldwork methods and reflections, as well as to collaboration with my research partner Primus Tazanu, and the concluding chapter will relate the findings more widely to discourses about social transformations.

2

Anxiety of mobility, New Media use, and imaginations of a "good life"

I have already highlighted the pervasiveness of New Media such as the internet and mobile phones in Bamenda. They play an important role in people's lives and their practices of social networking and their felt connection to "the world". Practices of New Media use and mobility are spatial practices, which have several dimensions: youths' seeking a position in society and their dreams of success are very much tied to imaginations of "greener pastures", related to migration in a sense of a physical mobility. Along the lines of youth's defining their social position and potential vis-à-vis others and pursuing projects of a "good life"[98], New Media offers opportunities to connect to physically present and absent others alike. Furthermore, New Media offers a range of imaginations of possible lives (Appadurai 1996), which in return serve as orientation for intentional practices of connecting to opportunities in the sense of a virtual mobility, in practices of media use. Not only that, however, being connected and engaging in the use of New Media is also taken by youth as an opportunity to shape their identity and an indicator to their potential for becoming successful and respected individuals in the sense of social mobility[99].

In this chapter I will explore how New Media and practices, and imaginations of mobility are interrelated. I will look at mobility as physical mobility, and of how dreams of social mobility intersect with it. This chapter thus deals with the first guiding question and related sub-questions: What role do the internet and mobile phones and their opportunities for "liveness" play in relation to practices and imaginaries of mobility? In order to explore the answers to the sub-questions, I intend to examine how New Media relates to mobility in particular, how

---

[98] I will come back to these notions of "good life", and „greener pastures" later in this chapter.

[99] Habitual, projective, and practical-evaluative agentic orientations are intertwined with the dimensions of mobility.

they are intentionally used for migration purposes respectively, and how they contribute to local imaginations of mobility.

Regarding imagination[100], this chapter furthermore relates to the second guiding question, regarding how New Media use impact on local social spaces, and thus, how imageries and imaginaries are produced and reproduced on a local level, as well as the production of imagined spaces. Appadurai calls migration and media the two major "forces" of transformation, and imagination their joint effect (Appadurai 1996:3). Specifically he calls it "work of the imagination", in the sense, as he explains, of both labour and culturally organized practice (compare Förster 2010). He thereby points to a negotiation between what he calls "sites of agency – the individual – and globally defined fields of possibility" (Appadurai 2009:42). He then rests his claim that imagination has become an important social fact or practice on three distinctions: first, that imagination has become part of the everyday mental "work" of people. Second, that – compared to fantasy – imagination is a basis for action, and third, that imagination is a collective activity, influencing and being influenced by "culture" and society, and thus the gear of social transformation (Appadurai 1996:4-7). Through mediation of diverse ranges of information and imageries, imagination has become even more pervasive in the everyday lives of people. According to Mbembe (2008:110), the "imaginaries of faraway" has undergone a revival in the era of globalization, with an increase of migratory practices and circuits of mediated communication between diaspora communities abroad and "at home". I will explore in the course of this chapter, how these mediations stimulate and lead to local negotiations of imaginaries and imageries, which cannot be viewed as

---

[100] I will need to distinguish between pictures in a material or visual sense as cultural representations, and mental images (Mitchell 1984), which often build on pictures. An image goes beyond a picture in a material sense, since it augments its depiction by individually, socially and culturally induced connotations and interpretations. "Imagery" or "imageries" could be understood as diverging aggregations of "images", which come about in individual's practices of imagining, or related to institutions' conveyance of imageries. These can be differing and even contradictory. An "imaginary" or "imaginaries" are then rather collective normative convictions shared in a society, or how these differ from such ideals. They are deeply embedded in social practices, and process oriented (Förster 2005:14, 2010). "Imagination" relates then to practices of reproducing or contesting societal "imageries" and "imaginaries".

merely a backdrop to the global on a local level, but need to be looked at in their own right (Förster 2010). Liveness as a potential is expressed in practices of reaching out and linking up, and imageries of a "good life" and "greener pastures". Of course, regarding the interrelations between New Media use and dreams of mobility, it is not always possible to separate New Media's influence from the overall context, and furthermore, mobility has a local history. I have concentrated on practices of mobililty, imagination, and interrelations with practices of New Media use especially during my second field stay in 2009, when I conducted interviews with internet and mobile phone users.

I will explore motivations for migration by relating to a sample of youth and their life projects. Consequently, I intend to unravel the notion of a "good life", and the reproduction of imageries through New Media practices, relating to specific appropriations of these technologies (compare Sey 2011). Later I will look into societal imaginaries and show how they consist of juxtapositions, and how a history of imaginaries in the setting might influence today's practices of imagination. Finally, I will examine the imageries related to in local narratives about migration and "white man's kontri", and what role media could play in their production and reproduction. As the issue of imagination is always interrelated with New Media use, imagination will continue to play a role throughout this book. In the next chapter 3, I will relate more specifically to how these imaginaries and imageries of faraway worlds materialize in the local sphere, and how these materializations in return inform imagination. The later chapters will then deal with practices of connecting through New Media, which are consequently guided by imagination.

New Media use and mobility – intentional practices and motivations for migration

Due to a framework of complex conditions, the question of whether the access to New Media can enhance opportunities of migration is difficult to answer. They certainly contribute to the importance rendered to the issue of migration and the impact on people's motivations for the use of these media. In a sense of a "virtual mobility", media can enhance migration or travelling through maintaining and creating contacts, providing the necessary information for such ventures and covering organizing aspects. New Media use, by providing opportunities to enhance and access different forms of capital, could thus lead to a certain "social mobility" – through physical mobility, or on a local level by achieving economic success through business practices, a job, education abroad, profiting from remittances and the status of being part of transnational social networks. Furthermore, as the pathway of migration has increasingly become a difficult endeavour, the use of New Media technologies and the assumed opportunities for connecting could also lead to a feeling of immobility in the sense of making lines of exclusion more perceivable. In this sense, New Media anchor mobility and transnational practices in the locality. In this subchapter I will concentrate on how users of New Media align their New Media practices with migratory ventures, and give examples of how New Media's uses are tied up with migration ventures and their imagination[101].

Connecting practices and an "anxiety of mobility"

I intend to see "being mobile" – and in most cases the imagination of mobility – as coping strategies, as well as dreams, desires, and aims of people to "change their lives for the better", in the context of economic and sometimes social constraints. Migration movements are not new, but might be stimulated further, when disadvantageous economic conditions, New Media technologies, and imaginaries about "faraway places" come together. Furthermore this needs to be seen in the context of striving for local notions of achievement, success, social position and prestige. Pink (2001:113) speaks of transnational intentions, and Horst and Miller (2006) about practices of New Media use as "expansive realization". In their words, with the help of these media, "desires that are historically well established" can be addressed more purposefully with the development of technological possibilities (Horst & Miller 2006:6,7, Massey et al. 1998:47). It is here that a projective dimension, as an orientation to the future, is constructed in line with existing imageries and life projects: the iterational dimension of agency is also important here, since the evaluation and possibility of migration is deeply

---

[101] The next two subchapters thus relate in particular to the first sub-question of guiding question 1, how New Media are intentionally adopted for migration purposes.

embedded in local social imaginaries and sets of action. However, these obtain a new meaning in transforming contexts, such as related to New Media uses. To be connected is an important issue here, to (re-)connect, to stay connected, and to negotiate connections. Practices of connecting are not new, but have become diversified and intensified. Desires for connectedness are accompanied by a prevalent dissatisfaction about one's environment and conditions, in view of parts of the world, where opportunities seem to be abundant. Reaching out and linking up while staying in one place could possibly be seen as a remedy from the feeling of immobility: to link up (compare Horst & Miller 2005, 2006, Miller & Slater 2001), in the sense of staying connected to people who are emotionally close but physically distant, or link up with people from all over the world, and reaching out to a global youth culture, such as music, films, sports, fashion and leisure – a "Western life-style" so to say. Or it means to inform oneself about one's opportunities to move, search for jobs and apply to universities and for visas online, or playing the DV lottery for an American green card.

For most people, moving or migrating are just imaginations, which figure in an array of dreams, or, at best, of plans and attempts. Underpinning the practices of "connecting" is the desire to move, even as a feeling of an "urge". Lindquist (2009) calls it the "anxiety of mobility", and Zymunt Baumann also writes about identity in the "globalizing world" which is increasingly defined by "anxiety". According to Baumann (2008:2), anxiety arises due to difficulties and constraints affecting everyday life, and similarly, due to the openness, freedom of choice and opportunities this interconnected world offers. Anxiety in general is related to insecurities and uncertainties in people's daily lives. However, anxiety does not simply signify "fear", but also has a purposeful aspect of aspiration and desire, to strive or make an effort to reach one's goals and pointing to a kind of courageous attitude. "(…) But the unsteadiness, softness and pliability of things may also trigger ambition and resolve: one can make things better than they are, and need not settle for what there is, since no verdict of nature is final, no resistance of reality is unbreakable. One can now dream of a different life – more decent, bearable and enjoyable." (Baumann 2008:2). I would argue, that there are cultural and social contexts, which are more conducive than others, in order to make people prioritize migration practices over other strategies, even in times and conditions, when opportunities become fewer. Nevertheless, it does not mean that migration is simply a conceivable outcome of investments, for example in New Media use, but youth refer to their life projects (compare Giddens 1984, Schütz 1967) to migration and mobility in differing ways[102].

Selected case studies – different life stories, different orientation towards mobility

At this point, I will introduce four young people, in order to provide examples of the ways in which New Media are adopted in order to pursue their life projects, with due consideration given to their differing life situations, backgrounds, their interests, motifs and imaginations of "the world", and in this sense, interrelations between New Media use, imagining, and dreams of a "good life".

At the time of the interview Lambert was 25 years old, he rented his own small room in town, even though his family stayed in Bamenda as well. He had opened his own barber studio at Savannah Street two years ago. It was doing well, more or less, even though he complained that "business in Bamenda is slow", and he thought about going to Yaoundé to join a female friend who had opened a salon for women. Since he was busy in his studio, he could only go to the cyber café about once or twice a week, for one or two hours. He mainly checked his two email addresses and browsed sites related to his fields of interest, which were, above all, bodybuilding, wrestling and football. He had specific sites related to these issues, where he informed himself. Furthermore, he said he liked to learn about foreign countries. He was also active on social networking sites such as Hi5 and Facebook for the purpose of, as he put it, "getting into contact with people from all over the world". His main motivation for going on the internet on a weekly basis were his friend in Belgium, a cousin in the US, and some classmates in Yaoundé. He said he would like to join his friend in Belgium, who had opened a barber studio there. He was also fond of the US, he had heard that there were many black people and good living standards – "one can really make it there" – but he knew that it would not be easy to obtain a visa, and also

---

[102] I want to state here that I am aware, that not everybody in Bamenda or Cameroon wishes or intends to migrate.

he did not know how to go about it, so it was all only a dream up to that point. His phone was especially important to him because he stored on it many pictures of himself, as well as football – and wrestling stars and bodybuilders. Most of these pictures he had exchanged via Bluetooth with friends, but he also regularly downloaded pictures from the internet. He mainly called within Cameroon, but from time to time he called his friend and cousin abroad for short calls. Nevertheless, even though he said he tried to economize, he would easily spend more than 10,000 FCFA per month on phone credit (Lambert. Interviews 2009/R). Lambert was a regular acquaintance of mine. We met often and I spent time with him in his barber studio.

Claudia was 23 years old, holder of a first degree (BA) from Dschang University –West Province – in linguistics, and at the time of the interview was unemployed, and living with her family in Bamenda. She was simultaneously preparing for exams for the teachers training college (ENS), and for the US Toeffl exam for American English. Therefore, during the period when our interview took place, she went to the cyber café on a daily basis for two to three hours, but she was not always online: she used a CD and headset in order to exercise filling out the test forms for the Toefl. She was very much determined to leave Cameroon, intended to foster her education and to see the world. "By next year, you will not find me anymore here in Bamenda", she said. She was at a turning point in her life, she expressed, and she followed up different options, in order that one might come to fruition. She had a sister and a friend, who lived and worked in the UK - the friend having studied in Sweden previously: They had recommended the universities where they had studied. She recently applied to both of these universities and her sister had filled out the application forms for her. She had also sent money for her passport and a deposit to the universities. In addition to this Claudia then applied to a university in the US. As she stated, she would prefer to go to the US: Her fiancée had been studying there since three years, and although he had just come to visit Cameroon for a few weeks, she was normally only in contact with him via email and phone. It was also her main motivation for using the internet, indeed, before her fiancée moved abroad, she did not go to the cyber café on a regular basis. Now they were sending emails to each other every day, and arranging to meet online, exchanging experiences of their daily lives. He also called her twice a week in order to hear each other's voice. They planned to marry next year, and if she were not be accepted at the university in the US, she would apply for family reunion in the US and study elsewhere in the meantime. She used her phone to communicate with relatives and friends in Cameroon. As she did not earn money, she limited her expenditures for phone credit to a minimum – not exceeding FCFA 5000 per month – and she often waited for others to call her (Claudia. Interviews 2009/F). I did not know Claudia before we met for the interview. She left Bamenda to Yaoundé soon after that, and our contact faded out.

Cletus was aged 26 years, he had studied law at the university of Buea, specializing in intellectual property rights. He followed the online study program of an NGO, which is affiliated to the UN. He was part of an international crew of students, who were working on papers, which were meant to be future directives for policies for the UN and national governments. He worked in a public cyber café on a daily basis for at least four to seven hours. All members of the online study group had to hand in their work once a week, and meet online in order to hold discussions. Furthermore, Cletus was active in discussion forums related to his field of work on an almost daily basis. He also maintained online friendships, most of which were also somehow related to his field of interest. When he had concluded his work for the day he was usually active on Facebook and on messenger chat. He then also allowed himself to browse on a few sites in some leisure related fields of interest, such as football, and the sites of BBC and CNN for news. He had been working for six months in Switzerland at the UN, and he had visited Italy, Malta, and a range of African countries. He said that he loved to live in Cameroon, as it was his home - his family and friends were there, and as long he could be connected to the world despite being based here, he did not see any reason to leave. Only a good job offer would make him to go abroad, he said. Phone calls were important for him rather in relation to local and contacts within Cameroon. He said he did not spend much on calls, but more for internet airtime. For contacts abroad - mostly friends and classmates, no relatives – as well as for friends in Cameroon, who were also often online, he used email and chat – the latter to a greater degree. He was currently living with his family, but after the conclusion of his studies in 2010, unless he would find a job, he would move out (Cletus. Interviews 2009/F). I met Cletus from time to time because he was part of my friends' network. However, he was very mobile within Cameroon, and stayed in Yaoundé for a while, so I did not see him often. He was one of the people I was in regular contact with when I was back in Switzerland however, through chat and Facebook.

Godwill was 27 years old, he was running a store on Commercial Avenue where he sold electronic equipment. The store was a family venture, and, as he explained, the business was going quite well. He had a wife and a small boy, and they lived in a family house in a better off districts of Bamenda. He was proud to be able to provide for his small family, he said. He thought that life could be good in Cameroon, when you have enough money to live without being continuously worried about the future. Since he was his parent's eldest son who stayed in Cameroon, he had a duty to maintain the family business. He hardly visited the cyber café, because he had no time to go there, and not much interest. He had however recently bought a laptop, because he thought he needed some experience of using computers, and his friend had shown him that he could inform himself about electronics equipment on the internet, so he was planning to have internet access at home. So far, his brother, who lived in the US and dealt with electronic equipment, provided him with everything he needed for the business, sending containers with goods from the US, and he also visited Cameroon twice a year. In this sense Godwill did not have an urge to travel himself. If he were to save some money, he would like to go for a business trip to the US, or to China. He stayed in contact with his brother, and friends abroad and in Cameroon only by phone. He spent a lot of money on phone credit, for private use and business issues alike - up to ten thousand FCFA per week (Godwill. Field notes 2009. 28.07.09). I regularly visited Godwill in his shop at Commercial Avenue. I never lead a formal interview with him, but talked to him on an almost daily basis[103].

Life situations, New Media use, and migration purposes

---

[103] The names of interviewees and other informants are changed throughout this book. See appendix for information.

These examples highlight the interrelations between the use of New Media and migration or mobility, which are highly diverse. New Media use does not necessarily need to be in a direct relation to peoples' life projects of their own migration. It is rather a complex arrangement within the framework of the personal, economic and familial situation, and individual preferences and inclinations. Related to this, imaginations of opportunities vary and have different references, such as for Godwill, "leading a good life" in Cameroon was very possible, others just dreamt of leaving, as Lambert for example, and again others were undertaking concrete steps[104], such as Claudia. In Cletus' example, he was already extremely mobile – which is of course an exception. Very often migrants – or others who are economically viable – contribute to people's personal satisfaction in one way or another. In Godwill's case, his brother was the driving force in the family business, whereas Cletus was involved in an international network of professional colleagues. Others are strongly involved in attempts to migrate and planning for their future, as in the case of Claudia, whose migrant sister supported her financially and in terms of assistance. For most, however, being connected to people abroad is related to just dreaming of a better life, as in Lambert's case.

F10: Taking passport pictures for the DV lottery at a cyber café
F11: Signboard relating to migration ventures, Bamenda

Social networking through the internet and mobile phones was for all interviewees the main motivation for using these forms of media, and for some, such as Claudia even a decisive reason to start using the

---

[104] See figures F10 and F11, related to undertaking concrete steps for migration ventures.

74

internet. Regarding maintaining personal social relationships the mobile phone is even more important in order to enhance emotional closeness, such as Claudia and her fiancé, or simply because communication partners prefer mobile phone as a technology which is more easily accessible for them, as in Godwill's case. Being less internet literate does not necessarily mean that people have smaller social networks – I have mentioned Godwill's vast business network. However, for some internet literates, using opportunities of online phone calls and chat enhances the maintenance of large social networks, as Cletus' example shows. Internet and mobile phone's communicative features are used for both, contacts on a local and translocal level. These networking strategies could, but have not necessarily to be related to migration purposes, as in Godwill's and Cletus' case. But, as we have seen in Claudia's and Lambert's examples, these contacts could also have the potential to enhance dreams of migration. The internet and mediated interpersonal ties also motivate reaching out to media contents, such as sites of global youth culture and information about countries enhanced Lambert's dreams of travelling abroad, and for Claudia the internet served as means to organize her potential migration venture. For Cletus, the internet provided opportunities in an educational and professional sense, and Godwill acknowledged the potential the internet could have for his business. In short, the use of New Media is not necessarily tied to one's own migration venture, but can be used as a means to support it.

Motivations for migration - Claiming a multi perspective approach

The question of why people move is of course central. Even though we should not see migration movements as exceptions, we have to be aware, that people do usually not migrate just out of a caprice. When taking into consideration local conditions of everyday lives of people in a certain context, various factors should be related to. Regarding building motivations, economic circumstances indeed play a central role, also in the context of my research.

Nevertheless, economic disparities alone are not sufficient enough to explain international migration. Why people migrate should be explained in multiple ways: the motivations can refer to different specific values and aims, such as material wellbeing, social status, comfort, autonomy, and security (Faist 1997:66). In this sense, Hannken refers to incentives for migration, be it "an opportunity" – mostly for a job – a risk to one's existence, and, most important for my work, a "graduell unterschiedliches "unbefriedigt sein" mit der eigenen Situation aus persönlichen und/oder strukturellen Gründen" (Hannken 2003:14,15, Braune 2008). Or, as Treibel (1990:44) puts it, the individual feels deprived vis-à-vis real or imaginary social reference groups. Shmuel Eisenstadt subsumes motivations for migration as "a feeling of frustration and inadequacy, and lack of gratification" (Eisenstadt 1954:2)[105]. Such individual motivations should be approached within frameworks of opportunities, as social networks, as well as drawbacks, such as financial capital, determining macro structures, visa regulations, and labour markets. So-called "meso" approaches, combining macro and micro, collective and individual approaches, are eligible (compare Hannken 2003:16, 17).

Interestingly, the motivations for youth migrating to work in the plantations, examined by Arderner and Warmington (1960) in the 1950's, seem to be relatively similar to the motivations stimulating youth today. Among motivations were reasons such as the desire for marriage, to maintain themselves, pay back investments made on their education, supporting their family, and to have life experiences, "to see places" (Ardener & Warmington 1960:249-253)[106]. Tardits states about young Bamiléké migrants in cities in the 1940ies: "Il est evident que les jeunes émigrés formulent leurs espoirs en termes vagues, il sera question de "voir la ville", "d'être libre", de "mieux vivre" (…)". (Tardits 1960:95). However, not only the possibility to gain economic independence, but also local conditions in the Grassfields have contributed to migration[107].

The endeavour for the better: the notion of a "good life"

In the Cameroonian context, motives for migration build a "package" of "hopes and desires", and could thus be examined by using

---

[105] The distinction of "enforced" and voluntary migration (compare Hannken 2003:16, Treibel 1990:20), the first relating to political, the second to economic circumstances, seems not to be useful for the purpose of differentiating between motivations for migration, in order to understand people's anxiety for mobility in the Cameroonian context.

[106] Ardener and Warmington (1960) made 1622 questionnaire interviews among workers in plantations in the 1950s.

[107] Compare to chapter 1.

the notion of a "good life" expressing an incentive for migration movements. It points to a situational and cultural interpretation of what a "good life" could be, and it emphasizes a migrant's agency and range of motivations (Levitt, Glick-Schiller 2007:185, Morawska 2003). Kaberry (1952:71,152) relates to the rural Nso', and their views of a "good life" in the 1940's, which contained good health, being strong in order to work and provide the family with basic necessities, including a "good" wife or husband. Central desires have not significantly changed, what might have changed are the opportunities of imagining a "larger range of possible lives", in Appadurai's words. I intend to unravel the notion of the "good life" in three dimensions, as it has occurred in informal conversations and interviews with youth, and I intend to relate the notion specifically to the option of migration. I have called the first dimension "taking over responsibility", the second "acquiring status", and the third "taking part in consumption".

The first dimension of the notion of a "good life" is to assume responsibility towards others. "My family comes first, I want my people to be fine, and I want to contribute to this" (Divine. Interviews 2009/F). The interviewees related to the motivation of "giving something back" in the sense of "succeeding in order to repay". It highlights a duty to reward the investments most families make in members in order to make their migration possible: "My mother sacrifices so much for me, she invests all she has in me, for that I can go abroad" (Christian. Interviews 2009/R). Migrant's duty for solidarity vis-à-vis the family is emphasized: "One cannot live abroad in luxury, and at home the family is suffering" (Christian. Interviews 2009/R). The notion of returning back to Cameroon is widespread: "I would always come back, I could not stay there forever" (Ivo. Interviews 2009/R). The truly good life is for most seen in Cameroon: While being abroad, migrants should not forget about their identity, where they come from, belong and originate. An issue here is to be able to start a family, however, due to the young age of the aspiring migrants, this seems not to be their primary priority at the time, but rather a plan for the future.

Regarding the second dimension of the notion of the "good life", acquiring status in migration is seen as an important motivation, especially for young people, as other opportunities to be successful are denied to them. As Comaroff and Comaroff (1999:291) state, movement adds value and status, however, a status, which is often highly contested[108]. A formation abroad is seen as loaded with prestige. Migration does not only serve to be able to earn money in order to support the family, but also increases the chances for a job when returning to Cameroon: „You are taken as a different person when you have studied abroad" (Ivo. Interviews 2009/R). In this context, status is achieved through taking part, and taking part means to play a role in redistribution in the sense of becoming a provider, which is for many related to taking over social responsibility (compare Goheen 1996, Argenti 2007, Fokwang 2008). "I have to play a role in my family, because I want to become somebody one day..." (Moussa. Interviews 2010/11). Status is achieved within the reference group, and thus, it is also based on the notion of return. Furthermore, since status needs to be demonstrated to others, it cannot be separated from consumerism.

The third dimension, taking part in consumption, points out to highly ambivalent valuations of individual success. Especially in relation to migrants and as well scammers, popular discourses point to attitudes of consumption in a rather ambiguous sense. Copying a "Western lifestyle" (compare Massey et al 1998:47) and respective behaviour as a marker of success, related to migration and transnational spaces, are seen by the interviewees as pointing to both the potential but also the dangers and decadency of the "West". "They want to copy these rap stars, and they want to go to the US as they think that everybody over there lives like this, in luxury" (Jack. Interviews 2009/FS). Consumer goods and demonstrating economic viability have always been expressions of status and capability, becoming potentially achievable, for example by investing in a migration venture. All the three dimensions are strongly related to and intersect with each other: status is acquired when taking over responsibility, and consumption is an expression of status.

Depending on the individual, the different aspects are more or less emphasized: the fulfilment of the desire to "lead a good life", is not for all necessarily tied to the option of a migration venture, as I have already tried to highlight when relating to the sample of interviewees. Whereas some clearly direct their energies towards migration, most follow various "threads of opportunities", in order to have various options while hoping that one of them "turns out favourable", as we have seen in the

---

[108] See later in this chapter and the next chapter.

example of Claudia. In this sense young people would apply for the few jobs available in Cameroon, which are mostly government jobs, as teachers or in the army or police force, or gather capital in order to start a business, but at the same time, they would probe prospects related to further education abroad. Additionally, the idea of „If you have money you can live a good life here in Cameroon, then there is no need to go abroad" (Ivo. Interviews 2009/R), is prevalent among youth, which is why Godwill did not intend to migrate.

In some cases interviewees addressed the great importance given to migration ventures in a mocking sense: „Many young people are just sitting and wondering why there is no money coming... they are just lazy. Instead of working they just dream about going abroad" (Clovis. Interviews 2009/R). Criticism of the excessive emphasis placed on migration by some was particularly highlighted by those who were in a more or less stable position, and did not pursue migration as anxiously as others. However, these people also tend to relate their life projects vis-à-vis migration and realms of potentiality. A young man, who had his own business, as well as a well off uncle in the US, stated: "I go make my own America here in Cameroon" (Mike. Field notes 2009. 20.07.09).

Aims, causes and intentional practice of New Media use

The desire to migrate often stands behind intentional practices (compare Giddens 1984) of the appropriation of New Media. Förster (2010) relates to intentionality as set up by habitual iteration, judgement and imagination, resulting in a creative process of negotiating sets of intentional practices (compare Emirbayer & Mische 1998). I have already pointed to the fact, that migration has a long history in the local context, and the symbolic evaluation of migration is given a great importance[109]. We could thus assume that practices oriented to migration ventures have become deeply engrained in society's framework for action, directed at a betterment of people's life conditions. Therefore, the motivations are often not referred to directly, but taken for granted. They are however also part of diverse narratives in the sense of "complaints", as they were expressed to me as an outsider. The causes underlying the desire to migrate are summed up by interviewees in the utterance of a general feeling of dissatisfaction with

---

[109] Compare to the previous chapter.

their life situation: "You have seen the conditions here in Cameroon…", or "What are my options here", in the sense of presupposing clear evidence of disadvantages. This generalized "feeling of dissatisfaction" could be differentiated into various and more specific reasons, why people are not satisfied with their current situation. It expresses an insecurity regarding income, continuously tapping resources through social networks and thus being dependent on others for even basic needs. It relates to a shameful feeling of being "useless" and involuntarily idle, and neither being in the position of supporting the family, nor taking part in consumption and competing with peers in that regard. It also relates to structural conditions: "The situation in Cameroon will never change"[110], or to a global disequilibrium. The afore mentioned causes which provoke the desire for migration do not only consist of conditions not met in youth's lives, but they are also related to imaginations of opportunities abroad.

When asking specifically for "motifs" or "aims" which are related to the will or desire to transcend these feelings of dissatisfaction, then youth's generalized statements could be subsumed in the notion of the desire to lead a "good life" or a "better life" than the one they are currently living. It seems here, that even though these imaginations are oriented to future trajectories of youth's life projects, these imaginations are also deeply rooted in socio-historical collective imaginaries. Despite the "fuzziness" of the aims, they are grounded on strong convictions about migrating as *the* possible solution to people's problems, and the "West" providing the respective conditions as a base or location for the fulfilment of the dream of this better life[111]. According to Castles (2007:37) in communities or societies, in which migration is rendered high importance, migration ventures are often not evaluated on a base of precise goals, but migrating is seen as self-evident under local conditions. The motivations could thus rather be captured in practices, which are implicitly directed towards them, in different degrees of determination

---

[110] This statement refers to the dissatisfying political situation in Cameroon. Fokwang points to attempts for migration as a kind of political protest, or an expression of dissatisfaction in the view of declining perspectives for youth, which are also related to an underdeveloped possibility of political participation (Fokwang 2008:247).

[111] I will relate to these underlying imaginaries in the coming subchapters of this chapter.

and intentionality – also continuously transforming and adapting to current life situations and evolving lines of sight. According to Husserl (compare McIntyre & Woodruff 1989) intentionality is interrelated with an interpretative approach to reality, intentions are not always clearly defined and rigidly pursued, and contain alternative interpretations. For most, as for Lambert, New Media use is not only seen as a means to an end, but an end in itself: many young people state that they like to have the feeling of being connected, informed, as well as "having fun" online and with peers in the cyber café. In this sense, New Media activities are not always related to clearly defined aims, but they are highly opportunistic. However, among those who adopt concrete steps towards the aim of migration, as Claudia, being determined to one's aims is important[112], thereby, the determination is itself valued as more decisive than a clear knowledge of how to go about one's aims. A strong orientation towards migration is nothing new in the context of Bamenda: with access to New Media, intentional practices oriented towards migration do not seem to have substantially transformed regarding the emphasis on social networking and accessing social capital, but they are integrated into a wider range of options and points of reference.

The (re-)production of imaginaries and narratives in the local setting

As I have already highlighted, New Media and related practices of their use provide a wide range of possibilities to pursue aims and dreams of migration and success. It seems that here, New Media do not substantially challenge societal imaginaries, but rather contribute to enhancing practices of imagining. In this subchapter I intend to show how media use and imagination are linked to mobility, and imageries of the "good life" and "the West". Most importantly, such collective imagination is produced and reproduced through circuits of communication and in transnational social networks. Another part is related to the imageries, which are conveyed through diverse media productions. The great importance of markers of status and success is also perceivable regarding consumption of "Western" consumer goods, and as well as being expressions of self-assertion.

---

[112] See later in this chapter.

The importance of social ties for initiating migration and providing information

According to Pries (1997), international migration flows are only understandable by integrating studies of social networks and ties and what he calls "kumulative Verursachungsdynamik" (cumulative causal dynamics), once migration movements are set: transnational ties reduce risks and lower the costs of further migrations. Massey (1994) speaks of communities, which are dependent on migration of at least some of their members, and particularly on emigrant's remittances. Migration becomes a self-perpetuating issue (compare Smith 1998:200). A migration venture is an investment in social networks, and also social networks are fostered by New Media of communication. A large part of the support rendered by families for one of their members venturing migration, is, however, effected on a local level, in particular when the pooling of finances is concerned[113]. Regarding the organizational aspects, migrants who are already abroad play a major role, in instances when aspiring migrants have to go through the same processes as they did when migrating. Assistance from migrants abroad, apart from providing financial means, could range from helping somebody with visa issues, writing invitation letters or providing assistance to fill out forms for applications for educational purposes, up to provide help to find jobs abroad. Thus, related to migration ventures, it seems to be crucial, whether people dispose over "social capital" in strong transnational social ties[114] (Nyamnjoh 2005, Fokwang 2008). In this sense, people's relation to kin abroad shapes their perspectives and options related to migration (Thorsen 2007:176).

An important way of rendering or receiving - non-material - support is providing or receiving information about migration ventures[115]. I often noticed that people used links obtained through assumingly

---

[113] I will come back to the role of social networking and information, as well as their role in migration decisions and pooling of financial means for migration ventures, in chapter 4.

[114] Not everybody has a close and trustworthy relative or friend abroad which points to the importance of "migration brokers", and "migration scams" on a local level. See later in this chapter.

[115] More than a third of the interviewees said that they had once or several times received such help. See chapter 4.

"knowledgeable others", to inform themselves about a certain topic rather than using a search engine. It is of course also a matter of not being experienced to use search engines on the internet, but, people seem to be inclined to ask somebody first for a valuable link. As an interviewee expressed: "I sometimes use Google to search for information. I then get hundreds of links. How can I find out which one of these links is really relevant? I'd rather then ask a friend if he knows a useful site" (Christian. Interviews 2009/R)[116]. In this sense, personal social ties impact on people's scope of information, despite holding dangers of being duped by apparently trustworthy persons, and the wide spread frauds which occurs on the internet. Many people decide on a specific destination for their migration venture on the basis of social contacts they have in the potential receiving country, or the country is specifically recommended to them by a trusted person rather than on the knowledge they have about the respective destination. The same is true for the migration process, which is often guided by knowledgeable migrants, leading to the tendency that there is not necessarily much concrete information pervading the local setting. Contributing to this obscurity is that migration is a highly sensitive topic, involving high levels of secrecy[117]. The organization of migration ventures is thus strongly supported by the availability of New Media, but their importance lays rather in their ability to provide opportunities to stay in contact with migrants abroad, than in their being a means to deliberately search for information.

---

[116] Searching through Google however also relates to the constraint of slow lines and connections; clicking on several links Google provides takes a lot of airtime. Having a direct – and useful – link is therefore an advantage.

[117] This related to jealousy and malevolent acts towards those who intend to migrate and returning migrants, in the sense of attacking and spiritually destroying their venture or success (Pelican 2010). See chapter 6.

F12: Posters in a [male] youth's home
F13: Signboard, Bamenda

The reproduction of prevalent imageries in popular mass media and publicity

Imageries and imaginaries of "the West", "modernity", and a "good life" are permeating the public sphere mediated through diverse media. In Bamenda, one can buy posters to beautify homes – of gardens, well-cultivated lawns and flowers, beautiful villas, of cars, film- and music-stars, and, often white couples in romantic situations[118]. Similar images can be chosen in photography studios. Prominent are landscapes and skylines of famous cities serving as backdrops. Film and video are an important part of visual popular culture in the public sphere, be they Nigerian, Hollywood or of other provenance. Regarding TV series, apart from Nigerian productions, series from South America, Venezuela, Brazil and the US are also popular, broadcast by African and Cameroonian TV stations[119]. In these films and series, imageries of a "good life" are accompanied by the disguise of material enchantment, houses, cars and well-dressed people. In regards to music, apart from globally popular hits and interpreters, African music industries from Nigeria and other popular African music styles[120] form the great part of the music consumed in the local context. Many local music productions – as visible in the respective music videos – allude to American Hip Hop and Rap stars and represent popular "Western" symbols of success and status, but they also include different elements, such as traditional

---

[118] See figure F12

[119] Anglophones prefer series in English language from Nigeria or Ghana, or those with an English overtone.

[120] As Zouk from Congo, Coupé Décallé from Ivory Coast and Makossa, Bikutsi and others from Cameroon.

authorities and scenes from the countryside. Copies of music and videos[121] on CDs and DVDs are sold in various places and by mobile peddlers: they are cheap, even though often of minor quality. Entertainment through music, video and film is constantly present in public spaces, such as in bars, drinking spots and shops, and in people's homes (Warnier 1993a:182, Fokwang 2008), as background entertainment, or actively consumed. Access to such media is easy, widely spread, and integrates all age and social groups.

Publicity is part of the visual imageries in the urban public sphere, competing for potential customers as part of marketing strategies. Thereby, local as well as international publicity propagate social imaginaries such as accessing a "good life" though the consumption of "modern" or "global" goods or services. Regarding local businesses, potential customers are attracted mainly through well visible painted signboards, catchy names and slogans. The names of the enterprises or their slogans often point out to connotations of success and modernity, alluding to potential realms of the Western world, as a notion of "German" suggest high quality goods, or the advertising of goods by calling them "American", or "European" goods[122]. The addition of "modern", "international", or "global" in names and slogans[123], or allusions to global places, from additions as "world" and "planet", up to famous metropolitan cities, such as London, Paris, or New York, serve as attraction. Similarly do the allusions to locations of power and formal institutions, such as business enterprises and government administration, names or slogans containing notions such as "enterprise", "corporate", "embassy", and "ministry", or they allude to transcending borders and taking part in a global communitas of consumerism, by adding the notion "connect", "linked", and "mobile". Another aspect is how New Media are attributed with the images they spread in the local context, in the sense that they are themselves seen as representations of imageries of a "good life" and faraway places. There are many examples of

---

[121] In an economy of pirate copies (compare Larkin 2008, Künzler 2009), there exists no differentiation between legal and illegal copying of music and film.

[122] As for example "German Mecanographic Werkstatt", "Swidish haircut", "Swiss shoes", "British library", and so on. See figure F13 which depicts a signboard with the addition: "global".

[123] For example "Modern Furniture", "Modern Electronic Equipment", "Modern Fashion", "Global Bizz", "Global Enterprise".

publicity for telecommunication, mainly by the providers MTN, Orange, and some of Camtel, visible in the public sphere, on posters and signboards, on every strategically useful corner, square, main streets in cities, and along interurban highways throughout the country, furthermore in advertisements in media as newspapers, local radio and TV (compare Caron & Caronia 2007)[124]. Technologies are associated with their ability to enhance and facilitate daily life, promising connectedness, pleasure, lifestyle and status[125]. On publicity posters, for example a young handsome man rests his elbow on the chassis of a sports car while manipulating his mobile phone, a happy and good looking couple enthusiastically enters using the internet on their laptop, Cameroonian football stars as icons of success appear in adverts for Orange, and young people seem to enjoy communality by profiting from the latest offers to enhance their mobile, fashionable and enchanting life style. In this sense, the local sphere is saturated with mediations of narratives and images – visible and audible – of "worlds of potential", and a "good life" of success, and sometimes of decadency. It is thereby difficult, or even impossible, to ascribe images to specific media, as they are spread through diverse media, and diverse media are enhancing and repeating, emulating and supplementing each other in the production of imageries, and thus also imaginaries within public spheres but also private realms, for individuals and collectives.

F14, 15, 16: Publicity for MTN and Orange, promising enchantment and unlimited connecting

---

[124] Apart from other publicities, such as alcoholic beverages, money transfer institutes or lotteries, the telecommunication providers were the main publicity makers in the public sphere in Bamenda.

[125] See figures F14, 15, and 16.

"Ordinary consumption" and blurring of symbolic capital

Neubert and Macamo point to the fact, that many African countries do not have to capacity to produce and maintain popular consumer products (Neubert & Macamo 2008:286). Thus, "products of modernity", as they are so called, are usually attributed to "the West" in the minds of people who consume them. "To many Cameroonians, imported goods have become crucial in the definition and expression of their identity" (Geschiere 1997:137). Warnier (1993:164) names it "l'ethos moderniste", which would be shared by the majority of Cameroonians. In pre-colonial and colonial times, such items became attributed with the notion of "white man's goods", as an expression of the prestige of those dealing and acquiring them as a privilege of the powerful (compare Geary 1996).

Consumer culture in Africa has become "ordinary" since some consumer items have been made widely accessible and more or less affordable to a wider group of people, especially since Chinese products have begun to flood the market, which are of cheap quality but affordable for people with scant resources. In addition, less economically viable individuals do now join into the "game of consumption" and status expression, which points to the lack of a distinctive "high culture" of the elites, rather differences are stated in degrees of "more" and "better" (compare Mbembe 2008). Mobile phones, for example, are seen as prestigious items, and continue to be so despite Chinese provenance. Despite their reputation for being low in quality, they instead offer access to diverse technological features - most of the recent phones range in the category of smart phones. Due to their wide distribution, the significance of many so far prestigious consumer items are also however increasingly doubted and brings about the impression of deception, since their consumption has grown to the extent that they have become "ordinary" and cannot necessarily contribute to distinguish social status groups. Nevertheless, being able to compete in spheres of conspicuous consumption is an important issue among youth, and explanations why some are successfully doing so are often related to imageries of realms of affluence of the "Western world", in one way or another.

The power of imaginaries and imageries – creating self-assurance and confidence

In view of the prevalent insecurity of a clear conception of these far away potential worlds and how to reach there, people adopt strategies in order to enhance self-assertion and confidence in their strengths and abilities in their pursuing imaginations of a good life[126]. Förster (2008:8) states, that the way to reach there could be seen as a passage, similar to "rites de passage" from one stage of the life-cycle to the other, as from youth to adulthood (compare Fokwang 2008). It is in an advanced stage, a development and achievement, or a realm where one could "become a man": this alludes to the images of the hunter or warrior, who proves his strength and abilities in a difficult terrain. It seems in this context that the prevalent insecurity, which is reflected in many ways in people's everyday life-worlds – the conditions of livelihood and prospects, the mistrust and duping practices and obstacles related to migration endeavours enhances a notion of a "strong self", also in the sense of circumventing the state, the powerful and the oppressive[127]. It points to the idea of the power of positive thinking (compare Nyamnjoh 2002:628). "One needs to bundle one's energy in order to reach one's aims", and a sense of relying on one's adaptability to changing circumstances, confidence, and fearlessness: "Things can only get better. This makes me strong, I have no fear" (Serge. Interviews 2009/O). Very important is anticipating success, the imagining of a future, when one would be leading a "good life" – it is seen to enhance decisiveness towards reaching one's aim. A striking example is that of Immaculate (see Interviews 2009/O), whose dream was to open her own business place, and she had already designed and printed business cards for her imaginative business venture, to support positive thinking as she explained. Ivo (see Interviews 2009/R) told me, that he had written in his diary that his personal aim was to be financially independent by the age of thirty: being at the age of twenty-eight, he believed himself to be

---

[126] This can be expressed in markers of success in the array of conspicuous consumption, as I have described.

[127] Who could be state representatives, obstructing youth from partaking, influential elders, jealous neighbours or friends – with whom one would not share one's aspirations due to the fear of witchcraft, or visa policies of Western countries, represented by employees in the embassies, who hold youth's destiny in their hands by their decisions.

well on track. Or a young aspiring businessman said frankly: "I am a very ambitious person, but I know I have the potential to succeed. I want to be the yardstick other people take their measurement" (Damian. Field notes 2009. 19.10.09). In this sense, youth seem to strongly integrate this "success-oriented" part in their negotiations of identity and self-evaluation, as comprised in a statement: "The size of your belief is the size of your success" (Cletus. Interviews 2009/F).

This self-assurance could also be extended to a group, as "we Anglophones", or "we Cameroonians". Damian, the business man I have mentioned previously, expressed: "We have this saying: L'impossible n'est pas Camerounais. Cameroonians never give up, that is why they can become successful. They are often not even as good as others, but they never give up, they continue and continue and in the end they achieve something. In that sense, Cameroonians are gifted, blessed by god, they are the chosen ones, the Jews of Africa" (Damian. Field notes 2009 22.10.09). Furthermore, hereby included and contributing to self-assertion are imaginaries of solidarity and social bonds of love and trust, and as well as faith in god, such as those which were expressed in statements as „my family will always be there for me", and "with the help of god I will make it" (Divine. Interviews 2009/F). These kinds of references reflect trust in the power of social relationships, and the lack of trust in formal institutions and state regulations. Narratives of dangers and failures are superimposed by the self-reference of individuals: "I will make it, I will take my chance" despite unfavourable conditions, as a personal achievement (compare Pelican 2010). In such a sense, narratives of failure only enhance the belief in success. Thus, a failure cannot be excused by the difficulties of life circumstances, but is attributed to a personal failure. Narratives of danger and failures do not serve to question the potential of greener pastures. In this sense, and in addition to the failure of migrants, such as not fulfilling their moral duties, it is taken as a motivation for achievement. "Because I cannot rely on them (the relatives abroad), I have to go and make it myself" (Elvis. Interviews 2009/FS).

Practices of New Media use, and the reproduction of imageries and imaginaries

We could assume that internet use relates directly to a differentiation of people's imaginations of "the West". Indeed, people with a wide range of practices in order to inform themselves in the realm of New Media, and also a wide network of social contacts pursue online in order to serve them to direct their practices and their life projects, are generally in a better position to pursue their ideas and notions of a "good life", in the sense of using opportunities for jobs, education, migration, and related information. Generally however, I was able to observe that, on the whole, among my interviewees media practices ranged above all within certain frameworks and not beyond their habitual paths[128], even when certain skills were given. I would therefore argue that New Media practices serve less to enlighten people about certain "truths", but are rather adopted in order to produce "their truths", in the sense that obtaining information occurs within a framework of personal interest, which is often based on previous knowledge: "You will only find what you are looking for" (Peter. Interviews 2009/FS). In this sense, some prevalent imaginations and public narratives are by no means only to be observed among those with limited education (Förster 2008:13). Regarding the media images conveyed, I do not intend to generate the idea that superimposed images of the West – for example in TV series and music videos – are not critically reflected upon and examined by viewers, but they contribute to shape imaginations: "Of course I know this is just film, but since they portray their culture in this way, there must be a certain truth about it. There must be some connection to reality here" (Julius. Interviews 2009/R). What seems to be of even higher importance is the (re-)production of imageries and imaginaries within circuits of communication – people are often more active in mediated social interaction, than they are engaged in using the internet as an information tool. Thus, due to pre-existing imageries, imaginaries and narratives, a persistent obscurity regarding information, and a certain

---

[128] This phenomena has also been observed regarding the use of internet in the West by psychologists and market researchers. See e.g. http://www.acrwebsite.org/volumes/display.asp?id=11855.

mistrust towards migrants' credibility[129], imaginaries are reproduced along the line.

Imagination of spaces and places

After I have referred to the reproduction of imageries and imaginaries, I intend to look into their contents, which are relevant above all for young potential migrants. As Massey (1994:154) states, a place is built on the consciousness of its relations to a wider world: Simone (2008) calls it "worlding". Hence, places are imaginatively co-constructed by imaginations of other places or spaces, be it as a memory or an anticipation of their qualities[130]. In both cases it is an overlapping of experiences, perceptions, hopes, dreams and fears of individuals and groups, shaped by cultural and social frameworks (Gupta & Ferguson 1992, Appadurai & Beckenridge 1989, Tsatsou 2009). A place is endowed with distinct potencies of connotative significance, which evolve in layers of intimacy with the place and the ability of the observer, to draw on these connotations (compare Förster 2005:33). Berutti (2008:9), Manovich (2004), and others call it "augmented space", in the sense that "information" is added to the experience of physical spaces[131]. Spaces are thus always perceived with distinct qualities contributing to people's "sense of place" (McIntyre 2004:168).

In order to examine this, I will have to delve further into the "driving forces" for the construction of imageries and imaginaries on a local level. Practices of imagination are influenced by a historically rooted base of a set of imaginaries, which is negotiated in contemporary frameworks of social change[132]. The sources and origins of prevalent narratives, imageries and imaginaries consist of historical remembrances and interpretations, and today's circumstances of life in a globalized

---

[129] See more in chapter 6.

[130] As for example, in memories of "the village" or of "home" when in diaspora, or as anticipation in the sense of "how will it be there?" and hopes and plans when one would get there, in dreams of migratory ventures.

[131] A social science's notion of space is thus relating to a subjective notion of space and creation of space in imaginations of human beings (Blotevogel 1995 in Pries, 1997:19-21). Compare to the conclusion chapter.

[132] In this sense, this subchapter relates to sub-question 2 of the first guiding question, of how local narratives and imaginations relate to mobility and how New Media possibly contribute to them.

world alike. When tracing the shared imageries of "the West", we have to look into how the first contacts between "the West", or the European invaders and the local population were probably experienced. From the point of view of the peoples in the Grassfields, during a long phase of initial contact between slave traders and merchants along the coastal region, they knew the Europeans only from hearsay, but nevertheless experienced the changes they had been bringing on their societies in their everyday lives (Chilver 1967, Ardener & Warmington 1960, Den Ouden 1987). The image of the "white man" was ambiguous, and linked to experiences of subjection and violence, but also to a powerful realm of wealth, prestige, and new opportunities in the quest for alternative life projects. Such experiences were integrated into existing institutions of power and cultural grids of meaning making (Argenti 2007). Subsequently, right up to the present day new imaginaries of success have emerged, contested along the lines of discourses about authentic culture and morality.

The promise of "modernity" as a promise of a "good life"?

Imageries and imaginaries of faraway places and of a good life are often tied to notions of "modernity", used here as an emic expression. I have often come across the notion as part of people's daily relating to life situations: local drawbacks versus an imagined better life elsewhere. In view of a sense of immobility perceived by many youth in the setting, John Tomlinson (1999:9) states, that the "paradigmatic modernity experience", for most people, rather consists in "(…) that of staying in one place where they experience what global modernity "brings to them", than in physical mobility". The discourses on "modernity", from a Western point of view of development and modernist theorists, are often understood in terms of worldwide capitalist systems of economy, commodification, and in the last decades, globalization, emerging from virtual finance worlds, a spread of Mass Media and technology, secularism, urbanization, new forms of social organization, work, institutionalization, and forms of state and governance. As experience shows, these features are not distributed equally in the world, and globalization and modernity don't necessary go hand in hand with each other (Geschiere 1997, Neubert and Macamo 2008:271,272).

On the contrary, these imaginaries of modernity end up rather as a disappointment for most people in the so-called "Third World". Geschiere, Meyer and Pels (2008:1) speak of a "gap between people's dreams of a better life, and their actual disconnection from the structures on which the materialization of this dream depends". (Geschiere, Meyer & Pels 2008:1, Ferguson 2008). In this sense, it is more a regulative political concept (Roitman 2008:215), than a benchmark of people's everyday experience. Tragically enough, "modernity's enchantment" (Comaroff & Comaroff 1992) is experienced as something external. In this sense, "disenchantment" (compare to Weber 1948) is doubled, by experiences of an everyday life reality of "underdevelopment", and experiences of exclusion by the colonial powers and today's elites (Warnier 1993a:194, Appadurai 2009), or through mediated experiences of such life worlds. The notion of modernity might have been criticized, unravelled and decomposed, or even abandoned – but nevertheless, it is influencing people's perception and images of "the other", in a wide range of social and cultural distinctiveness. I think we cannot deny that symbols and representations of modernity are closely related to imageries of "the West" in an African context, not only set up in admiration, but also, and increasingly, with critique, and in any case, in specific forms (compare Comaroff & Comaroff 1999). Modernities in plural – or "nuances" of modernity - may be used in order to point to the creative integration of differentiating forms of modernity in specific localities (Ferguson 2008, Nyamnjoh 1999, Page 2007, Geschiere, Meyer & Pels 2008:5).

The village and the city, "modernity" and "bush" – interdependent and symbolic notions

Narratives on "modernity" seem to be in the first instance based on a dichotomy between "tradition" and "modernity". Co-productive acts in the evolving of symbolic spaces of juxtaposition also occur on national, regional and local levels: The city and the village form such a pair of juxtaposed spaces: the dichotomies between the village, the realm of tradition, origin and past, and the city, the realm of modernity, modern lifestyles, technologies and cosmopolitanism (De Boeck 2008:128,129, Mbembe 2008). I introduce this pair because I intend to widen the scope of juxtaposing imaginaries to how people might perceive their current environment vis-à-vis their imaginations of "the West", and the social disparities that could occur with this differentiation. The juxtaposition of "civilization" and "barbarism" in Cameroon's national anthem, also indicates such distinctions: "Cameroun, berceau de nos ancêtres; autrefois, tu vecus dans la sauvagerie et peu à peu tu sors de ta barbarie". (Ndjio 2008:206).

I do not intend to delve deeply into the cosmology of Grassfields societies[133], but I want to make a few statements, which could be useful for the understanding of the notions of village and city. The most important core opposition is the wilderness and the realm of humans, forest and village. Both of these realms are ambivalent, dangerous and powerful at the same time. The realm of the forest is related to a realm of sorcery and occult powers, but also of mythical descent and origin. It is the realm of the king, the "fon", who is the master over these forces and institutions dealing with them, such as various secret societies. The human realm is the tamed realm of social relationships and regulated power, hierarchies of the palace, and institutions related to it. It contains the family, the land, and the ancestors. Fisiy & Goheen (1998:386) state that the conception of the home village as "(...) a physical, relational and metaphysical space. It is also the space that provides the spiritual links with the ancestors with access to land and with social security in retirement." However, from a migrants' perspective, the village is also the realm of witchcraft, jealousy and relatives, who compete with each other to "eat"[134] migrants' wealth (Page 2007:436). It often leads to tension between the spheres of the village and city or "the West", between kin and the migrant (Nyamnjoh 2005:244). "Tradition" of the village and the family can be used as a means of seizing on and controlling the attempts of individual pursuit of success, expressed in the saying of "kontri fashion go catch you" - tradition will indict you[135] (Nyamnjoh 2005:251).

Close links exist between rural and urban areas, in regard to the structural frameworks, the rural areas feeding the city, constant demographic exchange and migration, forms of political power reflected

---

[133] See for example Geschiere 1997, Ardener & Warmington 1960, Warnier 1993, Goheen 1996, and others.

[134] I will relate to the notion of "eating" in chapter 6.

[135] Strongly interrelated with „nyongo" or witchcraft believes, see chapter 6.

in the villages in realms of traditional hierarchies, and interrelations via social networks. In view of mutual interrelations, we cannot assume the urban as the node of transformation, and the rural or the village as a place of stagnancy and conservation. The city's identification with modernity constitutes itself by claiming difference, in the sense that the imaginative configuration of the village allows the city to define itself as the "city" (De Boeck 2008:128).

"Modernity" is an ambiguous term, likewise is "the city" and "white man's kontri". These notions are often related to an untamed, dangerous and powerful realm of "wilderness" (Hardin 1993:127,155). De Boeck states that many places and sites within the city are related to wilderness, and the model of the "hunter" becomes a notion of identification for youth, especially there where "the city most fully displays its "urbanity" and modernity", the nocturnal realms of bars and clubs, where those who can afford it "(…) capture, through ostensive consumption of beer, women and consumer goods, their interpretation of the "good life" as promised by and defined in their notion of "modernity"". (De Boeck 2008:129). It is no coincidence that the "West" is designated as "bush", and the term "bushfaller" is a migrant, who has gone abroad in search of game. In this sense, ambivalent images arise, the imaginary of "the West" being interlinked with imaginaries of "modernity" and enchantment, but also of "wilderness" and danger.

Betwixt and between: Bamenda as neither a village nor a city

Juxtaposing imageries and imaginaries of spaces and places are competing. Bamenda is embedded in global and local contexts, not only in a material sense of connections, but also related to subjective imaginations of people and their different normative references (compare Appadurai 1996). Bamenda could in a sense be seen as "betwixt and between" (compare to Turner 1969:95). Compared to "the village" it is a place of modernity, but compared to big Cameroonian cities, such as Yaoundé and Douala, or even seen on an international level, it is only a provincial town. I suppose that probably especially due to this state of "betwixt and between", practices of "reaching out" are strong in the local context. Bamenda could be considered a city, because of a certain level of centrality, which appoints certain urban characteristics. Regarding infrastructure, Bamenda is connected to streets leading to the capital and main industrial centres and airports. There is an availability of a wide range of commodities, of local as well as imported goods and services. There is access to education – primary, secondary and high schools, schools for professional education, and private computer training centres – businesses, communication, and media sites. Regarding its economic performance, Bamenda is feeding the region, and even the whole nation, through its channelling of agricultural products. Furthermore, many people state that some major developments have taken place in Bamenda in the last ten years, such as improvements in infrastructure and a more bustling economic life. In preparation for President Paul Biya's visit in Bamenda to celebrate the fiftieth anniversary of the armed forces in ending 2010, work had been done such as tarring roads, installing streetlights along the main streets, and reinstalling the neglected airport[136]. Many people were pleased with this work and statements like "Bamenda looks like a real city now", or "Bamenda has become modernized" were often heard. From this perspective, Bamenda can be seen as a hub for opportunities, it is a "window to the world", and a "stepping stone"[137].

On the other hand, Bamenda is just a town, which could be seen as rather insignificant, and "boring" as some inhabitants stated. Even though most do not deny that in Bamenda some developments and investments have taken place, it is – by many people – perceived as "backward", "rural", and "provincial". Influences of the rural area and nearby villages and fondoms[138] are strong in a cultural and social sense (Fokwang 2008, Awambeng 1991): The mutual interdependencies between the rural and the city could also be seen from a negatively biased perspective, in the sense of the town's dependency on agriculture

---

[136] The event took place beginning of December 2010: it can also be seen as a provocation or at least a demonstration of force and power towards the inhabitants in the region, which is infamous for being the hub of political opposition, in view of the coming elections which were previewed to take place in 2011.

[137] Fokwang (2008:104) relates to people's being proud to be Bamendian, "Bamenda is a symbolic marker of identity for most citizens from the Northwest Province, who, as migrants in other urban centres of Cameroon and abroad, often identify themselves as coming from "Bamenda" regardless of the specific location of their ancestral village".

[138] Kingdoms, or rather areas of influence of a fon, related to the land a specific ethnic group. The fon's influence is of symbolic nature, but not solely: since most of them are integrated into the ruling party, they also exert political power.

and lack of "modern" sources of income generation, industrial and service sectors. In this sense, Bamenda is "not a real city", which is providing alternative livelihood opportunities in the modern sector. Regarding the historical and political situation of Bamenda, a feeling of neglect and marginalization is prevalent, not only of the city itself, but the entire North West Region or the Anglophone region in general. Related to these discourses, people are aware of a lack of infrastructure, bad roads, continuous power failures, of disadvantages and difficulties regarding the economic situation, regarding accessibility to consumer items, purchasing power, corruption and high taxes. This was indicated in the statements of inhabitants concerning the latest investments in town: "The streetlights are great, they enhance security. But they will just allow them like this. As soon as the bulbs are down, they will not replace them, you will see" (Jim. Field notes 2010/11. 16.12.10)[139]. Or: "In one year from now, when elections are over, you will see potholes everywhere again. The tarring is not done properly, it was just supposed to look nice and the president should not feel any pothole when driving through in his car" (Killian. Field notes 2010/11. 15.12.10). Hence, Bamenda is in various ways connected and disconnected alike.

The notion of "greener pastures"

Imaginaries of promising lands, as the notion of "greener pastures" points to, seem to serve as a category to imagine other worlds, and at the same time criticize situational conditions. "There are no jobs here in Cameroon, that's why we Cameroonians are looking for greener pastures" (Patience. Interviews 2009/O). The notion of "greener pastures" indicates that the "grass on the other side is greener", indicating not only global interdependencies and the "hierarchies" in the scramble of global economies, but also to the fabric of interpersonal social relations. It is a notion, which is widely used in Anglophone Africa, in popular narratives of people in their everyday lives and media, and also in some cases related to in scholarly work (compare to Nyamnjoh 2005, Jua 2003, Förster 2009, 2010). The bible speaks of

---

[139] Unfortunately Jim was right: in outer quarters bulbs were taken out only a few weeks after the president's visit.

"green pastures"[140], which relates to a promised land, which can be enjoyed by those, who follow the right path. In the experience of a lack of the basis to fulfil the dreams of a better life, the images of these mythical lands relate to material abundance, and opportunities to access realms of status and social recognition. Here, various new churches[141] play a role, which pursue at times rigorous ideals and images of "greener pastures", using "faith" in order to achieve the wealth of the promised land, not only related to the nether world, but the worldly realm (Förster 2009, Knibbe 2009:139). It is also expressed in notions used in public discourses, of "money kingdom", or "lands of plenty".

Imagination as social practice – imageries and narratives about migration and "white man's kontri"

"The West" – and its symbols - has become the locus, the ideotopy of imaginations, where the fulfilment of desires for a "good life" could become possible. For Appadurai, even though imageries and imaginaries about distant worlds or other forms of lives are nothing new, imaginations have taken on a new intensity in today's globalized era, in the sense that these relate to veritable activities, to „the imagination as a social practice". Being so, social practices of imagining are also spatial practices, building virtual "scapes" of imageries[142], constituted of imaginations of persons and groups, and relations to physical spaces or locations, within specific historical settings. Imaginations consist of longstanding narratives and imageries conveyed by media and via transnational social relations. These imageries and imaginaries are thereby produced and reproduced,

---

[140] Expressed in the analogy of sheep grazing on green meadow, and refresh with cool water, guided by the saviour, where there is abundance and no lack, redemption of worldly constraints.

[141] These Pentecostal and charismatic churches are a way to partake in success and the promise of well-being regardless the social status and material preconditions of the followers. No wonder they are so successful (Devisch 1996).

[142] In order to grasp connotations of "fluidity" and spatiality of such practices of imagination, Appadurai describes dimensions of global cultural flows, which he terms ethnoscapes, mediascapes, technoscapes, financescapes and ideoscapes (Appadurai 1996:33). The suffix –scape also points out to that these scapes or not objectively given, but relational constructs – imaginaries themselves, with their impacts onto life reality in life-world contexts.

negotiated and sometimes transformed. Their power lays in their force as counter imaginaries vis-à-vis the deplorable situation in Cameroon: there are many different nuances of narratives of opportunities and dangers found in the local context. Nevertheless, the West as a locus is not particularly defined in its own right, but rather serves as a stage for the realization of life projects. Overall, the prevalent imaginaries are not easily challenged by New Media, but rather enforced by them, since they allow media users to actively take part in their production. The narratives and imageries I am looking at in the following subchapter, are related to notions of mobility and white man's kontri[143].

Spaces of potential, but difficult to enter

The narratives of the "West" are very much determined by the difficulty to attain access to this realm of potential. Possible migration ventures were one of the most common topics of conversation with me, since I was seen as potentially helpful regarding such matters. The most intriguing constraints for mobility in the field seems to be a lack of knowledge regarding opportunities and how to go about a migration venture, a lack of financial means to fund such a venture, a lack of reliable social relationships to support a migration venture, and a lack of accurate skills in order to fulfil formal conditions. Massey (et al 1998:13) states the importance of admission policies, as decisively conditioning the character and volume of international migration. "The legal admission policies of developed countries are now highly selective, and certain personal traits – education, skills, wealth, family connections – are essential in assuring success, (…)". Being young, unmarried, most often not yet having children, jobless, and financially not viable, create the highest drawbacks related to such attempts (Fokwang 2008). It thus seems that youth especially, who are most anxious in regard to mobility, are those facing most difficulties in order to obtain a visa for migration to a Western country. In many cases, as interviews and conversations have shown, the knowledge of migration issues as conditions, legal procedures and opportunities, was surprisingly poor, compared to the determination expressed towards migration ventures, and sometimes previous attempts to migrate. Even though legal frameworks influence mobility considerably, "(…) few migrants know much about law and

---

[143] I will thus relate to the 2nd sub-question of guiding question 1.

even fewer would point to law as a major factor in their migration decisions. However, law influences those decisions at every turn." (Brettell 2008:241, Castles 2007:27). Thus, policy related conditions have become opportunity structures to be compared and negotiated. The prevailing insecurities range from a choice of where to go to, purposes for migration, visa procedures, narratives about the journey, up to life circumstances abroad.

FOR THOSE INTENDING TO TRAVEL

- Photocopy of National Identity Card
- Photocopy of Birth Certificate
- Photocopy of Ordinary Level G.C.E
- Photocopy of Advanced Level G.C.E
- Photocopy of any other Diploma if available
- Photocopy of any Attestation of service if available
- Travelers passport
- 4 passport size photographs (4 x 4) white background
- Curriculum vitae
- International vaccination card (yellow card)
- Registration fee of 15.000frs
- First Deposit of 500.000frs

F17: Information available at a counselling office, offering support

F18: Locally distributed magazine (produced in South Africa) about migrating. Note the subtitle: "The bible of moving abroad and achieving success"

Knowledge is distributed unevenly among aspiring migrants, some are better informed than others. The main differences exist between those who are better educated, and often internet literate, and those with less education. Education levels can, but not necessarily make a difference regarding knowledge but also regarding opportunities to migrate. Those who have at least A-level or probably attained a first degree (BA) in studies, could migrate for the opportunity for education, an option, which is seen as the most promising. Others, working in the informal sector, youth with only O-level or school leavers, see themselves confronted by difficulties of availability of jobs and accurateness of their formation, which is not suitable for a migration venture.

Under such conditions, the option of clandestine migration is generally not completely denied. This also relates to narratives of people who "just went" and have not come back, which is interpreted in the sense that they have migrated successfully. Most interviewees however emphasized, that "One needs to know for what reason one is migrating. You cannot migrate just like this, without having a purpose" (Dennis. Interviews 2009/F). Many knew that they needed a valid passport, a birth certificate, an admission from a university or job provider, or an invitation letter, and possibly a bank statement in order to apply for a visa. Less spread is knowledge about conditions, which the applicants need to fulfil in order to have a viable chance of obtaining a visa. Here, the belief in the pliability of regulatory frameworks by competent and influential people is strong. Furthermore, general knowledge is lacking about immigration policies in potential receiving countries, opportunities for jobs and education, work permissions and living conditions[144].

A crucial issue is financial means, which is necessary in order to start procedures for visa acquiring. Just obtaining a passport, which people do not usually possess, costs up to 60,000 FCFA, and also integrates organizational work and time, not to mention the sums paid to intermediaries in the process of obtaining a visa, and of course money for the journey itself: Costs could total 2 to 4 million FCFA[145]. Some

---

[144] See figures F17 and F18. Media such as the depicted „bushfaller magazine" ought to inform people about migrating.

[145] Three million FCFA correspond to approximately 7000 Swiss Francs, which is enough money to start a business, to buy a piece of land and even start to build a house. These sums of money also relate to legal procedures of obtaining documents.

mentioned that they needed to work first and acquire a sound economic base. They knew that being financially viable is to a certain extent a prerequisite to be considered positively in the process of acquiring visas. However, many youth seemed not to be aware that money is also needed in order to start life abroad. In this sense, migration is not only a means to reach out to economic ends, but it also involves exorbitant expenditures, and could be seen as "big business": Migration ventures involve a whole economy of obtaining visas, passports and other necessary documents (Hahn & Klute 2007:18,19). Interrelations between evaluation and access to social capital and human capital as knowledge and education play a crucial role here. Many youth who want to migrate entrust huge sums of money to intermediaries, who would organize migration for them. These intermediaries could be a relative or friend, or professionals in this field. They would try, or promise to obtain visas for the aspiring migrants by faking their documents[146], or even if the visa application procedures are treated in a legally correct manner, clients take the risk of the high probability that the visa application would be turned down and the investment was in vain. The possibility of losing money on dubious intermediaries is generally high (Nyamnjoh 2005, Ndjio 2008, Jua 2007), as in the case of a young man, who had been duped for a large sum of money: "I lost 1.5 million FCFA for this man. He brought my cousin over to Dubai. He is there, that is why I had no doubt. But when I gave him the money he disappeared, neither my cousin in Dubai, nor the man's family in Bamenda know where he is. The police are looking for him. For me it means that I have to work for

---

According to Mbembe and Roitman (1996) there is a "fake parallel" within the official bureaucratic system.

[146] Ernest (see Field notes and interview 2009/O and 2010/11), an informant and friend in Bamenda told me that he was working in Douala for somebody's business in the "production of faked visas", in 2005. They were specialized on business visas for China, which at that time, as he stated, had been easier to issue than now. It contained faked documents as tax certificates, business registrations, invitation-letters of enterprises in China and either confirmations of employment or order of a registered enterprise in Cameroon. They had issued about 5-7 visas per week for prices ranging between 250 and 500,000 FCFA, it was a lucrative business; after he left, several people were condemned to prison in this matter. Fokwang (2008:243) writes about the scope of faked documents (faux dossiers) produced by the so called "doki men", ranging from visas, passports, driving licenses, up to marriage certificates and business attestations.

years, to pay back the money I have borrowed, and then try again…" (Richard. Field notes 2010/11. 29.10.10).

The difficulty in obtaining a visa is apart from a lack of finances, knowhow and trustworthy intermediaries also related to the arbitrariness of visa policies and embassies or consulates in Cameroon, since, as Nyamnjoh (2005:249) puts it ironically, the visa "(…) could always be denied for one "capricious" reason or another". Christian expressed his grief of having been sent back by the immigration police in Austria when he arrived there with a fake business visa: "It seems that it was a devilish intervention" (Christian. Field notes 2010/11). A female interviewee in her 40's complained, while showing the third rejection statement by the US embassy: "They did not give any reason, I fulfil all the conditions, I have work, I have money and I have relatives over there. It seems they will never allow me to enter their country" (Cynthia. Interviews 2009/O). Some are denied a visa, as happened to Dennis (interview 2009/F), even when they have an approval from a university, bank draft and a finance plan; in such cases, the disappointment is understandably particularly large. A denial of a visa is thus often not seen as the problem of individuals not fulfilling the required conditions to obtain a visa, or not having done everything possible, but as an obstruction. "How can they refuse someone a visa when he has all his papers?" is a statement, which demonstrates a different perspective on migration as personal as well as group related venture and risk taking, and not an issue of immigration policies (compare Alpes 2011). The belief, that "one could have more luck next time" is also expressed in many people's repeated attempts[147], and not giving up, even if they spend incredible sums of money for such a venture[148].

Images, which circulate related to the journey, are most often related to entering a plane. Migrating by crossing the Sahara is related to clandestine migration, which is an issue, but is often negatively evaluated. Rumours of people who crossed the Sahara desert relate to perilous circumstances of such a journey (Förster 2008:8), how

[147] See figures F19, 20. Playing the DV lottery in order to win an American green card is very common. However, winning a green card does not automatically authorize people to get the visa.

[148] Few expressed doubt about pooling huge amounts into insecure migration undertakings: "Sometimes I think that if I had invested into my business, I would be fine today, but the hope is always there" (Pride. Interviews 2009/R).

intermediaries would exploit clandestine migrants who had paid them exorbitantly to bring them North, that migrants would be sometimes left alone in the midst of the desert, and of people who disappeared or probably died. A young man described the return of a friend, who had been expelled in a Maghreb country[149]: "He was a fat man when he went, when he returned he was almost inexistent" (showing with the small finger of his hand, how that friend had looked like) (Hans. Field notes 2009. 24.08.09)[150].

F19: DV Lottery and passport photograph publicity in Yaoundé
F20: Publicity for DV lottery in a cyber café, Bamenda

Most often, migration is related to entering a plane as a symbol of departure. But also here, clandestine migration is not completely denied: Förster (2008) describes discourses of the opportunity of hiding in a plane, by holding onto its exterior, and I was seriously asked by one young man, if it was not possible to hide in a suitcase in order to be taken along. The lady who was repeatedly denied the visa for the US said jokingly, after she had asked me about opportunities for migration: "If I was a witch, I would turn into a bug, and creep into somebody's luggage" (Cynthia. Interviews 2009/O). Other narratives circulate, such as of the American national, who enters the scene each year and each

---

[149] Also an issue are the difficult living conditions in these Maghreb countries, and the danger "to get stuck there" and never attain the aim to cross over to Europe (compare Schapendonk 2010).

[150] Possibly, I suppose, people were also not open about such ventures with me. Moreover, because it is said that "whites do not like clandestine migrants", and the possibility of failure is considered as high: when being forcefully repatriated, the migrant could experience being judged that he had not known how to migrate "in the right way".

time takes with him several people[151]. Such adventurous imaginations of travel itself relates to the lack of knowledge regarding acquisition of visa and papers, and the fact, that the flight is not the difficult part of the journey. However, the cost for the flight could obstruct people in order to proceed in their venture. As in the case of Ernest (interview 2009/R), who won the DV lottery, got a visa and then could not raise sufficient funds for the flight within given time, which has left him deeply disappointed. Narrating stories and experiences of travelling could be used to express superiority towards others: I witnessed discussions among friends of whom some had been travelling: They related to experiences in airports, in planes, and in the respective countries. Narratives about travel, and especially travel to the West are expressed with excitement and fascination. To position oneself as a mobile individual and cosmopolitan is part of status creation. The insecurity related to the journey and opportunities of access enforces the image of the realm of "the West" as difficult to access, jealously guarded, alluding to the prevalent discrimination of Africans when it comes to their claim to a "good life". In the light of the narratives "how to reach there", the achievement of having entered these realms becomes even greater.

Narratives of "Western modernity" – or life in "white man's kontri"

Even though sometimes narratives of "reaching there" seems to attract greater attention, people have distinguished imaginations of a life "over there". These imageries are not always positive, but still seen as favourable compared to the circumstances prevalent in Cameroon. They are mainly related to narratives of opportunities, such as finding a job, earning money, getting access to good education, participating in consumer culture and enjoying a "nice environment" and facilities.

The West is seen as a place where economy is functioning and booming, where there exist opportunities for everybody to contribute

---

[151] Narratives about the disappearance of people allude to the stories of "zombies" in Europe. Zombies are bewitched kin, possessed by occult forces, who are traded to white man's kontri in exchange for western goods by their families. Or migrants or white men could take them over from Cameroon to white man's kontri, where they would work as slaves without having a will or a sense of themselves (Nyamnjoh 2005:242, Comaroff & Comaroff 1999:289), as the basis on which their masters build their success on.

105

with knowledge and skills. Many interviewees stated their assumptions, for example, that they heard that in Europe there would be "plenty of odd jobs, because their economy is strong"[152]. On the other hand, it is also stated "one needs to work hard in Europe. But I am willing to work hard, I have no problem with that" (Bertina. Interviews 2009/R). In the view of a lack of work on a local level and the complaint that youth are condemned "to be idle", the prospect of having a chance to work hard is seen as a great opportunity. "White man's work" is often related to as non-physical work, "book-work" and administration, contrasting "black man's work", hard physical work, "Work which they (the white man) do not want to do, we could do for them" (Nick. Interviews 2009/O) (compare Nyamnjoh & Page 2002). Or another interviewee stated: "Children are sent out to go and help the family. Not to sit and looking at white people passing. If I leave this Cameroon (…), I will work like a slave, I will do everything to help myself. Sweeping, cleaning people's shoes, I will do it. I will not be ashamed, as long as I can make money" (Valeria. Interviews 2010/11). Related to imagined chances in "white man's kontri", some stated that even though it was not easy, it was at least very possible "to make it" there, "If you work hard and try to make money, you can gradually make it over there, there is at least a chance for it, whereas in Cameroon it is very difficult" (Claudia. Interviews 2009/F).

Generally, in the Western world people are imagined to enjoy a high life standard. A central issue is the infrastructure in Western countries, which is related to a materially well-equipped public sphere, buildings, streets, highways, airports, train stations, and the transport system in general: "everything looks nice", and "there are no potholes in the streets", as well as a well-functioning communication infrastructure[153]. It is widely known that in the West, most people have computers and

---

[152] For example mobile phone or computer repairers, or taxi drivers were asking me about job opportunities, stating their belief that there would be "much to do for them" in countries with high economic and living standard. Some related to the impression, that "everybody has got a mobile phone", and "big cities where there are a lot of taxis needed". It points to an evaluation of the situation deriving from a local understanding of economy and lifestyle.

[153] In 2010, the year of the 50-year independence celebrations of Cameroon and other African countries, I repeatedly came across the puzzling narrative that "they are celebrating 50 years without power failure in the US this year".

internet access at their homes. Another dimension of infrastructure emphasized by the interviewees is public education, which is important for those who wish to migrate for higher education: Western education would be of high standard and sponsored by the government. Regarding consumerism, some expressed the idea that in the West everybody possessed a car, hereby taking the valuation of status symbols as cars as a starting point, in the sense that high status people must invest in certain status symbols. "They do not waste time, after school, young people are already wealthy and have their own flat and car" (Nick. Interviews 2009/O). In general, consumer items are said to be cheaper in Europe[154], for example electronics and cars, computers, and also costs for telecommunication. This derives from the impression that consumer goods originating from the West must be cheaper there, where "one is at the source". Many interviewees might not have been aware of the relations between income and expenditures for livelihood in Western countries. However, some highlighted this fact in particular: "Things are expensive over there, many people do not know. What is big money in Cameroon, is not more than a token over there" (Titus. Interviews 2009/F). Regarding financial means, some were well aware of conditions in Western countries: "You need money in white man's kontri. Many just go like this, they have no idea that you need money over there. They only think of going, not of being there" (Clovis. Interviews 2009/R).

Shifting hierarchies of notions of "a better place to be"

Even though the notion of "the West" was used in a very general way, when related to the imaginaries of "a better life" or as a "better place to be", there are also differentiating notions of these realms emerging, building up new hierarchies of places or spaces: For youth in Bamenda, the movement towards cities in Cameroon, as Douala and Yaoundé, is related to an attempt to better their situation, be it going there for education or in search of a job opportunity, and these cities are seen as destinations prior to finalizing migration. As one interviewee stated: "There you earn double compared to Bamenda, and there are many jobs for students. Compared to Yaoundé, or Douala, Bamenda is

---

[154] This is not untrue, as imported goods are expensive in Cameroon, not only in relative terms. Also outranged items from the West, such as second hand cloth (okrika), and electronics are sold for relatively high prices.

just empty" (Patience. Interviews 2009/O). Between Cameroon and "more developed countries" there is a big gap perceived. However, among countries, which are attributed as being "more developed", the superiority of the West has been undermined in a relative sense. Related to African countries, South Africa is seen in a positive light: "South Africa is in many aspects similar to a white man's country" (Dennis. Interviews 2009/F). Other – above all neighbouring – African countries, such as Nigeria or Gabon, are not top in their evaluation, but they serve as destinations in search of opportunities, above all for business or education. The advantage here is that these countries are relatively easy to enter for Cameroonians. Increasingly other countries, for example in Asia and the Arabic peninsula, most prominently Dubai, India and China[155] (compare Dobler 2007) are seen as highly attractive. China and Japan are seen as favourable when dealing in electronics and cars, China for teaching, since English teachers are recruited in Cameroon for China, Dubai is promising for business and construction or domestic work - (compare Pelican & Tatah 2009), and India is considered a good destination for ICT education. For Muslim youth, Arab countries such as Saudi Arabia and countries in the Near East are seen as destinations for work and education (compare Pelican & Tatah 2009). These "new destinations", which have not been ranging as mental images in the minds of people for a long time, have gained importance and are seen as a potential stepping stone (Pelican 2010)[156].

When it comes to imagining durable instalment and living abroad, however, European and North American countries are still seen as the highest of dreams. But "the West" is also differentiated, for example, distinguishing the US from Europe: in North America, the US and Canada, the "American dream" could be fulfilled, which is attributed to

---

[155] As a young business-man stated: "You cannot live in China for longer, it is just for business", and: "You cannot make much money over there, the salaries are small" (Elvis C. Field notes 2009. 20.08.09). Elvis had lived in China for a couple of years and had come back. He was also counselling Cameroonians, who intended to go to China.

[156] However, there are imageries, which indicate insecurities regarding new destinations. Among other factors, it is also related to the "mentality" of for example the Chinese, evaluations, which could derive from their increasing presence in Cameroon. In local discourses Chinese and Indian people are sometimes seen as "strange" and even potentially dangerous, as expressed in narratives about their "occult powers", and witchcraft relations, deriving from their "medicine power": Particularly "Chinese medicine" has become a notion in the last few years. See figure F21.

"the opportunity to work one's way up", as examples of migrants prove who came with nothing and are now successful and high ranking people. Since the taking office of president Obama, the US has gained an even more positive reputation, and the US DV lottery (compare to Fokwang 2008:248), which is entered annually by a wide range of people – a number of them winning American Green-cards – is seen as a sign "that in the US it is still possible". Canada[157] is seen as a kind of alternative to the US, since it is known as a country with high living standard, and a bilingual country, the same as Cameroon. "Europe" is sometimes described as "more civilized" compared to the US: some interviewees stated that they had heard of racism in the US, of poor people living in the streets, and some also mentioned the "warfare" of the US. Some people pointed out to "Europe" as "welfare states", as providing great security regarding livelihood. "Europe" is then sometimes differentiated. For example, "Scandinavian Countries" is a notion widely known, mainly related to the idea that these countries offer favourable conditions to foreign students. Countries of the former colonial rulers have a special importance in people's imagination: The UK, France, and Germany. Overall, evaluations are strongly related to economic opportunities, security, and issues of acceptance.

F21: The Chinese are very visible in Cameroon, here Bamenda
F22: Advertisement to "fall bush" to Canada, Buea

---

[157] See figure F22.

Imageries of whiteness

So-called "narratives of modernity" are also related to people, here the "white man" [158], and attributed mentalities. Such imaginaries are held up from colonial times by the powerful and successful: by migrants, politicians, businessmen. Den Ouden (1987:3) states that the notion of "white" relates to people holding certain positions or taking over certain social roles. People are sometimes called "white black man", in particular when successful or having high positions in government, or at times related to migrants (compare Fokwang 2008), in admiration but also to taunt. In this sense, bushfallers coming back from white man's kontri are often attributed certain physical traits, related to skin, body, and colour. They have a light colour because they are not "working under the harsh African sun". Their skin is soft because of their well-being, hands are soft because they do not work physically. They are well nourished – "fat" – because they eat good, nutritious food over there, and they are in a good physical condition, which is often related to that they look "shiny", and "bright" (compare Tazanu 2012:140ff).

The shaping of imaginaries of whiteness is implanted in the notions of difference of the opposing worlds of "Africa" and "the West", related to the difference of environments and also of race. As an interviewee quoted her sister's statement, upon the latter's arrival in the US and first impression of the apparently perfect environment: "My sister told me on the phone, calling from the US, that she was not aware that God indeed makes a difference between black and white, but now she knew better" (Valeria. Field notes 2009. 25.10.09). Such statements highlight the difference between the "white man" and "black man", in the sense of different opportunities of access to wealth and power: "The white man is ahead. They are better educated and know how to deal with life" (Christian. Interviews 2009/R). In this sense, reflecting the other makes up self-images, which are often placed in comparisons to people's everyday narratives. Such could also involve a negative self-representation of Africans, the "black man", or Cameroonians, in order to express dissatisfaction with Cameroon's treacherous elite and a corrupt system. Statements like: "I know us Blackman very well. Black man are wicked and dishonest" (Christian. Interviews 2009/R) then

---

[158] I mean here whites originating from Europe and North America. Asians are at times also called "whites".

lament the occurrence of deceit in Cameroonian society due to competition for scarce resources.

People's estimations could – apart from circulating narratives - partly stem from "white man" they perceive in the locality, and for a few from personal encounters. The "white's" character is in the Cameroonian context often attributed to their being "excited", "distanciated" vis-à-vis native people, always "moving in groups", and their high need for security and luxury. Generally, "whites" are seen as physically weak but usually well-educated and "civilized" in their conduct. Intelligence is attributed to some of them, however also relative to the strange environment – in the African context whites are seen as comporting in "stupid" and "naïve" ways (compare to Nyamnjoh & Page 2002). I had as well come across critique regarding the opposite gender, in juxtaposition to the setup of "whiteness": women attribute to white men that they are economically viable, determined and more faithful in marriage, whereas men evaluate white women as not looking out for economic gain in a relationship but being interested in true love.

In relation to the "white man's" attitude towards the "black man", there are ambiguous narratives circulating. Some interviewees stated that apparently people in the West were most often welcoming and open, from what they heard through friends who had migrated. "People are nice over there, because they are content with their lives. Hungry people are angry people, you know…" (Christian. Interviews 2009/R). On the other hand, some said they heard about white people being deprecatory and even racist: "Is it true that there are many racist people over there in Europe?" (Julius. Interviews 2009/R), pointing to images of "whites", who are closed and reserved, and who do not want to interfere with people from other cultures. A perceived ignorance of "whites" towards Africans is mocked upon, especially alluding to "white's" preconceptions of backwardness of Africans. It is also a topic, that people "in the West" have negative imageries, and no clear knowledge of "Africa" and people's lives in African countries. "They think we live in the stone-age, and then they are surprised when I write them on a friendship site in the internet." (Patience. Interviews 2009/O). The same interviewee also stated that she was surprised about such ignorance because she would have thought that people – in Canada in this example – are well educated in Western countries. Such evaluations relate to people's general impression of neglect and disinterest of "the world" in their fate. "The only moment when people in the West might hear of Cameroon is at the football world cup" (Xavier. Field notes 2008. 17.10.08).

People who have the chance to exchange online with others from all over the world, said they used the opportunity to involve themselves in such discourses to in a sense "correct" "the white's" attitudes and ignorance: "My online friend from China was very surprised when I contacted him the first time from Cameroon, Africa, through the internet. He asked many questions. By now he has become very knowledgeable when Africa is concerned" (Titus. Interviews 2009/F). Another interviewee wondered about the „strange kind of questions, the Americans (his online friends) ask about Africa": "They love pictures, they always ask for it. So I sent them a picture of myself. They were wondering if I was a real African, because they thought that they have to be very dark skinned (the interviewee has a fair complexion). Then they wondered about why I would be dressed in T-shirt and wearing a baseball cap. So the next time I sent them a picture in "kontri-dress", they loved it too much! I think to go now to my village and snaf (photograph) some scenes from traditional life there, and send the pictures to them. And also I want to record some traditional music, I am sure this will make them happy." (Peter. Interviews 2009/FS) The interviewee seemed in this case not to perceive the ignorance of his online communication partners as an offence, but expressed amusement, but he was also willing to serve their fantasies. In the internet, he stated, he could change their images, by sharing a common space online and chatting on an equal level[159].

The flip side of the West and counter-imaginaries

Discourses also point to ambivalences, as we have seen, and thus narratives of dangers – counter-images which positively evaluate "Cameroonian culture and values". They encompass manifold, rather negative imaginaries related to the West, of decadency and corruption of morality, concerning the accumulation of wealth and power, redistribution and solidarity. These imaginaries are by interviewees

---

[159] In comparison to Nyamnjoh & Page's (2002) examination of imageries of whiteness, the narratives I have come across are altogether more positive. Maybe people withhold criticism and "mocking whites" in my presence.

related to the "West's" emphasizing of individuality and individual achievement, without caring for other people, a lack of solidarity within families, of respect towards the elder generation, and of the capacity to discipline youths and children. Related to a disintegration of families, people would lack social contact and security: "People are sometimes lonely over there. They want to have a family. Sometimes they even come here to Cameroon because they want to take a child with them for adoption" (Patience. Interviews 2009/O). Moreover, loneliness is also seen as a threat for Cameroonians in these realms, far from their families and surrounded by people who are indifferent or even hostile. "At first when going there, I would try to find some friends, one cannot live without having friends around" (Miranda. Interviews 2009/R). That people in the West suffering from mental distress was mentioned by many interviewees, caused by isolation, too much work, and the lack of respect for moral and spiritual values: Another issue is the loss of religion, as a sign of moral decadency: "Is it true that in Europe people do not go to church anymore?" (Valeria. Interviews 2009/O). Regarding gender roles, some – not only men - deplored, that women would not fulfil certain tasks that are ascribed to typical female activities, such as cooking, domestic work, and rearing of children. Many women would not intend to have children, since they preferred professional careers. In general, sexual decadency is highlighted, related to representations of sexuality in films, as conveyed through images of rampant sex life, people kissing in public or homosexuality as a norm. This is the flip side of a "life in the West" – the price to pay for indulging in the comfort and affluence, which also endangers the bushfallers (Förster 2008:11). In popular narratives, such issues are related to the conduct of migrants, who are accused of adopting "white man's habits": Those who cut or treat social connections selectively and do not share their benefits, are suspected of having adopted the egoistic and greedy life style of the West – which is called the "bush disease" or " bush syndrome" (Tazanu 2012:165). Such imaginaries of the West contribute to the image of a realm of potentiality, but also dangers[160].

---

[160] It is not surprising that narratives of "nyongo" evolve around these realms. Witchcraft discourses are an important aspect concerning Western consumer goods, technologies, and migration practices. Geschiere (1997:4,5,10) speaks of ambiguous discourses about witchcraft as a levelling force and as a means for accumulation of wealth and power, and thus as both, a weapon for the weak and a resource for the

Adverse to the imaginaries of "the West" are those of Cameroon, ranging between criticism and affirmation. Valuations differentiate in view of the imaginaries people share regarding other countries, cultures, and social systems. Counter imaginaries stand in relation to what I have called the flip side of the West. People seem thereby to make a difference between what they refer to as "Cameroonian people, culture and society", and to the conditions imposed by "the system", government and governmental intermediaries and institutions. The latter, "the system", is assigned negatively by many people, related to corruption, exploitation and marginalization. The first aspect is what Cameroonians associate with "home" and "country of origin" tied to people and land. What is often deplored – among migrants, but also on a local level – is an understatement of origin: "One should never forget where one comes from", in the sense that if people lose roots in a community, place, and moral world, they lose grounding. "A root problem is that they are not self-confident. Many young people reject their culture and tradition without knowing what they are looking for instead." (Alexandre. Interviews Diaspora. 26.06.07). "Getting a wider picture", as migrants call it, regarding certain aspects which are in contrast to their home society, could also lead to a "mystification" and re-evaluation of "home" (Wiles 2008). Issues attributed to "home" or "feeling homely" are for example social characteristics of communities and everyday habits: positively valued are the tight social relationships and stronghold of families, and some traits of people's characters: their openness, friendliness, hospitability, generosity and religiousness – "We are god fearing people", the positively valued demarcation line between gender roles, and "everyday pleasures" as Cameroonian food, or going out with friends.

Conclusion – Transnational mobility, liveness as a potential, and the power of imagination

In this chapter I have been dealing with, overall, the first guiding question about how New Media contributes to imaginaries and imageries of mobility, and the sub-questions, how media is used

---

powerful (Geschiere 1997:16). In this sense, social, economic and power differences in general can be well explained by occult forces involved, be it differences among one's social group, or on a national, and even global level.

intentionally for mobility purposes, and how New Media possibly contribute to mobility practices. The chapter also relates to the second guiding question of how practices of New Media use and imagination shape social spaces on a local level, here rather concerning the part of mental imageries, and the third sub-question of what narratives of social change are provoked by them.

Practices of imagination have become important for my research, as I have emphasized with youth's "outward orientation". It is closely related to a projective dimension of self-realization, which is subsumed in youth's strategies to reach out for a better life, in the sense of social mobility. Thus, it is the feeling of being connected to opportunities, which induces liveness as a potential here. Such is expressed in alternative practices, for example in fraudulent businesses in the internet or the purposeful quest for migration and consumption. Similarly, I have highlighted that those practices of pursuing life projects are based on the local society's schematization of experience directed at the imagination of a "good life" to be achieved elsewhere, pursued through physical mobility, which relates to a habitual dimension of agentic orientation. In this sense, I have emphasized the high valuation of physical mobility, and sociality in the sense of maintaining connections. Migration or having fruitful relations beyond the locality is conditioned by – real and imagined – perceived disadvantages on a local level, set up in juxtaposition. Migration and imagination for a good life have a particular history in the local context, expressed in narratives, images and practices. Here, I have related New Media's role in the reproduction of imageries, as an explanation for the unceasing popularity of the notion of the "good life" to be fulfilled elsewhere, which is reaffirmed and not necessarily challenged by New Media. Similarly, as I have described, media's pervasiveness instead contributes to the reproduction of existing narratives and also opacities regarding migration ventures, rather than enlighten them[161]. Likewise however, New Media and modes of their use are important means to dream and to adopt purposeful practices in ways, which had not been possible before the advent of these media. We can perceive a creative

---

[161] In chapters 4 to 6, I will discuss the orientation to an ideal sociality and solidarity, which are seen as possibly on-going and coming to fruition on the basis of sensory liveness, achieved in mediation through New Media.

and reflexive dimension followed up by intentional practices of New Media use, as a kind of virtual mobility - although at the same time, their outward orientation may remain more or less stable. This then relates to liveness as work or effort, and to a practical-evaluative dimension of agency.

As I have described in this chapter, continuously produced and reproduced imaginaries and imageries are circulated across social networks in local and translocal communities, as collective imaginaries and as reflections of local conditions and interpretations of global phenomena, constituting reflexive social entities. Regarding the hegemonic force sometimes attributed to New Media – and their dissemination as technologies and media contents – even though they appear to "come from elsewhere", they simultaneously emerge from the local context and conditions concerning their materiality, as well as practices and imaginations related to them. The power of these imaginaries and imageries seems to lie in particular in their being undefined and "fuzzy"; they are to be seen as collective imaginations, but also as offering the space for individual purposes alike, and they seem to lie in the core of continuities of societal debates about morality, sociality and solidarity. Appadurai does not understand imagination as being an escape into "other, better worlds", but as a basis for genuine societal action and practice within a context of a larger range of opportunities (Appadurai 1996:7). It is thus not that access to New Media would replace mobility, but, at least on a large scale, they enhance imaginations of and about mobility and contribute in the inducing of it (compare Burrell 2009), enfolding in evaluative practices of New Media use by taking reference to both, habitual maps of action as orientation, and to the imagination of alternative pathways for a better future (compare Emirbayer & Mische 1998:970,971).

Whereas in this chapter, I have concentrated on the imaginaries and imageries, which lie behind intentional practices of media use, I will relate to those practices in the coming chapters of this book. In the next chapter, I intend to look in more detail at the materialization of New Media technologies and their meaning and use on a local level. Thereby I intend to examine how these technologies contribute and interrelate with local face-to-face social interaction and their user's "feeling of place".

3

New Media, their materiality, and their contribution to social spaces: between potential and local conditions

As I have tried to show in the previous chapter, societal imaginaries and the perception of space are closely interlinked with practices of the appropriation of physical space, such as practices of physical mobility as migration movements, and virtual mobility as the use of time and space transcending the media of communication and information. In this sense, spaces are according to Bourdieu (1977) frameworks for social practices, but also places for social integration, socialization and reproduction of society, social hierarchies and power relations. The appropriation of physical space is in its topology a reflection and representation of the social space: it is visible, in our example, in the urban topography of New Media – cyber cafés, and mobile phone and computer related businesses. In this chapter, I intend to approach notions and imaginaries of "modernity" related to their material representations, and the respective characteristics of this materiality in the locality. New Media practices are spatial practices, and they contribute to the constitution of local "social spaces". Such social spaces related to New media are adopted - not only - but in particular by youth, where they display lifestyles and "youth culture", practices, which point to negotiations on youth's position in society and their adoption of new coping strategies to overcome feelings of immobility and disconnection (Jua 2003, Burrell 2009). In such ways, New Media spaces are related and oriented to imaginaries and imageries of "faraway places" and potentialities. Furthermore, these spaces are social places where people interact on a face-to-face level and under specific local conditions. Thus, online and offline, mediated and face-to-face, but also public and private spheres intermingle in interesting ways. I will thus explore the contributions of liveness as a potential, in practices of media use and their framing by local conditions. This chapter refers to the second guiding question, which is related to how practices of New Media use impact on a local level, and consequently, how local and translocal sociality intersect. In this sense, New Media technologies contribute to people's "sense of place". The data on which this chapter is based has been collected throughout my different stays in Bamenda[162].

The materialization of cyber cafés and computer technology in the setting

When talking about the phenomenological qualities of places, practices of imaginations and the constitution of social spaces, we should not forget about the material endowments of places. As Larkin (2008:147) expresses, cities and urban space are produced through infrastructures: "(…) shipping, trains, fibre optic lines, warehouses, all exchange is based on such infrastructure. Space gets produced and networked. Infrastructure is the structural condition of the movement of commodities of whatever kind. New electronic communication has intensified these processes". As a part of urban infrastructure, a very important characteristic of a city is the availability and accessibility of media technologies[163]. Media differentiate in their materiality, and how they are perceived by media users regarding their materiality, technology, organization, and cognitive experience. Experiences of New Media are further shaped by the conventions, plus cultural and social modes of their use: shifts may take different forms due to historical developments of media use and practices of socializing. How these media influence social activities and cultural productions depends on how media are embedded into everyday life and practices in the local context. Thereby, according to Emirbayer & Mische (1998:1004), "The agentic orientations of actors (…) may vary in dialogue with the different situational contexts to which (and by means of which) they respond". It relates to the actors' dealing with unfolding temporal-relational conditions as contexts for social action, which are set up in mutual interdependencies between agents and structure. In the next two subchapters, I will shed light on New Media technologies' material

---

[162] I have concentrated especially on youth social spaces during my stay in 2010/11.

[163] See figures F3.1, and F3.2, maps and overviews of computer and mobile phone related businesses, as well as public cyber cafés in Bamenda.

features and their peculiarities, and the consequent influence on social and symbolic practices of their use as material objects[164].

Computer technologies and internet are spatially bound – which, unlike mobile phones, are not personal and mobile devices[165]. Apart from examining the materiality of internet technology in public cyber cafés, I will also relate to the use of the internet in people's homes, which has recently increased, and to computers as business items and status symbols.

Outer and inner appearance and location of public cyber cafés in the city

The customers in public cyber cafés and their practices and needs do not only enhance the spread and distribution of cyber cafés in general, but they also influence their appearance in the city and their internal set up. Overall, the labelling of the signboards most often include the terms "cyber", "internet", or just "–net". Most of the cyber cafés are found along the main axis and most busy streets. They are located within distinct buildings, rooms or floors, or less common in containers, with a clearly marked entrance and more often than not signboards, and therefore easily recognizable and accessible from the street[166]. Regarding different "categories" of cyber cafés, those with relatively quick lines and moderate prices, located in the core of the city, are those with the highest turnover of customers. Other cyber cafés are dispersed in outer quarters, on side streets, sometimes hidden. These cyber cafés depend on a stable circle of customers from the quarter. Apart from these two categories of cyber cafés, some others specialize in order to keep up with competition. Some almost entirely depend on scammers, who come there on a daily basis, whilst a few depend on other pillars, as for example vocational training courses, counselling sessions for those who want to inform themselves of study opportunities abroad, or they try to engage in business in a "niche", attracting another public – less scammer oriented – for example by asking higher prices and offering a convenient

---

[164] I am relating here to the first sub-question of the second guiding question, how the materiality of New Media technologies influence their use, and vice versa.

[165] Of course, laptops are mobile objects, but in this context they are not yet very common.

[166] See figure F23.

working environment. Other cyber cafés in the periphery of town are in the vicinity of educational offers such as higher professional schools.

Regarding the setup of the cyber café's interior, I would differentiate between a customer's section with computers and an employee's section with a counter, where customers could buy tickets for airtime. The counter is always placed near the entrance, either in a separated front room or in the same room separated from the tables with customer computers. Regarding the organization of the tables, chairs and computer screens, there are a few main styles, which depend on the size and shape of the room. Either the tables and computers are placed along the walls, the screens pointing to the centre of the room, or the other way round, the screens facing the walls and the customers sitting at the wall. In some cases tables are placed in double rows, the computer screens pointing in opposite directions (compare Braune 2008:156,157)[167]. There are a few cyber cafés, which have rows or small groupings of tables throughout the room, or such, or those which consist of several small rooms, where a few computers are placed. Between the computers there are either wooden partitioning walls or curtains, or no partition at all. Most cyber cafés are equipped with bulky second hand computer screens. There is however a tendency to replace these screens by newer flatter monitors, which need less space: in 2010 there were four cyber cafés in Bamenda equipped with laptops. The technological equipment is exposed to heavy usage by the customers. Insufficient technological maintenance or neglect can lead to rapid deterioration: regarding hardware this is especially true of keyboards or the external mouse. Additional electronic features, for example mobile web-cams or head sets are often only given out on demand, but most cyber cafés are not equipped any longer, since these web-cams are notoriously stolen.

The materialization of user's practices and habits in public cyber cafés

In the sense of a materialization of habits, practices, needs and expectations of the internet users, these have transformed compared to the year 2003, when the technological equipment and performance was far less advanced. Expectations towards technology have become

---

[167] See figure F24. Compare also to Tazanu's description of cyber cafés in Buea (2012:110ff).

comparatively higher and certain standards are demanded. Users are used to quicker lines, and as they could accomplish more work within a shorter time, some customers – especially scammers – are even ready to pay more for greater convenience and faster and more powerful machines with suitable keyboards and higher bandwidth. This development has led to a pressure for cyber café owners to update their equipment. Swift service by technicians is expected, especially in cyber cafés with a higher degree of capacity utilisation and turnover of customers.

The awareness of internet security has risen. People expect from cyber café owners and employees guarantees of a certain standard of security, including regular updates of anti-virus and anti-spyware on machines. Along with security, privacy was also an issue among interviewees. In general, internet practices, above all scamming practices, but also dating or interpersonal communication are seen to demand a certain degree of privacy, also a fear of scams is involved here. When investigating the reasons for this apparent higher demand for private spaces, or shifted consciousness for privacy, there are a few factors, which seem to be crucial. In 2008 it was unusual to see young people using a computer in a group, as I had often observed in 2003. That most people occupy a computer screen for themselves for their computer or internet work, is related to their level of computer- or internet literacy, more inexperienced internet users rely on the help of peers. Furthermore prices for airtime have fallen to the extent that people do not feel that they need to split the costs by sharing a computer. Some cyber cafés have policies of prohibiting computer sharing to make better gain and to prevent unrest and noise. It does not mean that people do not exchange, but generally one person per machine is the standard. Most cyber cafés provide a more private space for the internet users by blinds or curtains between the computers. In other cases, as I had seen in Yaoundé, computer monitors are built into the surface of the desks, in order that user's activities on the screen are not visible for others. Less common are "séparés" with one computer, or separate rooms for customers who pay more to work in a private space and quiet environment.

Regarding work effected in public cyber cafés, it is related to a "internet standard time"[168]. Apart from people of the category of frequent internet users, who spend several hours a day online, the "standard time" when buying airtime is one hour[169]. The time span of internet use and activities to be effected depend on the level of internet literacy, speed of typing and the quality of the internet connection. Regarding the use of one hour airtime, since the tickets are valid for two weeks after purchase in most cyber cafés, there is no need to finish the bought airtime at once. Expenditures for airtime vary widely related to the intensity of internet use[170].

Some cyber cafés have adapted their opening hours to the working practices of their customers – in this case the scammers – who work in two "shifts". Some of those need to work late in the evenings and nights, as they have to consider the time difference for North America, Australia or South East Asian countries, when preferring to meet their "clients" in their waking hours, in order to chat and email in real time, or making calls. The day shift is for European countries with only minor time differences. The prices for airtime in the night are lower than during the day, 1000 FCFA for eight hours. There are a couple of cyber cafés which are therefore open all night, closed and locked after regular opening hours at around 10 p.m. until 6 a.m. for security reasons in order that people can work inside without being disturbed. Usually it is

---

[168] Excluded here are those internet users who do not have a time limit in this sense, browsing at home, in the office or as employee in a cyber café, or those who buy airtime "in bulk", having their personal accounts.

[169] The airtime, which can be bought in cyber cafés is sold in categories of – most often – 30 minutes, 60 minutes, 90 minutes, 120 minutes, and so on. In some few cyber cafés the categories are slightly different, as 20 minutes, and 65 minutes for example, in order that users could log out properly. In this sense, internet use is related to "standard times".

[170] Whereas an occasional internet user could spend amounts which are in a range of 500-1000 FCFA – around 2 Euro – per month, regular internet users who frequent the cyber café twice or three times a week, spend about 2000-3000 per month. Most frequent internet users, such as scammers, who use the internet for their "profession", spend huge amounts when they work several hours in the cyber café on a daily basis. Some said they spent 30,000 up to 40,000 FCFA a month. This is of course only possible when they have an according "return" from their "work". Expenditures for airtime for the internet is easier to estimate by people, than expenditures for phone, since it is related to their habits of internet use, which are rather regular, and the fixed standard expenditure: 200-300 FCFA per hour.

announced on posters in cyber cafés, if night-browsing is offered, which is primarily the case in small cyber cafés in the outer quarters, as part of strategies for additional income.

A small range of different software for cybercafés is used, which serves to produce a customer's code to log in. In some cyber cafés the employees pre-print the codes, in others they just write the code on a piece of paper. Regular clients often have their personal accounts, which they enter with a password. Hacking software and "stealing" airtime from the cyber café is also a common issue and sometimes done by a few internet versed people – said to be mostly scammers. Due to these software systems the line is just cut off when airtime is finished, in this sense the user needs to be aware of the remaining airtime, which is indicated on the screen, in order to have enough time to logout from the mailbox. When time is about to finish, users could either call the employee to add time, or they need to go to the counter to buy a new airtime ticket.

Most cyber cafés offer other services, such as copying, scanning, documentation - typing documents for customers - laminating, binding and stationeries. Often there is also an opportunity to make international calls[171], perceivable most often by a separate small call booth, to provide privacy and obstruct noise from the cyber café, usually equipped with a chair to sit. Another service is playing the DV lottery, which is widely used also by people who usually do not frequent cyber cafés, consisting of filling out forms and taking passport pictures[172]. In many cyber cafés, there are opportunities to buy snacks and drinks, or they could be purchased outside, and in some there are also street vendors passing by. Usually it is not allowed to eat in front of the screens. In almost all cyber cafés there are opportunities to sit down on extra benches or chairs for those who need to wait or make a break. Hanging around in or in front

---

[171] The prices for international calls are lower than on call boxes in the streets, because the lines are maintained through the internet, such as by voip or skype programmes. Call boxes are most often only used for local calls. There are call centres, which specialize in transnational calls. However, during the period of my fieldwork in Bamenda there were not many of them. This is why I have not paid special attention to them in my research. There seems to be a difference in this regard compared to the situation in Buea, as it is described by Tazanu (2012:120ff). I explain this difference with the density of public cyber cafés in Bamenda, of which most offered services for international calls.

[172] Compare to figure F10.

of the cyber café is common for cyber café users, in particular among those who spend several hours online. Despite the slightly negative reputation of users of public cyber cafés – due to the scamming phenomenon – youth concerned stated that they were not bothered they were seen entering, leaving or "hanging around" a cyber café, as "nobody knows what we are really doing in the internet". Furthermore, using the internet could also contain a prestigious connotation – being internet savvy and thus potentially successful (compare Lee 1999, Miller & Slater 2001). In this sense, internet could play a more indirect role in the sense of possibly contributing to conspicuous consumption practices[173].

F23:Cyber café at Meta Quarter, Bamenda
F24: Interior of a (new) cyber café in Ntarinkon, Bamenda

Users' motivations for a personal computer and internet access at home

Motivations for having internet access at home are a private space for work, to prevent scams, to browse in a calm environment, not being bound to mix with "irresponsible youth", and conveniently logging in at any time. People who do not live centrally in town also calculate the costs of a taxi to and fro. Even though having a computer and internet access at home has become increasingly common, it still concerns a minority of internet users, and thus, according to cyber café owners, this trend is not yet felt as detrimental to the public cyber cafés.

Nevertheless, I discovered computers in a number of homes, mainly in family homes, where young people live together with parents and

---

[173] I will come back to socializing practices in public cyber cafés later in this chapter.

relatives. Having a home computer already contributes positively to people's use of the internet in public cyber cafés in view of saving costs for airtime, in the sense that they could prepare documents and even emails at home by prewriting them, and taking the documents along to the cyber café on a flash drive. However, it seems to be relatively often the case that those who possess a private computer also install internet access, since the computer is the main investment. The expenditure to purchase a modem, install an internet connection and the expenditures for airtime for domestic use have lowered considerably[174]. Additionally, youth who live alone could obtain a personal computer and internet access, or at least they consider this option as their next financial step and investment. In cases of less frequent internet use, a prepaid version could be chosen, for a family with more than one regular user a standing line would be the better option. It matters on the bandwidth people could afford to install: the pace of the lines seems to be a problem concerning internet at home. However, some of my interviewees who had switched from using the internet in public cyber cafés to their homes, calculated their expenditures for airtime in the cyber café as higher in the long run[175].

An additional reflection is the stance of the computer equipment and security aspects. I had come across a number of frequent internet users – not only scammers – who preferred to come along either with their personal laptop, or they installed their personal computer in the cybercafé they frequented, or they would regularly reserve a particular convenient machine. On their personal computer people are conscious about the security precautions such as the instalment of anti-virus programmes and regular updates. The fact that they could not socialize with friends on a daily basis, is deplored as a drawback of working at home. However, homes could, under the influence of an available computer and an internet connection also turn into a social space of

---

[174] Compare to chapter 1, regarding prices for private internet subscription.

[175] Derrick (Interviews 2009/FS), who had installed the Orange device "live box", including a cable to obtain an optimal connection to the Orange wireless network, said that he had spent up to 100,000 FCFA – approximately 220 Euro - for the installation. He had been saving for this expenditure for a while. Expenditures for airtime would be 30,000 FCFA per month. By working 6 hours a day, 6 days a week, he estimated his expenditures for airtime would be about the same.

interaction of friends and neighbours. At times, computers and internet access at home are shared among friends, who work in shifts. Similar reasons as far as convenience of security and privacy are concerned, is the need for computers and internet access for offices and businesses where it is crucial to be able to work online. It is furthermore also a matter of prestige – a well-functioning office increasingly includes computers and internet access as a standard.

The laptop as a status symbol and working tool

The mobility of a laptop is seen as its primary advantage, as some people expressed in statements such as: "My laptop is my mobile office", in the sense that investing in a laptop could give them the opportunity to work wherever they are located. It indicates people's mobility on a local level, especially young people who switch places, for example between the family home and their single home, going to Buea or Yaoundé for studies, temporarily staying with friends or relatives in another city, spending time in their family's village, and so on. In addition, they could use a laptop in public cyber cafés in order to access the internet, with the privilege of working on their personal machine. A laptop is also a mobile status symbol. They are easily recognizable as laptops when put in a laptop bag and carried around. However, security reasons usually limit people's travel with their laptops. Thus, the use of laptops is not comparable with the mobile phone as a mobile status symbol and multi-purpose gadgets in daily public use. A laptop could also be considered as a more personal device. Laptops are more often possessed by individuals than desktop computers, which are regularly used by a greater range of people. However, often laptops are shared by several people. Laptops are prone to being stolen, even when left at home, since burglaries and attacks on homes – especially unsecured single rooms, which young people rent – are not uncommon. As mobile items, which do not need as much space as a common computer, laptops are more easily transported – notably shipped - in greater bulk, and they are valuable objects, accumulating financial value[176] even in their spare parts. Dealing with laptops is thus considered to be a good

---

[176] Some business people only deal with mother-boards and scraps, the most valuable parts of computers.

business, and likewise laptops are seen as enhancing business opportunities[177].

Computer repair sites and their materialization in the city

Regarding the materialization of computers or their parts in the city it is possible to find most of the computer repair workshops in Bamenda in central town, however, not along the main roads. I am referring here to workshops, where computers are sold and hardware repairs done, which make it a space demanding business. Furthermore, what stands out for me in such computer repair sites, is that computers – no matter whether they are laptops or bulky screens - seem to be less regarded under the aspect of being highly sensitive technology items, as might be the case in a context of industrialized countries. The handling of computer materials, as well as the functionability of the technology in general under conditions of environmental influences, such as dust, humidity, sun and even rain, was sometimes surprising to me – the same is the case regarding mobile phones. Computer repairs, dismantling, de- and readjusting, and splitting an old computer up into valuable and waste spare parts, is a common and a flourishing business. Computers, especially the "outdated" ones with bulky screens, are imported in bulk in containers. They could contain computers that work perfectly, and others, which might not even be worth repairing, but could maybe serve as a reservoir for spare-parts. It is at the risk of the buyer who makes the evaluations, which computers he wants to repair and sell, which he uses for spare parts and which are useless to him. Even though most computer parts could be split up and re-used, computer waste is perceivable in every computer repair workshop and sometimes in the streets. However, in view of great business opportunities in computers and electronics, this is taken as a normal part of daily waste dumping in the locality[178].

---

[177] Software related computer updates, maintenance and instalments are only demanding according knowhow and a laptop with software programmes. Computer savvy people could earn money by offering such services.

[178] See figure F25.

F25: Computer waste in Buea
F26: Open air music download business in Bamenda

*Figure 3.1: Bamenda city centre; Overview situation computer-
and Mobile phone related businesses, 2008, Map: CAMGIS,
Bamenda*

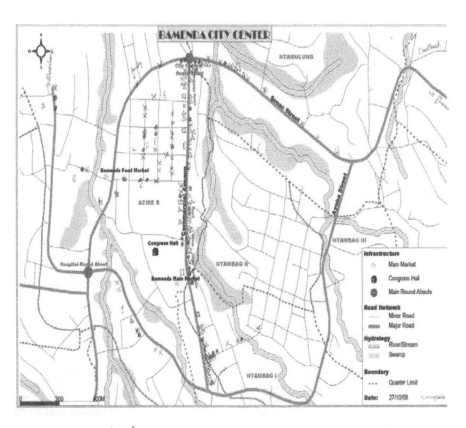

- ● Comp. Repair
- X Handy Repair
- ● Electronical Dev. Repair
- x Computer services + DOC   ✗ + Training
- △ money transfer
- ☐ Telefonanbieter
- ○ Brother Computer Verkauf
- ▮ Handy Verkaufs booths
- △ Mass Photos

Figure 3.2: Bamenda, overview public cyber cafés, centre and suburbs, 2008 (including update 2009), Map: CAMGIS, Bamenda

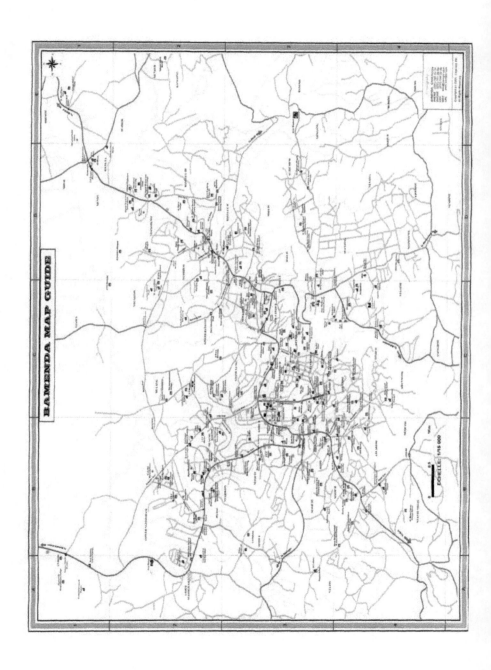

The materialization of mobile phones in the setting

Mobile phones are found virtually everywhere in the public sphere. This poses certain analytical difficulties in regards to how to go about them, but also makes it easy to observe their use. Thereby, I do not see the use of mobile phones as separated from the use of other media of communication, such as internet based communicative features: regarding technologies, they are intertwined, mobile phones have become mobile computers and likewise the internet offers opportunities for calls and writing text messages. Mobile phones are also very prominent in cyber cafés. However, the use of mobile phones, which is not dependent on computer literacy, has been adopted by all age groups and genders, not only in urban but rural areas too – given the relatively good network coverage. The increasingly wide spread use of this technology has also become possible as prices for mobile phone devices as well as costs for communication have dropped. When looking at the materiality of mobile phones, we should consider the following aspects: On the one hand, mobile phones are items of prestige and social gadgets as such, contributing to local face-to-face sociality in their materiality. On the other hand, the presence of absent social others influences sociality on a local face-to-face level, induced by their materiality as being a mobile and personal communication devices. I will concentrate here on the aspect of the material presence and handling of mobile phones[179].

The presence of mobile phone related businesses in the city

The selling sites for mobile phones range from shops up to mobile booths of wood and glass, especially along the Commercial Avenue[180]. Here, regarding some sections of the street, a range of showcases displays phones – almost exclusively new Chinese phones – and accessories[181]. Furthermore, there are various shops where mobile phone accessories and spare parts are sold: some big shops specialize in spare parts, and spare parts are also sold along with mobile phones in shops and the afore mentioned booths. Regarding spare parts or accessories

---

[179] Hereby my observations concern mainly the urban environment and young users.

[180] See figure F27.

[181] I will come back to the social interactions around these selling sites later on in this chapter.

widely in demand, such as chargers, cases, and protection covers for keypads, these are sold in many different places. Regarding the various places where mobile phones are repaired, some of them are well visible for passers-by on the street, including catchy signboards, whereas others are in more hidden places. These sites do not necessarily need much space and they could range from relatively spacious workshops up to tiny rooms[182], a niche in another shop, or simply a table in the open air. Thereby these places consequently differ regarding their stock of spare parts, the range and numbers of phones the repairers are working on at one time, and the technological equipment. Some of these businesses are one-man ventures, whereas others employ several apprentices.

Important in regard to the "materiality of mobile phones" in the city-scape are the innumerable public call boxes along the streets[183]. Selling credit is organized in hierarchically regulated operation systems of contractors, offices providing credit to small vending places, and on the lowest levels the call boxes, where phone cards and credit are sold. In almost every shop, where things of daily need are sold, phone cards and credit are offered as well. The reason for so many sellers is the policy of the providers, who distribute their licenses to sell, and the few investments that are needed in order to set up a call box as a small business[184]. Call boxes are usually made up of a wooden box, painted mostly yellow and labelled with laminated signboards, indicating the offers of the different providers, Orange and MTN and at times Camtel[185]. Most often, MTN and Orange credit is sold simultaneously, and often the selling of credit is coupled with the offer of all kinds of

---

[182] See figure F28.

[183] I will come back to social interaction at call boxes later in this chapter, page 116ff. Compare also Tazanu's description of call boxes and interaction in Buea (2012:124ff).

[184] What is needed is a vending box and equipment, at least two mobile phones and a special SIM card (evd) for distributors to transfer credit. The profit is small: the gain for selling credit for 10,000 FCFA was 700 FCFA in 2010. It is an occupation in order to bridge financial shortages, unemployment, or as occupation during school holidays. Very often girls and young women work at the call boxes, sometimes they are hired by the owner of the call box, or they are family members who contribute to the family's income, or given the chance to earn a bit of money for themselves. In a few cases the vendors own the call box themselves (compare Mbarika & Mbarika 2006).

[185] Camtel is government owned. It employs a slightly different system: the selling of credit is channelled through their main offices. Nevertheless, they are about to introduce new strategies, in an attempt to catch up with their private competitors.

imperishable foodstuffs and other small items of daily use: cigarettes, biscuits, handkerchiefs and sweets – and at some call boxes: books, newspapers, and magazines. Typically the call boxes have umbrellas to protect the vendors and equipment against rain or sun, mainly in yellow, provided by MTN. Yellow and increasingly orange are the dominating colours in the city-scape (Nkwi 2009, Brinkman, De Brujin & Bilal 2009).

Another business related to mobile phones is the "download business"[186]. In 2010 especially, these businesses seemed to multiply, discovered by many young people as a business, which does not require a great amount of investment to start. It is mainly made up of people working in the open air, especially along the Commercial Avenue, only equipped with a computer on a table, loudspeakers, and an umbrella to protect them from rain or sun. In their computers they have a range of different music uploaded according to popular taste. Clients can have music downloaded on their mobile phones, generally for 50 FCFA per song, but often the price is negotiable. Music is usually downloaded on the memory cards of client's mobile phones, in some cases people download music on their flash drives instead. Either people give the download of music to commission, or they spend time sitting, listening to music and telling the service provider what they want to have on their phones. Music is also converted into a specific format to be used as ringtones accordingly. This business is often combined with services such as burning and copying of CD's and DVD's. Acoustically these businesses are even more noticeable than visually: When clients express their wish for songs, the service providers start playing them, and quickly switch to the next one. This has led to a situation where different music can be heard from each loud speaker – the constant switching of music and songs thereby transforms the acoustic environment.

---

[186] See figure F26.

F27: Showcase along Commercial Avenue, Bamenda
F28: Mobile phone repair workshop, Bamenda

The mobile phone as an extension of the body

For most people the mobile phone is an indispensable item in their everyday life. An example is how people carry their mobile phones around. Whereas it is most common for men to have them in the pockets of their trousers, for women it is more common to put them in their handbags. It is also very common to see people carrying their mobile phone in their hands when walking around (compare Horst & Miller 2006:61). Having one's mobile phone at hand is an aspect of a bodily feeling, such as searching for it in the pockets – in a streak of anxiety – when the "feeling of having it there" is lacking, or just as a verification. "I feel amputated when my mobile phone is not there. It's like a part of my body is missing" (Ernest. Interviews 2009/O). Or: "I feel like naked without my mobile, when I have forgotten it in the house for instance. I will continue to search for it automatically, even though I know it is not there with me" (Linda. Interviews 2009/F). In Marshall McLuhan's (1999 (1964)) sense we could call the mobile phone an "extension of the body" (compare Horst & Miller 2006, Kriem 2009, Katz 2003). Another aspect of the mobile phone being symbolically looked at as a body part is the notion of the "phone doctor". Many mobile phone repairers deliberately call themselves "doctor" in their names or in slogans, and clients address the mobile phone repairers as "doctor". In a number of mobile phone repairer workshops the spatial organization of the premises is based on a division of workshop and waiting room, where clients wait for admission[187]. Most clients would prefer to wait rather than leave without their phones: since the phone is indispensable, they hope to have their phones repaired immediately[188]. When en route with mobile phone repairers, even when off duty, I was able to observe people addressing them and explaining their phone's problems. Mobile phone repairers are consulted for advice in regard to mobile phones, in the same way people take medical advice[189].

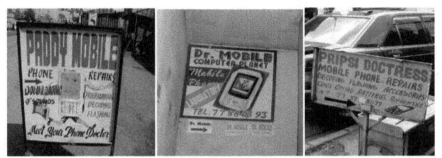

F29, 30, 31: Signboards of mobile phone repair workshops at Commercial Avenue, Sonac Street, and Fon Street, Bamenda

Mobile phones as entertainment devices and objects of attention when socializing

Mobile phones are objects of attention in a material sense – being attentive to them as an individual user or as object of attention for groups of people - and in narratives. A very popular use of the mobile phone is to take it out and hold it in the hand, as it is often observable for example in taxis, or when people are sitting down in a restaurant or any other form of socializing in public places. It is seen as a part of conspicuous consumption to demonstrate connectedness, but also the phone could assume functions for entertainment. In this sense it serves to "pass time" or to circumvent a situation in which one is without company[190] (Goffmann 1963, Humphrey 2005:813). The mobile phone

---

[187] One mobile phone repairer even had a secretary who was responsible to note down the name of the client and the problem of the phone, like a receptionist in a medical practice.

[188] After a quick look at the phone's problem – a diagnosis – the client is either told to wait, or to go and return later.

[189] See figures F29, F30, F31.

[190] This observation is probably a worldwide phenomenon.

is a gadget with multipurpose functions (Horst & Miller 2006, Caron & Caronia 2007), apart from their functions for calling and texting they serve as MP3 or MP4 devices, they are used as a play station, to listen to the radio and music, or to access the internet - however, this service is only used by a small minority[191]. Apart from using the phone for entertainment, the mobile phone's technological features could also contribute to the support of every day life's challenges. I observed that the phone's calculator or calendar was consulted, and many rely on their phones as a watch. Some phones have a torch, or people use the activated screen as source of light, which could be useful when a blackout occurs or when walking in a dark street at night. It is common to use mobile phones to take pictures, since most Chinese phones have an integrated camera[192].

Other aspects of entertainment are the ways in which mobile phone technologies have become an issue of social interaction and activities with others present. When young people socialize, the mobile phone is an object of attention. Thereby they could collectively concentrate on a phone, or each on their own phone, in the sense that exchanges or comparisons take place. Collectively listening to music and looking at pictures or watching music videos are common activities. The Bluetooth function makes it possible to exchange contents: Many interviewees had similar collections on their phones, stars from film, music and sports, erotic pictures, and the latest popular music. Chinese phones continue to be popular, because the need for additional features and technological gadgets is high, not least these socializing aspects are included in these considerations.

---

[191] MTN introduced internet browsing through mobile phones in Cameroon at the beginning of 2008. It was free, and just needed the respective configuring of the phone with the MTN SIM card. After two months MTN stopped the offer. They had underestimated the demand of this service, and that many people possessed phones, which were designated to accessing the internet, as the case with most of the new Chinese phones. Some of my interviewees said they still used the service, although not regularly, since it was expensive, about FCFA 600/hr, but prices coming gradually down in special offers. Since mobile phone screens are small, it was not seen as convenient to browse, but solely to check one's mailbox. It is also used by some scammers when they are en route, in order not to miss any important messages, or business people who are busy and do not often have the opportunity to go the cyber café.

[192] Also in the view that not many people have a digital camera at their disposal.

Mobile phones as objects of exchange and the need to be connected

Despite the impression that mobile phones are personal items, it seems that their personalized use – in the sense of individually being contactable and able to contact – should be considered as the most important aspect. The SIM card and number as personal contact coordinates are vital, but the phones themselves are also interchangeable. According to my impression there is a high turnover of phones among youth: I had come across different habits of lending and exchanging phones. One friend told me that he could not have a phone longer than one or two months, then "It becomes boring to always use the same phone. I will just take my SIM card and swap phones with some good friends of mine. At times we give the phone back after some time, at times we exchange them permanently. Of course we need to evaluate its monetary worth, and we might add some small amount of money to balance" (Richard. Interviews 2009/O). Swapping of phones has several aspects: one was trying out other brands and functions, and taking over music or picture collections – provided those are stored in the phone's memory – from the prior owner. Furthermore, having a new phone is regarded as prestigious. Another habit is to buy and sell mobile phones among friends. "After some time I want to have a new phone, so I ask a friend how much he would pay for it, and I buy another one for myself. I can buy a phone for 80,000 and sell it to a friend for 60,000 after four months, then I add money in order to buy a new phone for 100,000, and so on" (Manfred. Interviews 2009/F). In this sense I encountered several specific phones handed around within groups of friends. Borrowing phones is necessary when someone does temporarily not dispose over a phone. Providing a phone to a trustworthy person for a certain time when that person is in need of one, is a common strategy to make others maintain connections. Having access to phones thus depends not only on financial means, but on social networks[193].

Swapping or borrowing is enhanced due to the fact, that many people have more than one phone. A reason for having several phones

---

[193] Mobile phone sellers and repairers who have access to their client's phones, are at times using their clients' phones for a while to test their functions, while the clients try to raise money to pay the bill.

is the different phones' range of use. Original phones – due to their longer durability and resistance – are often guarded carefully and kept for a longer time: "They have become scarce, and they last for long. It is always good to have an original phone" (Emil. Interviews 2009/R). In turn, Chinese phones offer a large range of gadgets. Some people consider their Chinese phone as extra: it could be sold if there is an urgent need for money, whereas the simple phone covers the need for communication. One of the main reasons to have several phones concerns the urge to have several SIM cards (compare Mbarika & Mbarika 2006) for different communication purposes[194]. It is considered as a necessity to have both providers' SIM cards, in order to profit from respective offers – calls within the networks are considerably cheaper – and also to secure always being reachable and able to call in case people move out of one of the provider's networks, or if one network is disturbed.

Mobile phones as basic necessities and status symbols

Having a phone has become a necessity. "Now like this if I do not have a phone, its not normal and very disturbing, because I have got used to it" (Andrea. Interviews 2009/O). Not having a phone at disposal is regarded as a deficiency (Horst & Miller 2006:59). Also important is the type of phone, how many phones and how long one maintains the same phone. Shame could also be an issue when a phone does not correspond to certain conditions. A friend of mine used to leave the room when making a call, out of fear of being ridiculed[195]. Also some youth apologized when they only disposed over what they call an old-fashioned and "ugly" phone. "In earlier times having a phone was already sufficient, even when you had one of these big phones with the antenna – you remember those? – You were already considered a big man. Nowadays it depends on what phone you have, not just any phone" (Moussa. Field notes 2010/11. 13.11.10). The mobile phone is a means to link up not only to others, but also to "life-style" and prestige (Caron & Caronia 2007:103).

---

[194] However, since Chinese phones all have a double SIM card, this problem could be solved by only using one phone.

[195] I have come across the notion of the "Kumba bread", a notion for old fashioned phones, appealing to its size - a sign of not corresponding to the latest technological developments - alluding to the rectangular bread loaves from Kumba.

In general, original phones are highly valued regarding their high quality and durability. In this case, they could also be small and simple. Furthermore, Chinese phones contain different categories. Certain characteristics are regarded as typical for Chinese phones, such as their "shiny" appearance with a glossy surface and the ringtones, which sound different from original phone's ringtones, and the elaborate technological features – many of these Chinese phones figure in the range of the smart phone generation. The brand is not so important, since there is an endless and always changing array of brands – and they are "anyway all faked". In many cases Chinese brands copy original brands, some are slightly transformed regarding the name – Motocola instead of Motorola, or use a slightly different logo, or they look similar to the original, such as the latest Chinese I-phones. Regarding the appearance of phones, a marker is what is called phones for women and for men. However, these definitions are not rigid. As a tendency, men seem to often have bigger and more technologically advanced phones than women, and there are some specifically "female" phones (compare Horst & Miller 2006:62) with shiny and colourful keypad covers, or a ray of coloured lights shining up when the phone rings. In addition, phones are often decorated according to a gendered taste, with artificial jewels, stickers, strings, chains and key holders. Regarding ringtones, most commonly – apart from the ringtones offered in the programmes of mobile phones' software – are popular music titles[196].

The display of phones is partly limited, since they are prone to theft and loss: Hiding them in pockets and bags is seen as more adequate. In a secured and familiar environment, phones could be exposed – it is not uncommon to see somebody exposing two or three phones at once. Since almost everybody has a phone in the urban area, and phones with elaborate features are wide spread, it is not easy to impress with a mobile phone: it rather seems that the meaning of mobile phones as status symbols relates to their users' ability to keep them in use. In relation to being connected through mobile phones is more a matter of the phone's use. A possibility to demonstrate connectedness is using earplugs or ear-bugs, small wireless receptors worn in one ear, which are used by a

---

[196] At times I had come across "religious sounds", such as Christian church songs, or Muslim prayer calls (compare Horst & Miller 2009).

139

considerable number of youth. Or it could be publicly calling and being called, receiving calls from abroad, having more often different or new phones than others, or disposing over scarce original phones, which could point to connections to somebody abroad. In this sense it is the combination of material aspects and the phone's use, which makes mobile phones become part of conspicuous consumption practices.

The mobile phone as a reason for conflicts

The materiality of mobile phones, being handy and small, but valuable, also contribute to their being prone to theft, loss and disappearance. Since it is related to connectedness, losing a phone could have a catastrophic dimension, less because of the financial worth of the phone, but more so regarding the loss of social contacts[197]. Mobile phones are not only prone to being stolen, but they could be given as a deposit, for example when owing money to somebody. Or they could in case of debts be forcefully taken in order to urge the debtor to pay back. I have mentioned the possibility of selling a phone when their owners are short of financial means, but in these examples here, the importance of the mobile phone beyond its financial worth is apparent. I had witnessed situations, when tensions were directly related to mobile phones, be it that they were the reason, or a catalyst for provoking quarrels – quarrels over mobile phones is common, especially among groups of friends, where mobile phones are displayed, exchanged and handed around. For instance quarrel ensues when a mobile phone's ownership is not cleared, somebody has borrowed a phone and not given it back, or a phone has disappeared. Another issue are accusations of having caused a technological problem in somebody else's phone by playing around with it. Or tensions could occur when another person's phone has been destroyed accidentally, as expressed in unfortunate stories that somebody has driven over a friends phone with a car, or has dropped it in a gully or a toilet. Other points of contention include using another's phone without consent, finishing another's credit, or even "stealing" - transferring credit from a friend's to one's own phone.

---

[197] At least, the opportunity to retrieve SIM cards, possible through the new system of registering SIM cards, makes the issue less severe than it was before. However, most mobile phone owners have stored a considerable part of their addresses in the memory since the capacity of the SIM card is limited, which means that a part of the numbers is lost.

Keeping the mobile phone in use

Not disposing over a phone in the urban area is seen as a temporary, not a permanent state. Apart from theft or loss, technological problems are reason to borrow phones, or maybe to persevere without a phone for a certain time. "My phone is bad", could serve to explain when somebody is temporarily disconnected. People could then understand that the person needs to repair the phone and that it takes time to raise the money for it. In order to be able to stay connected, it is then important for "the disconnected", to let, for example, friends know that their phone number is still valid, just temporarily not reachable. It depends on peoples being connected to others, how easily they could access a new phone or borrow one.

Whereas in urban areas "people are supposed to have a phone", the sharing of phones is of greater importance in rural areas, where people have less chance to acquire or repair a mobile phone. To keep the mobile phone functioning and in use in a very basic sense, for example, is to make sure that the battery is charged. A problem with the Chinese phones is that the battery only lasts a short time, in the sense that daily charging of the battery is necessary. It means also that people often need to move around with their mobile phone charger. Mobile phones are charged in many places in the public sphere, such as restaurants, business places, car parks and travel agencies[198]. In rural areas uploading a phone could pose a problem, since power supply is not available everywhere. However, electricity infrastructure in urban areas is also often deficient[199], and power failures could be a problem: In the summer of 2009 when they lasted for days, sociality was severely disrupted. Thus statements such as "my battery was flat", serves as an explanation why somebody could not call or be called. To keep mobile phones in use also means that it needs to be constantly switched on, in order to "never miss a call" – or not miss opportunities. "Now you feel so secured with your mobile phone with full battery, ... you know that information can come through (any time), that will change your life positively (...)"

---

[198] See figure F32.
[199] See figure F33.

(Festus. Interviews 2009/F). The main problem of keeping the mobile phone in use, are however financial means for credit[200].

F32: Charging a mobile phone in public (Primus Tazanu, Buea)
F33: Inconsistencies of electric power

New Media's influences in the public sphere - materiality, discourses and critique

Not only are the media present in their materiality, but mediated experiences intermingle with what I intend to call "the immediate", which build together the experience of the life-world and meaning making (Geurts 2002, Krotz 2001, Rydin & Sjöberg 2008:40). Contributing to such intersections are a high medialization and accessibility of New Media in the urban sphere. They form part of overall different information and communication media on a local level. The result is a continuous negotiating of qualities and properties of spaces on the local level and beyond (compare Cresswell 2006). When using the notion of the "city space", I relate in the first instance to a public space or sphere[201], where people come together, interact and communicate. A space can be experienced as public, and at the same time it also encompasses private spaces. Public and private space can

---

[200] I will come back to this difficulty and its influence on practices of mobile phone use in chapter 5.

[201] The English notion of the "public sphere" or -space is underlining the "spatial" in a more differentiated way than the German notion "Öffentlichkeit", that is why I would prefer the notion for these examinations (compare Iveson 2007). I relate here to the notion of the "public sphere", rather as multiple overlapping public spheres, which are related to specific groups of common interest, than to "the public" in Habermas' sense.

exist in tangible material or in "virtual" form within the minds of people and within media. These relate to modi of social practices, which are visible or invisible to others, and involve individual or collective intentions and actions (Weintraub 1997, Förster 1999). From a material, locational point of view, public and private spaces blur: they cannot be attributed clearly either to a set of specific collectives – groups of people – nor to collective activities of these groups. Which activities or collectives are to be considered as private or public[202] becomes an issue of continuous negotiation, related to social and cultural norms. Media contribute especially to the transformation of private-public distinctions[203] and lead to an altered view upon the materiality of private and public spaces (Habermas 1990, Iveson 2007) and related practices. The public sphere must also to be examined regarding resources and diverse interests of different agents, on a more global scale and rather locally induced. First, I refer to social practices of New Media use regarding their users' different ways of connecting, and respective conditions, which are constitutive for the shaping of social spaces. Secondly, I relate to narratives about New Media and their attributed qualities for sociality, and thirdly, to pre-requisites of availability and access to them[204].

Notions of private and public regarding New Media practices and sites

I have already related to how media in general permeates public and private spheres in the previous chapter. Regarding the media sites themselves, for example the public cyber café could be seen as a "hybrid" place (Iveson 2007), where private and public intermingle in complex ways, as spaces in a material sense as well as regarding online and offline social interaction. On the one hand, it is indeed a public space, a gathering place, where people interact and exchange socially[205].

---

[202] The notion of privacy needs to be evaluated differently in the context, as compared to a "Western" notion. Privacy in the sense of withdrawal, of invisibility towards others, has not had a stance in many forms of co-residence in an African context so far. However, this has also greatly transformed, in particular in urban settings.

[203] See figure F34.

[204] This sub-chapter relates to the 1st sub-question of question 2, how technology's materiality influences their use and vice-versa, and the 3rd sub-question, which narratives of social change are provoked by the presence of New Media.

[205] See figure F35.

Nevertheless, regarding the collectives within this setting, even though not exclusive and generally open "to everybody", it is a place of belonging for certain individuals and groups of people, people who are computer and internet literate, able to afford airtime, and therefore excluding others who do not meet these preconditions. Moreover I also refer to more inclusive groups of people who know each other and meet in the cyber café on a daily, or at least regular basis - as for example the scammers. Cyber cafés could thus be seen as semi-public spaces. Regarding activities, which are effected within the space of the cyber café, public and private are equally blurred. On the one hand regarding the offline setting, which contains both a social space of exchange, and as well as a private working space. Then, also the blurring of public and private relates to online activities: Online private spaces or activities contain personal emailing whereas contributing in a public chat-room can be considered as a more public activity. In general, most forms of the consumption of digital contents are consumed publicly and privately (Darley 2000:182,183). However, in the example of Bamenda, the consumption or use of digital media – internet in public cyber cafés – mainly takes place publicly. Limited privacy could have implications on categories of – from a Western point of view – private and public contents, which may be blurred by respective practices, for example looking at porn – sometimes collectively – in a public cyber café[206]. The need for privacy regarding interpersonal communication and also business practices has arisen with the differentiations regarding the use and accessibility of media technologies. As such, cyber cafés are spaces, which support both a room shaped for collective public interaction as well as private activities on the screen. It is therefore possible to speak of a "double sociality".

---

[206] Looking at porn was more common in public cyber cafés in 2003 before the spread of mobile phones, whereas nowadays such activities are mostly effected on mobile phones. This is however not an issue of examination in my research.

F34: Signboard in church, asking the congregation to switch off mobile phones during church service
F35: Social interaction in a cyber café

Whereas public cyber cafés are separate spaces as part of public spheres, mobile phones permeate the public sphere since they are mobile devices. Moreover, mobile phones transcend the boundaries of private-public as they can be used privately and publicly alike. Phone calls in public are perceived as common in the local context, there is no specific attention given to a person speaking on a mobile phone, which means that a certain civil inattention (Goffmann 1963, 1972) is given. Before the advent of the mobile phone, there were few public land lines, a caller was most probably in the presence of many others waiting to call, attentively listening to what he or she said, also depending on a societal framework of mutual knowing and relatedness (compare Nkwi 2009). However, mobile phones also allow the withdrawal to more private spaces in a society where social issues are often discussed in public spaces, or within larger social configurations, such as extended families. The mobile phone offers opportunities – since it is a personal device – to regulate private and intimate matters and relationships in more concealed ways (Archambault 2009). Mobile phones also contribute to the local public sphere and face-to-face social interaction due to their presence – be it through their materiality, as described in the previous subchapter, or be it through their bringing together mediated and immediate social interaction, as I will look at in the next subchapter. We could here also speak of a "double sociality". It seems that the use of the internet – as activities in a virtual sphere performed in public cyber cafés or at home - as well as the use of mobile phones – here regarding the difference of using the personal phone in privacy or at the public call box - indicates discourses on social morality[207].

Media discourses, and discourses about New Media

Certain issues are mediated through various media in the public sphere (compare to Rajewsky 2002, Beisswenger 2002, Sturken & Cartwright 2004) and therefore influence a wider audience or public[208]. Nevertheless, media are not only reflecting societal discourses, but they are objects of address themselves. Part of these discourses was "talking about" media in co-presence. When sitting together, the technological features and outer appearance of phones are often a topic of discourse. Furthermore, in public cyber cafés, technology itself is an issue. Above all media contents are issues to talk about – pictures, music videos and games on people's phones, or diverse media contents in the internet (compare Caron & Caronia 2007:108). The other part of such discourses relates to a meta-personal and societal level of discourses. Discourses about how media shape the public sphere have often been ambiguous. In general internet and mobile phones are ascribed undoubtedly as being positive influences, for many people they are indispensable in everyday life. However, negative impacts of media – in particular on youth - are widely debated, also in Africa. These discourses are also informed by shifting cultural and social norms of private and public.

The internet has a special stance in the array of media. This might be interrelated with the fact that it is a medium, which many people do not have much knowledge of: it is known as an access to a realm of potential and dangers, an access to "another world". The "dark side" of technology is often portrayed in regard to the internet. By far the most popular discourses regarding the harmful influence of the internet on youth in recent years, is the "scamming issue", since it is a phenomena

[207] This is related to the fact, that privacy is also a matter of economic prosperity: Having a private internet access at home, or being able to afford the use of a private phone. It points to youth's social position in society and their activities slipping out of control and out of the comprehension of elders and "the public".

[208] In Cameroon media has been liberalized since 1998, and are therefore quite free to publish critical reviews, also in the local print press. Nevertheless, it is common, that newspapers and private radio stations for example, are closed down repeatedly by the state authorities, measures, which are not exercised on the publishing of homepages, which is more difficult to control. However, sites, which address political issues, are mainly hosted abroad (Nyamnjoh 2005a).

widely perceivable in the local context. The possibilities for concealed and incomprehensible activities in the realm of the internet in the scamming field have raised discourses on morality, youth, and their position in society[209]. This is again fuelled by the rise of distinct media spaces such as the public cyber cafés as youth realms, where youth could be seen as slipping out of the control of the public, in the sense of building sub-publics or sub-cultures. Furthermore, the discourse on scamming has been escalating, serving as an explanation for many "negative tendencies" related to youth and youth's conduct and appearance in the local context. The internet is also seen as an insecure realm for internet users, due to its interactive nature, an inconceivable range of contents, and sometimes the lack of internet experience of the users. When using search engines, activating a link, or by windows popping up unexpectedly, and as well, in public forums and chat rooms, and on dating and friendship sites, users are often confronted with non-expected or non-wanted – and sometimes confusing - information. Very often, such kind of information is then seen as potentially dangerous, be it by their immorality, their unknown source, or the seductive offers that they make. An example of this was given by a young woman: "I never use links, or open an email of which I do not exactly know who has sent it" (Clothilde. Interviews 2009/F). She related to her fear of being led to a "potentially dangerous site", and not of a computer virus or Trojan horse. The fear of occult sites, which could harm when browsing on them and even more so when engaging in communication with somebody on them, is widespread especially in the field of scamming. Another issue are immoral sites related to porn, or the links to porn videos, which are sent through chat and dating sites. "When I see something like this, I immediately delete it" (Bertina. Interviews 2009/R), as expressed by a young woman, and as a young man said, he once followed such a link: "There was a white girl on the webcam, showing me her breasts. I was shocked" (Manfred. Interviews 2009/F). Some interviewees related to the immorality of the internet as an "invasion of the immorality of the West into the African sphere", "It contributes to spoil youth. Even homosexuality spreads like this"

---

[209] However, such discourses have always been an issue, for example concerning social youth movements, occult spheres, or new realms of youth empowerment, in particular since the colonial era.

(Herbert. Interviews 2009/R). Narratives circulate of young Cameroonian women – and increasingly men – who chat with men/women in white man's kontri, by using webcams in various cyber cafés in – mainly – Yaoundé: "Have you seen these séparés in cybers in Yaoundé? They need some privacy because they undress themselves in front of the screen" (Jack. Field notes 2009. 18.09.09). Another issue is scams widely spread in different fields on the internet. It could be false information on faked sites, such as those related to immigration and visa information or job offers in the field of "migration scams" for example. Furthermore, notorious are "personal scams" on dating sites (Smith 2007:79,81), as experienced by a female friend, when she was looking for white men on a particular dating site. These scams consist in very personal emails, based on communication exchange, promises, and in the end a fee to pay. Related to these various narratives circulate concerning warnings of specific sites, and stories of those who have been scammed on these sites respectively.

In a similar sense of negative allusions to technology and its use, mobile phones are related to the disruption of social cohesion in the sense that they are seen to enhance deceit at the level of interpersonal social relationships and business practices. In addition, mobile phones are seen to provoke a transformation of sexual comportment (compare Nyamnjoh 2005a) on a local level in presence availability. This issue is in particular related to female sexuality, as women could more easily slip beyond social control (compare Nyamnjoh 2005a, Archambault 2009). Mobile phones are seen to enhance the possibility of deceit in relationships between husbands and wives, and lovers, since it is easier to maintain a large social network with the help of the mobile phone. The issue of maintaining a network of potential sexual partners is also mentioned by Horst and Miller (2005) for youth in Jamaica (compare Jua 2003:20). Resulting are continuous rumours about marriages for example, which split up, and the mobile phone enhancing the spread of HIV Aids. The use of different SIM cards provides a possibility to keep social networks apart: when – in particular men – use different SIM cards on a level of interpersonal social relations of intimacy, it is said that one is for the wife, and the other for the "deuxième bureau" or the secret lover (Nyamnjoh 2005a:318). Keeping an extra SIM card for secret relations or calling instead of writing text messages, are security measures in order to conceal the contact. The popular narrative hereby is that of the jealous wife, who would control the phone of her husband, to see with whom he is communicating, and who would even try to call the numbers in his phone. Also often related to mobile phones, is the issue of lying as a threat for morality of conduct, concerning relations between opposite gender but also beyond (Horst & Miller 2006:101). Lying through mediated communication is seen as having become a habit, which has become public. When somebody conveys for example wrong spatial coordinates in a mobile phone conversation, it is often related to cheating since the communicator seems to have concealed reasons for lying about his or her current location. Lacking social clues could lead to the enhancement of opportunities for deceit (Horst & Miller 2006:98): "Deceit is always normal with communication, but especially when you cannot see the person face to face. If I want to tell a lie on the phone, it might be easy, because I know you are not there to judge my moves and facial expression" (Festus. Interviews 2009/F).

These discourses, which range around deceit and mistrust in mediated social relationships, are related to incomprehensible and hidden activities in an online realm, but also to social interaction in presence availability[210]. Morality in social conduct – related to intimate relationships, sexuality and porn, but also notions of a morally conducive sociality and solidarity in general – have become a matter of public concern, because these issues seem to slip out of public control.

Perceived drawbacks of technologies – serving as a catalyst for critique

New Media technologies and their all-pervading importance in the public sphere, specifically in the urban context, are also susceptible to what could be called "technological failure". Technologies are seen as part of infrastructure, which ought to be provided by the state authorities, or indirectly, as the state gives out service mandates to private companies. Other players[211] are the global telecommunication companies in the market in Cameroon, their price and assessment politics, which influence the access and use of New Media as internet

---

[210] The idea that New Media foster deceit is related to an overall perception of deceits taking place in face-to-face everyday social interaction. Compare to chapter 6 and chapter 2. Compare also to Tazanu (2012:149ff).

[211] I have already related to these players in chapter 1.

and the mobile phone on the local level. I will try to show in this subchapter how inconsistencies in technologies are dealt with, but also valued and interpreted in a context, where these technologies and being connected is considered to be of vital importance. The local use of technology is shaped by assumptions and expectations related to their performance in different contexts and situations.

Within the array of internet and mobile phone performance could be measured by network coverage. The telecommunication providers use this for their publicity, for example by quoting: "Already 80per cent covered by Orange network", or "4.000.000 Customers, and still counting..." by MTN, and so on. In an urban area it is assumed that network coverage is given, and internet and mobile phones were supposed to work. Even in the countryside the expectations of network coverage is high, and having network or not refers to shifting levels of "bush", a village where network is there, and where this is not the case, where "one still has to climb a hill". Regarding functioning and materiality of technologies within urban areas, as in Bamenda for example regarding the cyber cafés, apart from slow lines, emerging from a congestion and not enough bandwidth, drawbacks mainly constitute frequent power failures. According to the season different "types" of power failures occur – in the rainy season lines might be destroyed by landslides and heavy rain, and this season is known for long absences of power, for hours or sometimes even days. The dry season is known especially for the low voltage problem, and thus short but repeated power failures. When longer power failures and shortages occur, as was the case in August and September 2009, when I was in the field, when it lasted for days, or electricity was only available for a few hours daily, the whole city was paralysed. In the case of cyber cafés, many did not even open as the situation was too unreliable and customers would not even come. Many cyber cafés had lost a part of their equipment, such as computer systems, which were heavily damaged by continuous and sudden power failures, since most of them could not afford back up power systems to allow a safe shutdown of the computers. Generators were widely bought in these days, but in the case of cyber cafés, generators could not balance the power needed for operation. When "lights went off", as it is expressed in Bamenda, people in cyber cafés used to scold and swear, and depending on the estimation of the situation, whether a rather short or lengthier power failure could be expected, they kept on sitting in front of their screens, or they left the cyber café. Power failures were commented with exclamations such as "ah, Sonel", or "AES"[212], which is an abbreviation of "Always Expect Shortage" – instead of American Electricity Services (compare Fokwang 2008:30), related to the electricity company AES Sonel, in a kind of fatalistic expression. The power failures are attributed to an assumed non-interest of the government in enhanced power systems in the North West region, interpreting it as an example of discrimination and marginalization by the Francophone dominated government in the country: "Bamenda is the only bigger city in Cameroon, where you have so many power failures" (Bertina. Interviews 2009/R). A further topic is the technological equipment in most public cyber cafés in Bamenda – most have old and slow computers, including bad keyboards and a high sensitivity for failures as well as soft- and hardware problems. Even though people are used to it, some have higher expectations regarding the functioning of technology. Internet users in Bamenda are also aware of the fact that they use the old "worn out computers people in the West want to get rid of".

Another interesting aspect is that technological inconsistencies often also serve as an excuse for failed communication[213] (compare Caron & Caronia 2007), such as the lines are not passing, the network is unstable, power failures, messages lost or not received – which could be seen as "soft" explanations, which everybody uses, which could never be verified, and which are thus also accepted as an excuse to a certain extent. Or connection could be there, but not corresponding to expectations, as one interviewee expressed: "At times lines worry too much. They are not always clear. I will talk and after five minutes she (the sister in the US) says aaahh? And that's all about it… we stay for some time as if it has stopped. Wetti? (what?) The conversation does not

---

[212] AES Sonel is the strategic partner of the Cameroonian government for electricity production, importation and distribution in Cameroon, emerging from the privatization of the electricity sector in 1996, originating from the American AES-Corporation. However, there is virtually no competition in this sector in Cameroon. AES Sonel is so far the only company in this sector, which may also have an impact on the service. Compare to: http://www.aes.com/sonel

[213] I will relate to this in more detail when I will discuss mediated communication practices, see chapters 4 to 6.

flow" (Valeria. Interviews 2010/11). As I have tried to describe, such drawbacks are often used as a catalyst for criticise of the government for example, the embezzling politics, and global injustices (Larkin 2008:152,153). Failures of technology adding to difficult circumstances, have an impact on people's perceptions of "inconsistencies of modernity" and feelings of disconnection.

In this sense, the consciousness for the public and private in uses of New Media, narratives about "the negative side" of technology regarding enhancing deceit in general and within personal social ties, as well as the perception of failures of communication technologies, are in the same vein of as a perception of insecurity or uncertainty regarding being connected.

Different loci of sociality – from face-to-face to mediated social interaction

Media of communication and information as the internet and mobile phone encompass complex configurations of, on the one hand reaching out beyond the local context in diverse scopes, and on the other hand creating a specific "design of locality" shaped by related media practices. It means that the local is co-produced by the specific imaginations of these spaces beyond the locality. However, also those not physically present influence the local life-world and practices. Appadurai goes beyond viewing locality in a spatial sense, but relates to "locality" rather as a "phenomenological quality", as relational and contextual, produced and reproduced by "(…) social immediacy, the technologies of interaction, and the relativity of contexts" (Appadurai 1995, 1996:178, compare Crow & Allan 1994), in the sense of "a structure of feelings, a property of social life" (Appadurai 1996:179, compare Tufte 2002:241). I will relate to three examples of social spaces as frameworks for social interaction, where New Media play a role - as communication tools, but also as symbols and materializations of imaginations, as well as in narratives about them. These examples are the "home", the public cyber café, and social spaces in the street. Thus, in this subchapter, in continuation of the observations regarding internet's – computer's – and mobile phone's materiality, I want to explore the relationship of these medias' materiality and presence on a local level in materially – and symbolically – confined social spaces, in order to look at how their presence contributes to social situations in these spaces, and how mediated social ties ad to face-to-face social interaction (Tufte 2002:241)[214].

The home – representing and relating to absent family members

The first social space I intend to describe is the "home": such as the family which could be seen as a primary node of sociality, the home could also be considered as the primary location for sociality within the family, depending on physical co-presence and localization in time and space of the family members[215]. It includes a notion of privacy, even though rather on a wider family's than on an individual's level. I want to relate to privacy more in the sense of a safety provided by a home, a place of intimacy for members of a household (Miztal 1996:165), being co-resident or not – as for example some youth live apart from their family's compound, but consider the latter as their "home" (compare Mayrhofer 2003). The highly considered importance of family and solidarity is also reflected for example in the interior of homes: Living rooms are representative spaces, they are mainly set up to receive guests, emphasizing prestige and comfort and thus success of the residents. It is also a space to represent family members even those in absentia: on walls or sills, pictures of family members have their prominent place, and also, photo albums are kept in a cupboard or under the tea table for example[216]. It is always the first act when welcoming a visitor, in particular when the visitor had not visited before, to show him or her the photo album and comment on relatives and friends depicted, by explaining their life situation, where they are located and what they are

---

[214] I thus specifically relate to the 2nd sub-question of guiding question 2, how face-to-face and mediated sociality ad to each other on the local level.

[215] A broad meaning of "home" encompasses the "home" in the sense of a place of origin, of a place of co-residence of family members, from the level of the house or compound of the individual's family, up to the village or wider chiefdom, and up to the level of a nation state (compare Wiles 2008). "Home" is transitional, multi-layered, and one could have several homes with different meanings.

[216] Young people most often have their own photo albums – some of them dedicated to their friends. It is a common habit of young people in particular to go in pairs or groups of friends to a photographer's studio in order to picture themselves together, as a representation or symbol of their friendship and relatedness (compare Egloff forthcoming).

doing. "This is my sister Linda, she studies in Yaoundé and will make her final exams next month", or "This is my cousin Divine, he is married to a Belgian and lives in Belgium (Cynthia. Interviews 2009/O), and so on. By relating to the success of physically absent relatives – but also the ones within presence availability, the source of comfort in one's home is emphasized, as people in question are taken into account to support the family.

Regarding being connected from home, better off homes sometimes have a landline phone, however, in most cases communication from home is provided by using mobile phones. Mobile phones are rather privately owned by family members, but sometimes people leave their phones at home because a member of the family has none and needs one for a particular time. In some cases people have a "house mobile phone" which is left in the house most of the time for the members of the family, and thus is a phone, which is collectively used. In a few homes, a computer with internet access is available. A motivation for having an internet connection at home is the fact that some families have children or other relatives abroad, with whom they wish to be in frequent contact. Most often in family homes, the internet users share their log in details. However, mailboxes, similar to mobile phones, are considered to be rather "private" spaces for communication. Sharing New Media communication gadgets – in the sense of a "house phone" and a "family computer", could have differentiating aspects of opportunities for privacy or collective communication with physically absent others. This is given due to opportunities to talk privately – leaving the room with the mobile phone, or to correspond through a personal mailbox. At the same time New Media at home could also be used as a means to socialize in the sense of a collective "family-event". Several youth said that they played computer games with family members or friends at home, browsed together, or they chatted collectively with absent family members. An elderly man whose sons and daughters were in different locations abroad, told me how they had united the family through a conference call including webcam on Christmas eve: "We were all gathered at home, and all of them out there were also in front of their computer screens. Everybody could talk to everybody, we interacted as if we were present there with each other, we even ate together. Our gathering lasted the entire evening, it was an unforgettable event!" (Alfred. Field notes 2010/11. 28.12.10).

The public cyber café – intermingling of online and offline social interaction

The second social space is the public cyber café. Cyber cafés are interesting spaces concerning their specific kind of sociality, a place to work, but as well for socializing and for "hanging out", in offline but also in online spaces (Miller & Slater 2009, Lee 1999): I have already described the cyber café as a space for youth social interaction. It is therefore an important site in some youth's socio-spatial geography (Tufte 2000:256, Miller & Slater 2009:362), especially the ones involved in scamming activities. In some of these cyber cafés, people interact in very lively ways, they sometimes comment loudly on their activities, standing behind each other's computer screens or talking in a loud voice across tables, discussing or assisting each other with technological problems. This interaction is based on a certain level of familiarity with each other[217]. They are also however, carefully watching over the non-integration of others when it comes to business, which is about to be completed, in order not to fall prey to being scammed by co-scammers[218]. Additionally, when related to pursuing private social relationships, such as chatting with friends or writing and reading emails not related to work, these activities are often regarded as "private issues". However, some activities are shared with others, as in the example of looking at pictures on the internet as a collective pleasure which is common, thereby calling each other: "Come over, you have to see this!", or also forwarding them to each other. In this sense, online contents could be a subject of face-to-face conversation. While at work not only scammers, but also other frequent and skilled internet users have opened different sites, email boxes and chat at the same time, in

---

[217] Scammers often utter mistrust towards „strangers" in their cyber cafés, being suspicious that they could install spy programmes on the computers. This is a reason why scammers are usually very stable users of cyber cafés, in order to know the conditions of peer internet users, technological standard and security. Often, if switching cyber cafés, for example when trying out a newly opened one, they do so by dislocating as a whole group.

[218] Spying is mainly directed at taking up on-going businesses when they are about to be finalized, and pocket the gain instead of another person.

this sense they are continuously receiving and sending messages, or chatting simultaneously with different people, and doing other tasks on the internet, such as browsing, writing a text for an ad, and so on – simultaneously concerning private and "business" issues alike. Caron & Caronia (2007:23) have called this a simultaneous polyvalence in media use, as part of the media habits of frequent internet users. Internet related features of communication are used for social ties ranging from a local to a translocal level. I often witnessed people chatting online with friends who were in spatial proximity. Mediated communication could also be used in a situation of co-presence, to keep something secret, as in the example of people working in the same cyber café could use chat – as they are online and have each other in their "friends lists" in messenger chat – in order to make a statement that they do not want to share with others present. The mobile phone is also present in cyber cafés. Sometimes, scammers receive international calls from their potential business partners. They could leave the cyber café when receiving such calls, in order not to disclose information to their colleagues. In other cases, however, phones might also be exchanged, and given to a friend to answer the call, depending on how far a business has advanced. Once I witnessed a scene when a "dupe"[219] from Spain called. The receiver of the call put the caller on loud speaker, apparently to create some fun (compare Smith 2007:41). This mixing of online and offline communication with absent and present others through different media, and thereby the mixing of public and open, or private and hidden activities, lead to interesting situations of interaction in cyber cafés.

Street life – mediated communication adding to face-to-face social interaction in a public sphere

The third social space I will describe is social spaces in the streets in lively quarters of Bamenda, such as along the Commercial Avenue for example, which is a hub for business and small shops and vending places of various goods. During the week there is a high turnover and movement of people, which ceases after the shops close around 6 or 7 p.m. It is a highly mixed public in the street, old and young, locally based and those coming to Bamenda from the village. It contains the whole range of social connectedness, meeting of strangers as well as of intimate

---

[219] A potential fraud victim, to be duped, or scammed, also called a "mugu".

156

friends, thereby including mostly face-to-face interaction but also mediated contacts. It is an urban space, in the sense that even though many people interacting know each other, it is also characterized by highly purposive interaction[220] among vendors and potential buyers. There are also a high concentration of mobile phone related services and goods, such as the selling of mobile phones and accessories, mobile phone repairer's workshops, many call boxes and places to download music on mobile phones. In front of the shops, which are located in one-storey buildings or in containers[221], there are various additional mobile showcases and booths, which are removed for the night. Mobile phones are not only popular as devices informing social interaction among people along the street, they are also the main means for communication to keep the business going. I often spent time with a friend, John, who was in his mid-twenties and a successful dealer in mobile phones. He always had a wide range of different models in his showcase, and he often called his business partners in between conversations with clients, to place orders for phones from China and Dubai. A very common scene on Commercial Avenue is youth congregating in front of shops or the mobile stalls, whether they work there, or they are just acquaintances or even just passers-by. A group of mobile phone sellers, who were my acquaintances, often sat casually nearby their stalls in order not to miss a potential customer, they also had a lot of time for chat and pleasure. Often they looked at the mobile phones, which had just arrived from Douala seaport and demonstrated the latest models to each other[222]. They also used their mobile phones for entertainment, most often collectively, such as listening to music or playing games.

Another dimension of social interactions observable in this space is the receiving and making of calls in public[223]. This concerns people's

---

[220] Nevertheless, it is also common that a seller or provider and a customer know each other, or often a seller might have been personally recommended to the client by somebody known.

[221] These street scenes with the containers had changed very much in 2010, when all containers – overall Bamenda town – had been removed by public decree, "to make the city look more nice", also in view of the elections previewed for 2011. I am relating with my description here to the state of the Commercial Avenue before these transformations.

[222] See figure F36.

[223] See figure F37.

private phones, as well as calls done at the various call boxes - call boxes are more places for calling out. Or people buy credit to be transferred to their personal phones at the call boxes, usually rather small amounts not exceeding 2000FCFA, at times also very small amounts for just a few local calls, from 100 FCFA onwards. People passing by or being around could hear what a calling person is saying, since the spatial scope to depart from the callbox for the caller using the call box phone is limited. However, when the decision is made to make a call from a callbox these conditions have already been taken into consideration. Calls in public are perceived as something common and are usually not given much attention to (Goffman 1963, Humphrey 2005). The interaction between clients and the seller of credit at the call box is often very short and concise and concentrated on the call and its costs. Of course, some people have "their call boxes", and when buying credit or doing calls, it could also be used as an opportunity to chat. As credit is usually bought in the instant one is in need of it, however, interactions are often more purposeful. I observed relatively frequently that clients treated the call box vendors, most often young women, in a condescending manner, corresponding to social status and gender hierarchies[224]. When somebody wants to make a call, the conversation is often limited to indicating the provider by for example "Give me Orange", taking the phone reserved for the provider, making the call, and then silently paying a 50 FCFA, since most people are making short calls and are conscious about the length of their calls. Or they ask the price or start to complain when it exceeds 50 FCFA, when exceeding 59 seconds. When buying credit, customers first say the amount they want to expend for the credit, then type their personal number into the phone, or dictate the number to the call box worker. Some clients then await the credit transfer indicated by SMS on the spot, or they just move on, since at times it takes a bit longer until the credit is sent to their number. Call box vendors are at times addressed verbally as "callbox!" by customers, to attract their attention, especially when they stop by in a car or on a bike, calling to the vendor from the roadside (compare Tazanu 2012:120).

---

[224] Even though call box vendors often have a higher education, or are students earning money in the semester break.

In short, these examples concern different ways of managing space in specific localities, including distinct conditions. The social interactions of various kinds, which occur here, span up between co-presence and absence, social closeness and distanciation, between inclusion and exclusion of others, visibility and invisibility of practices, demonstrating and concealing, between face-to-face and mediated interaction. We could hereby speak of a doubling of sociality, the face-to-face realm where communication acts take place, and the communicative interaction with physically absent others. Thereby, practices of mediated sociality could more easily be detached from the public sphere[225]. Privacy, which could be obtained by certain uses of New Media, seems here to contradict the notion of a morality, which presuppose social control over, in particular, youth sociality, which could now be more easily withdrawn from a public sphere.

F36: Newly delivered mobile phone models, Commercial Avenue
F37: Call box, Ntarinkon, Bamenda

Interestingly, it seems that in a society where a clear private-public distinction regarding socializing and communication has never played a significant role, these media raise fears of concealment, whereas in Western societies with a distinct idea of privacy – where a private sphere for sociality is highly sanctioned and presumed – the same media are attributed as having "dragged private issues into the public", where there ought to be no room for them. I have also related to media realms as spheres of potential and dangers, which could be compared to insecurities about realms of the West, where migrants reside. Thus,

---

[225] I relate here to the 3rd sub-question of guiding question 2, how narratives of social change are provoked by the presence of New Media.

anxieties about laxities in social control are not new, for example, as I have described in chapter one, non-migrants have always struggled – at times successfully and in other instances in vain – to keep migrants at their disposition. Now diverse media create more possibilities to pursue individual life projects, by reaching out to social networks and resources, possibly without others knowing. Such opportunities lead to increasing confusions and anxieties about other's social networks and activities: who is doing what with whom. Furthermore, activities can be performed simultaneously on different levels of social reach: in presence availability and on a mediated, local and transnational level. Multiple layers of sociality intersect with face-to-face sociality, and practices of connecting in different realms are interdependent. It also raises insecurities about scams and deceit in New Media spaces, but also in interconnection with face-to-face sociality. Notions of privacy regarding New Media use are of increasing importance here, and are rather to be seen in the light of these fears of deceit, than in the sense of à priori defining private practices. Narratives about public-private transformations point to anxieties of being connected and connecting.

Youth and embodied imaginaries –performances and feelings of being (dis)-connected

So far in this chapter I have addressed the materiality of technologies, which is shaped by and contributes to the shaping of media user's practices. Regarding medias' materiality, and how they are integrated into practices of connecting, we could also see them as representations of social imaginaries of sociality, solidarity, or simply, connectedness. In this sense, these media and related practices also allude to absent social others and spaces of potential. In this subchapter, I intend to take a closer look at bodily performances or embodiment of imaginaries by youth. This is related to discourses on legitimacies of conspicuous consumption and youth's social position and status in society (compare Mbembe 2008:110, Cruise O'Brien 2003:155, Fokwang 2008:8). Youth's orientation towards mobility and an emphasis on the material culture of success could be seen as a strategy to deal with disadvantageous life situations (Jua 2003:16). "We have this saying: Life begins with forty. But this is not true anymore, why wait? Young people want to become successful now, not only when they are older. Like in the west where young people are successful and economically well off at an early age" (Peter. Interviews 2009/FS). Although such practices are based on historically rooted social imaginaries of success, in correlation with New Media use, youth also reach out to alternative pathways for social mobility, in the sense of a projective dimension of agency.

In this subchapter, I would like to emphasize narratives in the local sphere evoked by youth's demonstrations of life-styles. Youth are generally suspected of engaging in "subversive practices", which escape public control, which is also strongly related to practices of the use of New Media and mobility[226]. I intend to broaden the view on New Media and their use in the public sphere as being a means for the demonstration of one's connectedness, literally and symbolically: The demonstration of status[227] in social spaces by practices of conspicuous consumption occur on the basis of connections, and ensures connectedness in terms of reputation and enjoyment of a – sometimes only temporary – rise in status. Likewise, not being able to take part due to conditions such as, most often, a lack of financial means can be experienced as an exclusion and disconnection. Apart from pointing to liveness as a potential, such practices of demonstrating success refer to liveness as continuous work and effort in order to stay connected.

Life-styles, staging success, or embodying the subversive?

Life-style alluding to success can include houses and comfort, a car and other status symbols, but also the body (Warnier 1993a:167). The body is for most people the most important locus of identity and statement regarding personal success, when house and car are beyond reach (compare Fuh 2010). Appadurai (1996:179) speaks of an inscription of locality – as the specific conditions and connotations with the immediate life-world – onto bodies. The body is thus, as Merleau-Ponty (1966) has said, a phenomena of expression and instrument of understanding. It is a media of communication to connect to the individual's environment. The conception of the self in the body,

---

[226] This relates then to the third sub-question of guiding question 2, what narratives of social change are provoked by materializations and performances related to imaginaries prevalent in the setting.

[227] Status is according to Goffman (1963) a position in a system of reciprocal relations, dependent on social roles.

practices, comportment and style, also includes anticipation of the perception of others, evoked intentionally or unintentionally by the physical appearance. In this sense, style reflects or opposes evaluated body standards. Argenti describes bodily practices constituting memory, but also including a vision of the future, which transcends officially sanctioned realms (Argenti 2007:248, Fuh 2010, Cruise O'Brien 2003), allowing for the expression of multiplied influences on identity construction. These are new figures of success[228], some of which are highly contested, expressed in particular life-styles, status symbols and bodily practices. Giddens' (1991) "life-style" refers to individuals' choice, opportunities, context and conditions; "life-style sectors" refer to the overall activities of individuals: the more an individuals' experience is influenced by media presence, they rise in numbers and provoke the formation of life-style sectors shared with absent others.

In the local context, emulating a "global youth culture" relates to an identification with certain life-styles and their local interpretations, represented by bodily styles. Regarding youth culture, it is stated that emulating is helpful in order to "think big", by taking successful and famous people as models, international stars from music, sports and film. Thereby, often blacks and Africans are taken as models, but also Cameroonians, such as most prominently Eto'o Fils, the football star, or singers or artists, such as Manu Dibango, Petit Pays, and others. These styles are made up by style elements, such as faked branded clothing, such as Gucci, Dolce & Gabbana and Armani, and the displaying of luxury items such as nice watches, golden necklaces, fancy sunglasses and mobile phones, or emulating the "Gangsta style" of famous black US Hiphop stars such as 50cent, Snoop Dog, Notorious BIG, but also of Nigerian stars (compare to Malaquais 2001, Bucholtz 2002). Furthermore, individual style elements are important and carefully chosen[229]. Orientations through styles are also pursued by young women, by fashionable clothes, shoes, jewellery and branded handbags, such as Gucci and Armani, stylish hair styles, styled foot and finger nails, and other "nyanga" – vanities. Among young men, such displays of

---

[228] For Banegas and Warnier (2001) "success" is a polymorphous and dynamic notion. They propose to study success by analysing material culture and practices emerging from the notion (compare Künzler 2007), but also by integrating values, moralities, status and social and economic resources related to it.

[229] See figures F38 and F39.

styles is tied to certain behaviour, such as handshakes, ways of walking, talking and verbal expressions, music and dance, which are part of performing identity. Even though young people, who express themselves in conspicuous ways only form relatively small groups, they are highly visible and contribute to the image of youth, who are opposed to society's norms of morality. Some youths' bodily performances and styles seem to point to the subversive. Youth life-styles, conduct and conspicuous consumption inform discourses of their accurate status in society. These discourses have gained intensity due to a rise in consumer culture, which allows more individuals to partake, and of pathways to material success with transnational references, such as scamming, profiting from remittances and migration.

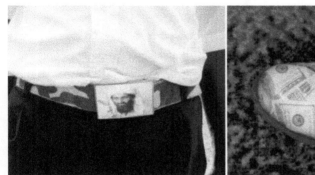

F 38, 39: Elements of youth styles

Between mocking and admiration: some analogies of status of scammers and bushfallers

Such negotiating of life-styles depend on local images of "what is worthwhile or desirable", to images of "success" and likewise to scammer's and bushfaller's – the migrants – images. Local narratives related to scammers and bushfallers show certain similarities, but also a few crucial differences. Both scammers and bushfallers are notorious for their staging of success, as Nyamnjoh (2005:296) describes it, as the production of "consumers without affordability and affordability without the typical signs of effort". It indicates an "irresponsible" way, of spending money, on oneself and on prestige goods, and in this sense not meeting moral prerequisites of redistribution. Furthermore practices of scamming and bushfalling both relate to a translocal and transnational space. One important aspect, which contributes to

163

negative attributions to scammers and bushfallers alike, is that the circumstances, in which wealth is accumulated, to a certain extent remain a "mystery" to outsiders, due to a lack of know-how about the internet or a life abroad. As a representation of an opaque way of accumulating wealth, scamming practices but also the ways how bushfallers achieve success, are often related to practices of "nyongo" or witchcraft practices, instead of "hard work" (Nyamnjoh 2005:248, Jua 2003, Ndjio 2008, Malaquais 2001:33). Scamming as well as bushfalling point to distorting reality, hiding the truth, telling lies, and playing with identities, from the perspective of non-internet users or non-migrants. Scammers, but also to an extent bushfallers, are said to contribute to negative developments and to a corruption of morality in society. Therefore there exists a great deal of suspicion towards bushfallers and scammers regarding their trespassing of societal hierarchies and moral norms. That said however, images of bushfallers and scammers are not only rejected, but also admired, especially by peers. Basically, these ambivalences highlight new and opaque spaces, of "bush" and the internet, opportunities but also dangers. Scamming seems thereby not to be a substitution for migration but a temporary strategy to "survive". Most of these youth still dream of going abroad "for greener pastures", and to become bushfallers (compare Frei 2012).

Youth social spaces are consequently viewed with suspicion by some people. For example the cyber cafés are often linked to the scamming issue and non-conformity with societal values in general. Correspondingly young media users - being scammers or not, and even outside of the space of the cyber cafés – are associated with respective the characteristics and attributes. Thereby, many people think they are able to identify scammers from their outer looks, appearance and comportment[230]. Regarding the bushfallers, they are also said to be recognizable by their styles, and in particular by prestigious status symbols (compare Pink 2001)[231]. For example, most of the cars, which are brought to Cameroon by returning young migrants "are Mercedes

---

[230] These styles are not confined to scammers. But style and scamming are very much equated, even though not all „fashionable" youth are scamming, similarly many youth who are scamming do not invest much in their style.

[231] Only a small part of the bushfallers are comporting in such ways, but local narratives are strongly influenced by those who do so. I guess that most try to be as inconspicuous as possible, also to control expectations.

and luxury four-wheel SUVs, which they sell before returning to the West at the end of the holidays. Ownership of these vehicles in Cameroon, a privilege of the ruling class, is seen as a reification of power." (Jua 2003:23). Above all however, locally, "One knows who has gone abroad, and when such a person is seen again in town, maybe driving around in a car, these news spread quickly" (Valeria. Interviews 2010/11). Bushfallers are, as Jua (2003) puts it, actively involved in practices of subversion, by demonstrating their new status in public. Furthermore, they are often seen to circumvent their obligations, regarding redistribution of wealth and to meet their attributed role as providers. An exaggerated display of luxury is attributed to a mockery on those who are less fortunate.

Additionally social conduct is part of expressions of conspicuous consumption – scammers are generally described as behaving irresponsibly and opposing societal rules of morality, lacking politeness and respect for other people and especially elders. According to popular discourses, scammers shout, fight, drink and smoke, go to nightclubs and have irresponsible sexual relationships. Bushfallers sometimes try to distinguish themselves through their performance of "behaving in a civilized manner". In this sense, they would be seen in "responsible and expensive places", eat in expensive restaurants, enjoy leisure activities, speak "good English", move to a certain extent in circles of other "big people", and show a "(...) multiple cultural competence to cope with the needs to respond to different cultural reference systems" (Martin 2007:231). An important part of performance is related to demonstrative consumption in the sense of showing generosity. The successful invite their friends and let them take part in their success. Among scammers, it is announced in bars and places where young people meet, who has made money, how much and by what means[232]. The act of spending or "eating" the money itself is at times extensively celebrated and performed, as a means acquiring status among peers (Malaquais 2001, Bucholtz 2002, Simone 2005)[233]. It is said that scammers and bushfallers

---

[232] For example stories like: a particular friend had just picked up 1.2 million FCFA from Western Union having sold a German shepherd to a "dupe" in Germany, and many other similar stories.

[233] I have been told stories of scammers who poured money over people in nightclubs, and I witnessed scenes when scammers were dancing collectively, their hands full of money and showing their full purses.

alike tend to spend huge amounts of money in places of leisure, for pleasure and status. In many of these local spaces of youth culture scammers and bushfallers compete - there are many narratives about altercations between scammers and bushfallers, in nightclubs around Christmas and New Year, the time when bushfallers return home to spend their holidays, when scammers and bushfallers try to surpass each other by demonstrating their economic viability. "They (the bushfallers) think they are better than us. Then they are surprised when they realize that some young guys here also have money" (Peter. Field notes 2009, 01.08.09). Such means of demonstrating success lack differentiation among social groups, however, to differentiate oneself is of great importance, and bushfallers would try to demarcate themselves from other groups of youth, and in particular the scammers, who are also involved in practices of conspicuous consumption but seen as inferior in status. The pathways to success are in this sense distinguished and evaluated in regard to their legitimization and moral integrity. In local discourses, whereas for bushfallers, the chance for acknowledged status when following the lines of moral duties, remains intact, and their conduct is judged against their estimated fulfilment of this duty, scammers could never obtain a consolidated status beyond their immediate peer groups[234].

In as much as such displays of success evoke connotations of illegitimacies: the acquiring of certain elements of status is also regarded as achievement - demonstrative spending is also admired. Somebody who is powerful must distinguish himself as an individual (Hardin 1993:208): "one needs to show who one is". The appearance and conduct of these youth enhance ambivalent feelings among elders, but also among peers, who do not want or are not able to take part in the competitive staging of success. In this sense, as Tazanu (2012a) relates, it also includes mocking of success. It has gone so far that the

---

[234] Being one of the few who are in the position to celebrate success more regularly, one could solidify a reputation of being a "good scammer". This is also expressed in circulating stories about one or another "famous scammer" in town, and calling one another by nicknames, relating most often to images of success in this field, such as highlighting their economic viability, their mastering of technology, status, or in general "global references". Such nicknames could be Euro, Dollar, Technique, Tension, Tresor, the American, and others. Not only scammers have their nicknames.

slightest insecurity about the provenance of somebody's material success has become a base from which to declare somebody "a scammer". Bushfaller's wealth is easier to explain since the notion of easy ways to accumulate abroad is prevalent - thus, youth who display success but are not bushfallers, must be scammers – in this sense understating other means of acquiring success in the local context. By evaluating every deviation of the norm, which would be considering youth as being economically deprived and dependent, youths' discourses take on the line of those discourse pervaded by elders and the powerful in society.

Competition and conspicuous consumption in places of leisure and youth culture

I have highlighted an atmosphere of competitiveness, in particular among youth, and most conspicuously among male youth (compare Fuh 2010, Fokwang 2008). On a local level, typically male activities are scamming, playing football or sports in general, and "hanging around" in peer groups. In such ways male youth are visible in different spaces and settings, such as bars, clubs, or other meeting places for young people and for diverse collective activities. Such places are to be considered arenas for competitiveness, which could be expressed by both, positive assertions or negative statements, as for example loudly commenting on others' abilities or disabilities, mocking and expressing loyalties, in the sense of equalizing social positions or showing superiority. Superiority or the competitive aspect in interaction is most often related to the display of wealth, be it in style and status symbols or as acts of generosity, to accentuate masculinity and virility (compare Fuh 2010). There are furthermore differentiations of levels of legitimacies of such display of wealth. This levelling also depends on the location, where display of success occurs, and who is doing so and to whom the demonstration is primarily intended. Social spaces of recreation and leisure serve as a kind of stage where appearance can impress and be judged at first stance.

There are those places of leisure, which are known in local narratives for the demonstrative consumption of bushfallers and business-people – but also increasingly scammers. In Bamenda, these places might be restaurants, hotels, bars, and nightclubs, where the average age of people and prices – for entrance fee and drinks – is

167

higher, in the sense of only allowing economically viable individuals' entry. The ambiance is described as "soft" – not too noisy and "civilized" – and the interior elegant. Often in these places, be it cabarets, nightclubs or bars, there are sections differentiating the guests into "VIP's" or "commoners". In the VIP sectors the drinks cost – sometimes much – more, and guests enjoy other conveniences there, be it better service, more space, more elegant interior – or also no difference at all, just having the feeling of being on the "right side". Some places also differentiate in temporal terms: For example Azam was the newest club in town in 2011. Going to Azam on a Saturday is seen as prestigious, and more for "the big people", when the entrance fee is 5000 FCFA. It is different on a Thursday, at "lady's night", when the entrance fee for women is less, which makes it more affordable for young men to invite their girlfriends, than on a Saturday[235]. These prestigious and luxurious "responsible" places are also stages for social events for the "better off", or prestigious parties - around Christmas, these locations could be rented for weddings[236]. To attend parties at special events – for "big people" with high entrance fees – is seen as prestigious. For example for the New Years' party 2011 at Ayaba - a hotel and nightclub - one had to pay 30,000 FCFA as entrance fee[237]. In order to afford high expenditures to go out, for entrance fees to the clubs and also the compulsory consumption of drinks, most young people go to such places in a group to enjoy, but also to share the costs. For those having money and wanting to demonstrate this, it is however a possibility to take friends out to the nightclub at their own expense. Other places in turn are notorious for their reputation of being rather "irresponsible places", and thus attracting youth that are considered to be noisy, rough, and in this sense "not responsible people". These places have a tendency then of attracting younger people, because the entrance fee is lower, or drinks less expensive than in other places. The other famous nightclub in town belongs to this category, Tropicana at Mondial hotel. There the entrance fee is 2000

---

[235] See figure F40, Azam on a thursday night.

[236] The narratives about the luxurious "bushfaller marriages" are very present at the time around Christmas and New Year, when bushfallers come home and marry, and in general the time when most social events take place.

[237] As a comparison: a room or a small house in Bamenda could be rented from 10-20,000 FCFA per month.

FCFA per person and it operates only on Fridays. Mondial is said to also attract "noise-makers", and that too much drinking, quarrelling and fighting is common there. Other places attracting youth are sports and leisure related places, which have a "touch of special", as these are for example the swimming pool and tennis courts at Ayaba hotel[238]. Being there means "to have a good time together", a space of demonstrating oneself and one's bodily appearance, competing with each other in terms of one's abilities in swimming, or showing elegant jumps into the water, or just sitting there and watching, a social space for flirting, young women and men admiring each other. Just to enjoy the leisure of being there, when bathing in the pool, one has to pay 1000 FCFA, an amount, which is relatively high for young people. Additionally having a drink or even eating a salad, which is very expensive at Ayaba, could also here draw lines of differentiation among those present.

F40: Azam Hotel nightclub, Bamenda
F41: At Ayaba Hotel pool, Bamenda

In short, places – which correspond to differing connotations of status – differentiate people, and people used places to differentiate themselves from others going to other places, which are considered less prestigious. Going out for leisure purposes is doubly rewarding, to have fun, and at the same time to demonstrate viability to peers. "Responsible places" represent a certain status, which also youth, who are not financially viable, desire to enter. For example scammers, as epitomes for irresponsible comportment, as well enter these

---

[238] See figure F41.

169

"responsible spaces": they also want to impress, and invite peers to these prestigious places[239]. In this sense social spaces do not clearly demarcate social groups, conspicuous consumption practices permeate different status groups. Since scammers permeate non-scamming youth, other youth could try to demarcate difference between them and scammers. As it was explained: "The problem is that when you interact with them, everybody will take you as a scammer too. If you do not want to spoil your reputation, you should better not mix up with them" (Simon. Interviews 2009/F). It has also to do however, with the anxiety of not being able to compete in terms of spending: "You should not too much involve with friends of this category, when you do not want to spend your money for drinks and pleasure. If you are serious, you should find friends of your own kind" (Ernest. Interviews 2009/O). There is also a feeling of dignity involved here, in the sense of not wanting to be involved with others who would constantly demonstrate their superiority: "By paying for your drinks, they make you feel bad. And they also expect something from you then, but I do not have money for things like this." (Ronald. Field notes 2010/11. 24.12.10).

Not having money at hand is the principal reason why youth are excluded from social life. Going out with friends involves the urge to be able to spend on them, if "one is not up to the task one should better stay home". The general complaint of not having money and thus being forced "to stay indoors" is widely heard. When I asked a friend[240], what he had been doing at New Year's eve, he responded: "I was only in the house, because I had no money. I could hear the fireworks from Upstation (a better off quarter), and I loved to be there, and I knew that many of my friends were there. So I just went to bed and covered my ears and tried to sleep. I felt very frustrated" (Eric. Field notes 2010/11. 02.01.11). These narratives reflect the high

---

[239] A whole group of scammers - from a specific cyber café in town - had collectively applied for VIP membership at Pen Pan Pacific hotel - including bar and nightclub. I witnessed when the hotel's car picked them up from the cyber café and guiding them to the hotel, where they negotiated about their membership fees with the manager.

[240] The friend even had a regular income: he worked as a secretary in a lawyer's office. However he was also complied to constantly feel financial shortage. His lawyer friends were financially better off than himself.

competition amongst youth. Female youths are of course also competing - regarding physical beauty, "sexy" appearance, dresses and status symbols related to their style, and often for the attention of those who are viable for spending (compare Nyamnjoh 2005b:304). However, the negative images of youth, of bushfallers and scammers are male images, even though the image of, for example the morality/immorality of young women, is also part of these discourses. Of course, gaining status among peers is not only defined by status symbols emulating youth culture, or demonstrative consumption and spending, but could be realized in other arenas, as sports, business, studies, and so on. I have nevertheless placed emphasis on conspicuous consumption practices as practices of demonstrating success, which are attributed to bushfallers and scammers. Undoubtedly, both of these categories have stimulated popular discourses about youth, success and moral legitimacies.

Conclusion – New Media and social spaces: liveness and a "sense of place"

In this chapter I have addressed the "materialization" of imaginations of potential realms of success and a "good life" in the local sphere. I have specifically concentrated on the materiality of technologies, being shaped by imaginations and related practices of New Media use. Media have transformed the sense of "being in the world" or "the sense of place" considerably. I have related in particular to the 1st sub-question, regarding the interrelations of media practices and materiality of technologies: I have described how New Media practices shape spaces in a material sense, as in the example of the cyber cafés or media use in a public sphere. I have furthermore related to the 2nd and 3rd sub-questions about the intersecting of mediated and face-to-face social interaction, and narratives of social change provoked by the materialization of imaginaries. Internet and mobile phones create layers of spaces for action, personal and collective, some depending on co-locality and physical presence, but at the same time alluding to faraway realms and absent social others, to which also a demonstration of success refers to: I have related to this in the subchapters about the public sphere, different loci of sociality, and youth's staging of success.

171

In this context New Media serve to create urban, mobile and "connected" identities, and in this sense status. In view of youth staging success and connectedness to worlds of leisure, pleasure and consumption, we have seen that the displaying of wealth and status raises extremely ambivalent discourses, rendering their "staging" viability, but not necessarily legitimacy. These narratives are also about youth's adequate status in society, and about places and how such displays of success would occur. As we have seen, scamming as well as bushfalling seem to serve as nodes of discourses, which address anxieties of social change and are related to underlying discourses on notions of "authentic culture" and moral values.

Overall, being connected is in itself prestigious, and it ranges from face-to-face to mediated interaction, from personal social others up to a connectedness to spaces of potentiality, expressed by conspicuous consumption, sociality and the use of New Media in youth social spaces. Likewise, being disconnected has connotations of not being able to "keep the mobile phone in use", or in the sense of not being financially viable to take part in social life and socializing practices based on consumption. Thereby, as I have pointed out, being connected is seen as subversive. I have mentioned the increasing confusion about social networks and activities: who is doing what with whom. This is closely related to shifting cultural norms of private and public. I have mentioned moral aspects here, for example in narratives of opportunities for social networking through mobile phones as a personal communication device, and the possibilities for concealed and incomprehensible activities in the realm of the internet, for example in the field of scamming.

In this chapter liveness consists of a perceived potential, imaginations are "coming to life" in materializations in the local sphere. A kind of liveness of success emerges, rendered visible and tangible in symbols of success and associated demonstrating, also as a sensory experience. At the same time, practices of New Media use are related to liveness as a continuous effort and work in order to maintain a sense of accessibility, choice, and connectedness, in the sense of social networking, habits of media use and performing. Thus the dimensions of liveness are part of a "feeling of place", and they contribute and pervade the perceptions of the public sphere and the city, which is not just a place, but a temporal-relational context for action.

In this chapter, I have related to the situation of New Media and described the thereby evolving practices of New Media use under local conditions. I will relate to youth's practices of connecting as evaluative and reflexive activities in unfolding temporal-relational contexts (Emirbayer & Mische 1998:969) further in the following chapter. I intend to look at practices of social networking – in order to create and maintain social connections and access to social capital – and how these are transformed and negotiated on a translocal and mediated level.

4

Practices of social networking – face-to-face and mediated social ties and support

In this chapter, I will look in particular into practices of social networking and access to social capital and support through these social networks, which are pursued on both a local face-to-face and a translocal mediated level. I will introduce different categories of social ties, and examine how notions of sociality and solidarity transform from local face-to-face to mediated social ties. This chapter thus relates to guiding questions number two and three. New Media seem to have a high potential in order to contribute to the transformation of New Media users' agentic possibilities, in intentional practices of social networking, in the sense of a practical-evaluative adapting of these media. At the same time, these practices are strongly tied to blueprints for an ideal sociality or connectedness, as well as to alternative pathways for social mobility through New Media of communication. In this chapter, I intend to adapt a predominantly local perspective from the view of non-migrants in the context of Bamenda, before I will integrate views of migrants more specifically at a later stage in this book, in chapter 6. In this and the coming chapters, I will describe the increasing interconnections supported by New Media of communication, how they have transformed sociality, and how related social imaginaries of roles and norms, and thus expectations and claims on sociality and solidarity are negotiated, revised or enforced. Thereby, new opportunities seem to be integrated into existing norms and patterns of sociality, evaluated on their premises and pursued along the lines of acknowledged practices, within altered frameworks of reference, where they are prone to negotiation (compare Giddens 1976). I thereby understand the term negotiation as an on-going coordination of social practices of inter-subjective meaning making, understanding and articulation by social groups and individuals. The concept of "liveness", in the sense of potential and work may here be adopted in order to describe New Media users' efforts to establish and maintain enduring social relationships and networks, and articulating[241] claims on social capital and notions of solidarity.

Social units and networks based on co-presence – families, friends and lovers

In this subchapter I will concentrate on primary social ties of connectedness, in varying degrees of closeness. In today's globalizing world, communities are prone to being dispersed in spatial terms. Nevertheless, physical co-presence of individuals is the base on which basic sociality draws on. Mediated – or indirect - social interaction and relations are usually based on a relationship which has been built during face-to-face interaction prior to physical dislocation, based on a shared life-world as a base for the valuation of other's comportment. In this sense, mediated social relationships should be examined contrasted with face-to-face social relationships of direct interaction.

Kinship, families and households

Kaberry (1952:15)[242] distinguishes different categories of kinship[243]: the "group of the house", which is the elementary family, in polygamous households several wives and their children of the same father, the "group of the compound", which also includes the co-resident patrilineage, and the "great group" of the patrilineal clan[244] (compare Den Ouden 1987:7). Social network analysis has shifted the unit of analysis from the family to the household, placing emphasis on spatial proximity, as families are increasingly transnational. In this sense, the unit of the household relates to a community living together in the same

---

[241] According to Förster (compare 2010), articulation is a part of practices of mutual negotiation, consisting in claim making, distanciation, connecting and dislocation. Such practices influence interpersonal communication processes and social relations, and are thus useful for my work.

[242] Relating to the Nsaw, or Nso' (compare Goheen 1996), an ethnic group in Cameroon's North West province.

[243] Studies of kinship and social organization lie at the core of migration studies in Anthropology. They evolve around the concept of the social network, which gained importance as anthropologists turned to the studies of complex societies and urban populations (Brettell 2008:124).

[244] Or that of the father's mother, or mother, varying between groups and situation in matrilineal or patrilineal tribes.

house or compound, in different interrelations, differentiation of roles, and division of work. Thus, a household is defined as an economic unit. However, members of a household who contribute to its functioning, do not necessarily live in co-presence: Men often live in separation from women (compare Kaberry 1952, Goheen 1996), where traditionally women occupied houses for themselves with their youngest children, today this can still be the case, particularly in polygamist marriages. It has also to do with the fact, that men are often located elsewhere in order to earn money (compare Kaberry 1952, Warnier 1993a). Den Ouden (1987:7) relates to this form of residence as the "matrifocal household". In an urban context, a considerable number of my young interviewees lived apart from their family's compound, in particular, youth, who study in Yaoundé or Buea, are renting there and live temporarily on their own, and return "home" during semester break. Some interviewees contributed to a certain extent to the livelihood of their families, whilst others stated that they struggled just to support maintain themselves and not be a burden to their kin, and again others depended greatly on their family's support.

Parents - fathers and mothers

Being ideally the provider of the family, the father has to care for his children, is responsible for them and should serve as a role model (Kaberry 1952:150,151). He is, as the head of the household or compound, in control of resources such as land, and in this sense active in acts of redistribution[245]. This is also related to access to opportunities and the chains of inheritance and rights to power positions within social networks. In this sense, the relation to the father is also characterized by the access to opportunities, though his – sometimes selective – support (compare Warnier 1993a:77). The father is the regulatory element within the family, the node of authority, and he has the right to exert power (compare Hansen 2003:203). In these ties of strong hierarchies and patrimonial power, relations can, in the case of the father son relationship for example, become strained and tense sometimes. I met

---

[245] The position of the father also serves as a metaphor for "natural power orders and hierarchies" and their legitimization, a power based on seniority and acts of generosity. It thus reflects not only familial relations but also relations of hierarchy and power in patron-client relationships and in networks of political power.

many young men who expressed their delineation from their father's views and comportment towards his dependents. Mostly, such complaints were related to the moral duties a father is ideally supposed to exert[246], which is more difficult to fulfil in today's economic circumstances. The father also seems to play an important role in determining his offspring in terms of encouraging migratory ventures. A young man said: "My father puts me under pressure, he says that either me or my elder brother have to migrate" (Hans-Derrick. Field notes 2010/11. 28.12.10). In many cases it is the father, who demands support in terms of remittances from the migrant. From the view of - in particular male - migrants, the father is also an important figure in terms of opportunities and migrant's aspirations to maintain their influence at home, laying claim to their land, or entering a "traditional society". In some cases, migration can however be also considered as a kind of flight from the father's authority: "My father wants to influence me all the time, he wants to tell me whom I have to marry and interferes regarding my education. I do not intend to migrate because I am supposed to contribute to the family's income. Rather I want to leave to avoid such tensions in my family" (Clive. Field notes 2010/11. 10.12.10). Or a young woman, who was jobless and living with her family, said that she was suffering from tight control, and that her aim was to get a job or start studying elsewhere in order that she could leave the compound and loosen the grip of her father on her life (Yvette. Field notes 2010/11. 25.12.10).

The relation to the mother is to a lesser extent connected to rights and power, as in patrilinear societies, paternal ties and attributes of rights and duties are considered primordial, and maternal ties subsidiary (Kaberry 1952:49). However, the role of women should not be underestimated[247] in terms of their influence on access and

---

[246] I met several young men and women who experienced problems with their fathers, involving struggles and quarrels regarding inheritance. In one case, the father had realized that his son was earning money through scamming, and had forbidden his son to come to the compound. Apart from these severe cases, a number of young people were critical of their fathers, saying that they were not responsible, "My father is a lad", was a commonly heard statement.

[247] Den Ouden states that – in Bamiléké kinship – women pass on their social position and influence, as for example access to land, rights to usufruct and inheritance of the land she is cultivating to her daughters (Den Ouden 1987:8).

opportunities, as social ties can always be strategically emphasized and used as a spring board, and are not only defined qua nominal rights. The maternal side of kinship is considered as important, for example for a woman, who marries into another kinship group, but as well for the woman's offspring (Ifeka 1992:151). Furthermore, for female youth in urban areas, mothers play a role in relation to pregnancy and child rearing of their daughters. Many young women in Bamenda told me that they spent some months with the mother or the mother-in-law, after they had given birth (compare Warnier 1993a:94). For unmarried youth, maternal kin have at least some symbolic meaning. Very often the maternal uncle – the mother's brother – or sometimes the mother's father (compare Pradelles de Latour 1994:27) plays an important role. According to Warnier (1993a:117) this maternal male relative is not in the position of authority, but he offers a friendly and supportive relationship (Kaberry 1952, Pradelles de Latour 1994, Ifeka 1992:150). Most of my interviewees uttered a strong feeling of responsibility and affection in particular for their mothers, whether they lived in closer proximity, or abroad[248]. An important motivation for migration is to "benefit especially the mother". Often the mother is the one providing for her children, when means of the father are not sufficient, when he does not take responsibility, or is absent. In urban areas, patterns of descent and virilocality and the influence of the agnates of one's patri- and matrilineage, can transform substantially. Reasons for this include, among others, the commodification of land, changing familial residential patterns – in the sense that increasingly core families build units of co-residence, in accordance with decreasing numbers of polygamous families – and residence beyond the territory of the wider kin group. However, these ties, and the notion of "my father's village" and "my mother's village", were considered just as important by youth in town.

Siblings – sisters and brothers

The terms "sister" and "brother" express closeness and they were used very broadly in popular expression. I usually needed to ask for the specific relation: In the cases of a biological relation it could be specified by speaking of a "direct brother" or distinctively of a "step sister". Apart

---

[248] In polygamous households the children's solidarity with the mother is strong – whilst relations are not always smooth with the father's co-wives and steps.

179

from blood ties, co-residence is often the defining node of "brother" or "sister", as an interviewee stated relating to her intimate relationship to a socially distant cousin: "She grew up with me in the same compound, so she is a real sister to me" (Valeria. Interviews 2009/O). Thus, "brother" or "sister" can also be somebody who is emotionally very close, such as a good friend. Siblings have been and still are competitors regarding inheritance and access to resources in the family's hierarchy. However, as Warnier (1993a:118) relates to an ideal of brotherhood: "Brothers are not always equal, but they are always contracted in a mutual relationship of help and loyalty." This goes at times so far that somebody "sacrifices" his or her own life project in order to support the one of a sibling. For example in the case of a young man, whose father had decided that he had to go to China to work for two years, in order to contribute to set up his younger brother's career as a footballer in Germany[249]. Or in several cases I had come across elder siblings who emphasized their duty to work and earning money so that younger siblings could pursue higher education. "It is important that some or even one of us can make it one day... we are a family and have to look after each other. We need to always stick together" (Minette. Interviews 2010/11/EC). Generally, hierarchies of age are important among siblings. Elder brothers and sisters often assume responsibility towards their "juniors"[250]. It often means providing for younger siblings for livelihood, food, school materials and -fees, to help them out when they are in need of something, to give them "pocket money", show care towards them, or let them profit from connections and relations. Sometimes elder siblings accommodate younger siblings who are sent to town by the parents from the village, for example when they have to go to school. Furthermore, being senior means the right to exert authority over younger siblings and claim on their responsibility towards the family. It seemed to depend very much on the parents' economic situation, the location of the family either in the village or in town, and the age and economic viability of the interviewees, in conjunction with the extent to which they contributed to their siblings' well-being.

---

[249] In this sense, hierarchies of age could also be reversed.

[250] These hierarchies could also change, for example when a younger brother migrates, and supports the elder siblings.

Beyond kinship – the notion of "friendship"

Friendship is based on a social relation out of free will: it is not confined to a kin relation, and in this sense tied to a "natural" connection within networks of mutual claims and duties. Friends are chosen and – often – share many social and emotional characteristics, sometimes more than family members do[251]. Simmel (1999 (1908)) relates to the notion of friendship being based on a range of mutual involvement of the personality as a whole (compare Dreher 2008:295).

There is a differentiation between "friends" as more superficial social ties, and "friends" in an ideal sense. When investigating beyond a casual use of the term "friend" and "friendship", the picture is differentiated, ordered in ranges of closeness. Whereas friends – in the sense of "real friends" – range on the level of greatest closeness, mates are considered to be more distanciated. The term "mate" often denominates somebody with whom people have shared a certain time of formal education at school or university: A "school mate", a "bench mate" or a "roommate", with whom one has shared bench and accommodation. Due to the sharing of such formative times a mate could also become a close friend. Many people said they kept contact with many former classmates over long periods of time, even if only by occasional emails or calls. A further term of a social relationship of decreasing closeness is an "acquaintance", which denominates a rather casual contact, related to co-presence in a locality; it could be somebody from the neighbourhood who one meets regularly in the course of daily routines. Degrees of friendship are furthermore related to as "real" and "true" friends, in the sense of a qualification related to the approval of validity of the relationship, proved especially in difficult circumstances, which also implies different degrees of intimacy and trust (Miztal 1996:180,189, compare Fokwang 2008). A "real friend"[252] is somebody who "never disappoints", and with whom one could "share everything". Here loyalty is the core of the proof of somebody's friendship and

---

[251] Studies of friendship in sociological examinations mainly refer to an occidental cultural context. It might be that the concept of a "friend" has different connotations in the Cameroonian context (compare Allan 1979).

[252] Such ideals of friendship are also related to in very popular forwarding messages by email, as they are frequently sent around by many internet users. I also received dozens of them. They idealize the normative traits and characteristics of a "friend" and of "true friendship".

concern out of free will: "A friend is somebody who will always be with me, no matter the circumstances, in good and in bad times" (Richard. Interviews 2009/O). In this more confined sense of the term, many people stated that they only had a very small number of "real friends". In this sense, many people seem to have high expectations of an ideal friendship, and conditions under which such a friendship could be developed.

Depending on attributions of status, achievement, personal relation and affection and most importantly, age, an individual could relate to a friend in further social distinctions. Hierarchies and competition are important aspects among groups of friends[253], who interact in different places of leisure and work. Among some – especially young men – there are also serious tensions observable. One such example was a young man, a scammer, who complained: "My friend has spied out my work and took away one of my most important businesses. I cannot trust him any longer" (Jack. Interviews 2009/FS). It also relates to a common differentiation of "good friends" and "bad friends". When I was wondering why they called a "bad friend" still a "friend", people emphasized the connection, which was also pertinent when strained. It seems that the valuation made is related to expectations and duties, which are called on in ties of friendship, by offering a denomination to express disappointment. A further notion is a "big friend" or "small friend". It could express relatedness in the sense of a potential access to opportunities, such as "big friends" are related to certain positions and contacts, exerting influence and responsibilities towards "small friends". It is thereby mentioned that "big friends" or "small friends" could not be considered as very close and intimate, because "We are on different levels. I cannot go too close to some of my elder friends. They have already families and made it to a certain extent. I envy them. I look up to them. But they are not the friends I would tell my problems or invite to my home" (Ernest. Interviews 2010/11). Close and intimate friends are often age mates, with whom interests and problems are shared related to a certain age category and social group, often commonalities also concern education and formation. People emphasized a stable

---

[253] Males and females also consider each other as "friends", but it is often stated that it was difficult for men and women to be "just friends". When a woman is married she would be less inclined to say that she has male friends.

integration of friendship within time-space coordinates: long-term and intimate friendships – "childhood-" or "old friends" - built on lengthier periods of physical proximity as a prerequisite. However, a friendship in this sense would not always relate to spatial proximity, but could also be maintained by occasional mediated contact (compare Miztal 1996:189). An interviewee stated, relating to his close friend abroad: "We only call about every two months, since he is busy and money is not enough – but we know each other so well... when we talk on phone, it is like we only talked yesterday" (Clovis. Interviews 2009/R).

Friends are – the same as kin – involved in networks of opportunities. The expectations someone has of a friend, such as in the example of a "true" friend, many young people stated that reciprocity was important, since the relation was based on free will and not on obligation. Mutual solidarity was considered as important: "A friend in need is a friend in deed", in the sense that friendship and "profiting from each other" was not seen as opposition. An interviewee stated: "Everybody has something special, a special quality – so as a friend, you should also be profiting from that. It is a mutual benefit, both should be able to develop and to be brought to another level through the friendship" (Anastasia. Interviews 2009/R). On the level of a rather casual denominating of "friends", the strategic use of the term comes out clearly, in the sense of "claiming relationship"[254]. An example of high expectations and high potential for disappointment is a friend who has migrated. Overall, disposing over a network of friends in different degrees of closeness is important, also related to the social capital involved with it, and a person with many friends is seen as socially successful: friends are an important part of self-esteem and identity (Miztal 1996:191).

Love relationships, sex and support

The difference between friendship and lovers are opposite gender ties, the physical attraction and involvement of sex. Whereas in friendships spatial proximity does not always have much importance once a sound base of emotional closeness and loyalty has been built, in love relationships between opposite gender spatial proximity it is seen as crucial. However, love relationships also need sometimes to be

---

[254] See later in this chapter.

maintained over distance (compare Giddens 1992, Gross & Simmons 2002), the possibility for a certain level of intimacy to be upheld through New Media is important then[255].

On a local level, love relationships are very much evaluated by societal notions of morality, also including differentiations regarding gender. In today's urban context, many young women are relatively free to choose to have boyfriends before marriage[256] (Nyamnjoh 2005b). Nevertheless, I think they bear the greater risk of being morally devaluated compared to their male peers. Discourses about youth's comportment regarding sexuality and immorality, are attributed to both sexes (Diouf 2003:3), however, young men's being "rude" and "making noise", in the sense of being rebellious as a rather male attribute, is to an extent "naturalized". In this sense, female youth are doubly judged, not only by elders and women, but in particular also by male peers who interact with them at the same time. Young women taking the freedom to go to a nightclub independently without a male escort are quickly regarded as prostituting themselves and comporting inappropriately. A young man stated that he ended his relationship because his girlfriend had been seen in the nightclub without having asked him for allowance. Moreover, a young woman told me that her boyfriend had accused her of drinking alcohol even when he was not around, and ended the relationship. Subordination of females is in such ways reproduced in youth discourses, as a male view of the need to control female sexuality, expressed in statements such as: "A man needs to control one's girlfriend", or "If you cannot control your girlfriend, how can you control your wife later on?" (Peter. Interviews 2009/FS). These discourses among male youth relate to their notion of male superiority, but also to economic despair and feeling powerless: since they are not able to marry, they have ultimately no claim over their temporary girlfriends.

---

[255] See Tazanu's description of love relationships over distance, in particular involving migrants abroad (2012:167ff).

[256] This kind of freedom for youth is related to in discussions about sex education and also in discourses about HIV Aids and its prevention in numerous local youth magazines, such as "Among Youth", "Entre Jeunes", "Jeunesse", "Youth Vibes", and so on, as these tasks are usually not taken seriously into account in schools, churches or family homes, where in many cases these topics are primarily addressed from a moralistic perspective.

The material base to have girlfriends is considered vital by young men, and it renders proof of "being a man" in regards of financial viability (compare Fuh 2010, Fokwang 2008). It was often stated by young men that women were "demanding"[257], and at times they explained not having a girlfriend, because they "could not afford it", or that they were tired of investing money into an insecure relationship. This insecurity is also expressed in statements about feelings of inferiority vis-à-vis more economically viable men, who seem to exert a greater attraction for young women. It is said that young women prefer bushfallers – migrants returned from abroad – and elderly or married men – so-called "sugar daddies" or "mbomas" (Fuh 2010:170, Nyamnjoh 2005b) – or lately, scammers, all of those who are seen as more or less economically viable. Some young women are actively on the lookout for men who would support them as an economic strategy (compare to Nyamnjoh 2005b:302). However, I do not want to generalize here, since I met many young women whose first aim was to stay independent from men. Also young women use to complain about men, who are seen as unreliable and never faithful (Johnson-Hanks 2007:651). In this sense, both sexes accuse each other of being deceitful and immoral[258] (Tazanu 2011:167,168).

However, apart from this, there is also the notion of romantic and true love relationships (compare Giddens 1992), based on equality and understanding, which could sometimes lead to marriage. As opposed to the apparent openness to deal with temporary as well as sexually and economically motivated relationships, youth refer to marriage as a morally legitimized and binding relationship. Overall, the main transformations related to marriage concern the decrease of polygamous marriages, marriage is increasingly postponed to higher age (compare

---

[257] According to Johnson-Hanks (2007:651), for women, men's engagement in the relationship is also measured by their gift giving, also as a potential future spouse, which would be a reasoning deriving from „the social structure of marriage exchanges centred on bride-wealth, translated into different circumstances of urban lifestyle."

[258] Johnson-Hanks (2007) writes about single, childless and educated young women in Yaoundé, who search for potential husbands from abroad in the internet. These women's requirements of men have changed, they dream of a relationship based on mutual trust and the husband's fidelity, which, as it is said by the women, can hardly be found among Cameroonian men. Also men, in the sense of providers, reliable and capable, seem not to be easily found in contemporary economic conditions in Cameroon (Fokwang 2008:255).

Johnson-Hanks 2007), and love marriages have become more and more common (Awambeng 1991). Marriage is "naturally" connected with having children[259], whereas being married when having children is seen as a normative ideal. Overall, especially young women's but also men's ideas about having children tend towards smaller family sizes. The great majority of my interviewees did not yet themselves have children. A few unmarried young men had children, but they were in all cases no longer together with the mothers of their children. Their assumption of responsibility varied according to their abilities. Three young women had one child, and two had two children each. Only a small percentage of my interviewees were married – two females, and three males with children.

In this subchapter, I have related to primary social ties of greatest importance. They relate to co-presence in order to develop a base of mutual knowing. Also mediated social ties base on such notions of responsibility and obligations and social and emotional closeness[260]. I assume that it is crucial to understand qualities and moral evaluations of social relations in a local context, in order to understand the scope in which they tend to be negotiated when taken on to a mediated level[261].

F42: Gallery of families and friend's pictures at a photo-studio
F43: Telecommunication publicity, new offer, 2009

---

[259] Children belong to society, and owe respect, obedience and solidarity to all elders in society, who could be seen as their mothers and fathers (compare Fuh 2010).

[260] I will relate to these notions of social and emotional closeness in the next chapter 5.

[261] See figures F42, and F43. In figure F42 one can see that family portraits are often „designed" in order that also those who were not present when the photograph was taken, appear in the picture, highlighting the strong kin relation also – and especially - under conditions of physical dislocation. In figure F43 close ties among family and lovers are pointed to in a telecommunication publicity ad, that these social ties could be smoothly maintained by phone.

Social networks and practices of networking through New Media of communication

Social networks are no fixed entities, rather they are situational decisions of people to relate and connect to others under certain circumstances and situations. Nevertheless, existing social connections in the categories of social ties on different levels of closeness, which I have described in the previous subchapter, are decisive. Does their basic core change, or are moral normative notions of these social categories rather taken on to a mediated level? I will explore answers to this question in this, but particularly also in chapters 5 and 6. The coming subchapters are related to the 2nd sub-question of guiding question 2, about how mediated and face-to-face social interaction and networking add to each other, here specifically how practices of New Media use are related to negotiations of social norms and moralities, in face-to-face and as well mediated social interaction[262]. Also, I will relate to the 1st sub-question of guiding question 3, how sociality and solidarity are transforming from a local to a translocal level. In the background of these examinations, I keep two dimensions in mind: physical presence or absence, and social and emotional closeness or relatedness[263]. In this subchapter, I will introduce different categories of New Media users. Furthermore, I will relate to uses of internet and mobile phone as media of communication, which are generally used on levels of local and translocal social networking, however, their usages point to differing tendencies in social networking practices and characteristics of social networks. I collected the data for this subchapter mainly during my field stay in 2009.

Face-to-face and mediated social networks and practices of networking

Social networks maintained through media are not the same as groups or communities, but they must be seen as ego-centred networks, or personal networks of individuals. They are in that sense sparsely knit,

---

[262] Compare to the previous chapter, which relates to social (New Media) spaces as arena of negotiations of social norms and moralities.

[263] Both of these dimensions have an impact on a perceived sense of liveness in social interaction.

which means that not all members of a social network interact with each other directly, or are aware of each other. Within personal networks, the nature of single relationships may be more important than the overall network (Wellman 1999:18, 24). Furthermore, a characteristic of such mediated social networks is that they are rather loosely bound, in the sense that they go beyond group boundaries of kinship, neighbourhoods or immediate community, „instead they ramify outward" (Wellman 1999:xiii). Social networks are furthermore valued by their range, size, and heterogeneity or specialization of social ties. Personal social ties can include different networks, such as different social and interest groups (Wellman 1999:7). Since personal social networks do not have clear boundaries[264] and transcend local and translocal levels, we need to be satisfied with a section of an individual's social network, and networks are also transient, they change over time according to changing circumstances in one's life (Wellman, Carrington & Hall 1988:137). It shows that mediated social ties cannot be separated from face-to-face social ties. They are interdependent, since – as I have tried to show through the introduction of different categories of social ties induced under conditions of co-presence - mediated social networks depend on local social networking practices and moral considerations of being connected.

Horst & Miller (2005, 2006) emphasize that practices of establishing extensive social networks and connecting to others in order to re-establish and validate a social tie, have not emerged with the mobile phone or the internet, but correspond to an ideal of sociality. I am following here Horst and Miller's (2005, 2006) notion of "link-up" as they describe its importance for youth in Jamaica. Related to practices of New Media use, the authors (2006:89) define link-up as social interaction in which the content of communication is subordinated to the act of connecting itself. They see link-up as a way of communication in its own right, which I would agree to, by, however, highlighting the interdependence of link-up practices and general access to New Media, as well as constraints of lacking financial means and media literacy. Due to such limitations specific medialities come to the fore. Calling just to

---

[264] Looking at overall networks only makes sense when we deal with distinguished and confined social networks, as for example on an institutional level, organizations, or in media newsgroups, internet communities, and so on.

"say Hi" (Caron & Caronia 2007) and to greet, is common in the local context, sending "beeps" by phone without speaking, or writing a short email, are expressions of local conditions of mediated social networking and adopting of a specific mediality for communication purposes. Furthermore, I would emphasize, that New Media of communication boost the considered importance of practices of link up, since internet and mobile phones offer various opportunities, also for a greater number of people. Likewise these practices are negotiated in narratives of social transformations. In this sense, they transform the ways, of how practices of social networking occur and what effects they have on sociality.

Samples of interviewees related to their use of New Media for sociality

In this part of the chapter I intend to examine social networking practices of young New Media users by building samples of interviewees. Thereby I assume certain conditions of New Media use: In an urban context the possession – or availability - of mobile phones is to a large extent given; also, I assume the availability or access to cyber cafés and thus internet in a basic sense[265]. Nevertheless, in their New Media uses, youth generally weight up conditions of scarce financial means and differing levels of internet literacy against the great need to connect.

I assume a strong contribution of the internet as media of connecting where translocal social relations are concerned. Since I was particularly interested in transnational social relations and networking practices, I was looking for and found my interviewees mainly among – to a greater or lesser extent – internet literate youth. I assume that everybody, no matter if they only use internet occasionally or frequently, combines the use of internet with the mobile phone, regarding their overall mediated social contacts – internet and mobile phone are most often used alternatively concerning the same social relationships. For this reason I do not differentiate between internet and mobile phone

---

[265] However, access to cyber cafés is of course not given in rural areas, and other constraints regarding the access to internet are also prevalent, mainly related to computer literacy.

users. Apart from that the mediality of these media also blurs[266]. I am of course aware that many mobile phone users do not use the internet. I emphasize the intensity of youth's internet use, because I found that their – in particular transnational - networking practices transform with the use of this media, and it also allows conclusions related to the use of other communication media such as the mobile phone. Regarding New Media users, I differentiate them into three categories: the "occasional", "regular" and "frequent internet users". These categories can serve as analytical tools to examine different characteristics of New Media use.

I conducted 52 formal interviews with New Media users during the field stay in 2009. Among these 52 interviewees were 17 females, which is about one third. The low share of females is especially remarkable in the category of the frequent internet users, which can be explained by the subcategory of the scammers – who are all male – I have integrated here. The lower share of interviews with women in general can be explained by a relatively low share of women in cyber cafés[267] in Bamenda. The interviewed New Media users in all categories were below thirty-five, most between the age of twenty and thirty years old. Of these 52 interviewees, 12 were students at the university or in school, 3 doing volunteer internships, another large part of 25 individuals were jobless or doing irregular jobs, and another 12 were employed, or self-employed and in business. Regarding education, some had O-, a majority had A-levels, a few had dropped out from school[268], and some were studying or had concluded their studies. I have included a list of the interviewees of the three categories at the end of this book in an appendix.

Categories of New Media users – occasional, regular, and frequent internet users

The first category of users I will call the "occasional internet users". They either do not use internet in a regular way: "I go there just once in a while" (Felix. Interviews 2009/O), or their regularity is confined to few occasions, such as once a month or every two weeks. I interviewed

---

[266] Due to such reasons, Tazanu (2012) has concentrated on „the call" in his analysis.

[267] Compare to the introduction chapter for explanations.

[268] Although education correlates with internet literacy, many internet users with a high internet literacy – for example the scammers – have learnt to use the internet by themselves or through friends, rather than in school.

sixteen users who used the internet occasionally, of whom seven users were female. People in this category differ in two subgroups: one consists of youth with low internet literacy, scarce financial means, and only limited mediated social networks. In this range, a limited internet literacy correlates with only an occasional use of the internet to a large extent. For this category of people, money is also an important determinant. As people in this category have "not much to do" in the internet, they are likely to evaluate whether it is worth it going to the cyber café and spend money for airtime. Few social ties pursued online and irregular contacts are correlated with each other, influencing the motivation for using the internet: their online networks could range from only two to five persons. These contacts are likely to be kin – if having "available" kin abroad – or also friends, as former classmates who study abroad[269]. Although not everybody in this category has a contact abroad, more than two thirds of the interviewees in this category of occasional internet users have at least one transnational contact, and others within Cameroon. Most of the internet users in this category consider the phone as more effective and their preferred communication tool: the reason why they additionally use the internet occasionally is because it is less expensive, in particular concerning transnational communication. I estimate that this category generally forms the greatest part of internet users, and also the share of females is greatest here. The other subgroup of this category consists of people, whose internet use depends on time constraints. These are mostly business people or employees - busy at their job sites with no internet access, and people with other responsibilities, as in the cases of two females who had children. Their level of internet literacy differs, some of them have a low level of internet and computer literacy, whereas others have basic skills. These people use the mobile phone for private and for business contacts, hereby most of them are more indifferent in regard to financial constraints, when they have a more or less regular income. Some are in this way connected with people in Cameroon and also abroad by regular phone calls. Concerning both of these subcategories, people use the

---

[269] It is less important for people in the category of occasional internet use, if it is kin or a weaker tie, also a class mate could be a motivation for them to go to the cyber café in order to maintain this social relationship.

internet mainly for communication purposes and hardly browse, they are limited regarding their computer literacy and slow at typing.

The second category which I call the "regular internet users" visit the cyber café at least once or twice a week, sometimes more, depending on their endeavours, on a more or less regular basis. I conducted seventeen interviews in this category, of which seven were female users. However, in this category, people's abilities regarding their use of the internet diverge greatly, and I would also allot them into two subcategories. The first subcategory consists of people who are not highly internet literate, for example they use the internet in a limited way mainly for communication, such as writing emails, or regularly visiting a certain friendship or information site. Some have only a few online communication partners, in some cases only one or two, in other up to ten. The proportion of social ties abroad are not necessarily higher here than in the category of the occasional internet users. The reason why regular users do not frequent the cyber café more often, is due to their limited need concerning internet activities: once or twice a week is seen as sufficient, then there is always an email to answer, another to write, and certain sites are updated. Others say that they do not have more time, being occupied at their job site, with their families, and so on. Making their evaluation of expenditures versus need, they limit their internet use to a certain amount of time – and money – per week. Some of the internet users in the category of "regular internet users", belong to another subcategory: they are internet literate to a relatively high extent, and also use the internet regularly to do research in their fields of interest, but they restrict their internet in accordance to their needs or time constraints. There are considerable differences between what those with higher internet literacy and others in this category could achieve within one hour online. Those of the latter subcategory are more apt at maintaining a greater number of social relationships in the internet: friends, classmates and kin, within Cameroon and also abroad, with whom they are in regular contact. Thus they also include social relationships in their networking, which are socially and emotionally more distant, whereas the less internet literate users in this category only write emails to their main contacts, mainly the ones abroad, and use the phone in order to stay in contact with those within Cameroon. Out of the seventeen in the category of regular internet users, thirteen had regular contact with online friends on the occasion of exchange in chat-rooms and on friendship sites, and two young men tried dating. Even though in this category, people use the internet to a higher or lesser extent also as an information tool, maintaining social contacts seems to be their main motivation and activity in the internet.

In the third category, which I call the "frequent internet users", I relate to youth who usually use the internet on a daily basis: I interviewed nineteen people here, of which only three were females. Certain activities of people who belong in this category are relevant for their frequent internet use: In three cases, people had become frequent internet users because they had acquired a computer with internet access in their homes. Other frequent internet users used the internet in relation to their job, such as one young man who worked for an NGO. Two others were involved in online studies. Six of the internet users in this category were employed in a cyber café, which gave them the opportunity to be online daily. Five of the interviewees were scamming, one was a former scammer, and another was cohabiting with a group of scamming youth. In this sense, using the internet on a daily basis is exceptional, but I found it highly interesting for my work. Regarding finances, a daily internet use implies considerable expenditures: either the internet is used as working tool, as in the case of the scammers, and those doing online studies, or the cyber café employees use the internet for free during working hours. In some other cases the internet is used in people's spare time, and they have an income from other sources. People in this category use the internet in a varied manner for research, business, scamming activities and pursuing of interests, which absorbs a major part of their time spent online. Regarding kin and close friends, they are not necessarily many more in number compared to the category of the "regular internet users", but as there people are apt at maintaining a higher number of weak social ties, with mates and friends – the number of overall social contacts increases, be they private or business related. In this sense, they often also have a larger number of social contacts with people abroad. The people in the category of "frequent internet users" are apt at having a large range of more or less active online relations in messenger chat and on various friendship sites. In general "frequent internet users" are in contact with others who are as often online as themselves – also in Bamenda, and thus their social networks also contains large proportions of mediated contacts on a local face-to-face level.

By introducing the three categories of internet users, I have highlighted characteristics of networking practices pursued by New Media of communication. Furthermore, I have shown how certain conditions interrelate with each other: the intensity of internet use could render insights into internet literacy, experience, and financial issues, and the combinations of media could be related to the size, spatial range, and the social or emotional closeness in mediated social ties[270].

F44, 45: Signboards of cyber cafés, Bamenda, indicating prices for airtime and different services

Interdependencies between intensity of internet use and media skills

Regarding frequency and regularity of internet use, which have served as start point, I have mentioned shifts in people's habits of internet use regarding time, which is spent online. I have also shown the differing levels of internet literacy. Undoubtedly, in many cases, the intensity of internet use is interrelated with the level of internet literacy; inversions could be explained with a temporarily shifted intensity of internet use. It could be due to time constraints and other responsibilities that people have to limit internet time and maintenance of contacts, or internet use intensifies because the conditions of internet access have changed, as for example when having acquired a home computer, or a new work place with internet access. There could also be a person and issue related "boost" in communication for the duration of a certain time span. For example some people said they communicated very often to a friend, when they had recently gone abroad, or a business

---

[270] See figures F44 and F45, sign boards of cyber cafés indicate airtime prices, as well as offers and services.

venture had to be organized in cooperation with an uncle abroad. Such shifts also involve a re-combining of media – for example adding weekly updates about a specific issue by email to the monthly phone call with the uncle abroad. In general, however, the different categories of New Media users cannot simply be attributed a certain level of internet literacy. In the category of regular internet users, a low level of internet literacy would not obstruct a person going to the cyber café three times a week to read and write emails to a few friends. In the category of frequent internet users, the level of internet literacy differs, ranging from highly internet literate scammers to averagely skilled users with their stable media habits[271]. Additionally, the interrelation of internet literacy and experience are relative. We could state that most frequent internet users have been using the internet for a longer time, and are thus more experienced. However, there are also occasional internet users, who have been using the internet sporadically over a longer period of time. This suggests that the period of time and regularity of internet use do not necessarily enhance technological skills to a large extent. Rather it seems that it depends on people's motivation, whether they adopt new skills or they continue using the internet in some limited habitual patterns of use. Overall, communication and the maintaining of social ties are given the priority in all categories of internet users. In this sense, by choosing a perspective placing emphasis on the temporal structuring of internet use, I intend to emphasize user's intentionality and purposive practices.

Interdependencies between internet literacy, size and spatial range of mediated social networks

I have suggested that internet literacy is not always decisive when the maintaining of regular social exchange with a close person is concerned, such as Claudia (Interviews 2009F[272]) who did not know how to use messenger chat, but was exchanging a great number of emails with her fiancé on a daily basis. However, correlated with the intensity of internet use and internet literacy is the size and range of mediated social

---

[271] Even scammers, who could be considered as possessing technological skills on a rather high level, are in some cases very much confined to „their area" of use, and do not go beyond it. As an example, there were young and not very experienced scammers, who knew how to change IP addresses, but they hardly used search engines and browsers.

[272] See selected case study in chapter 2.

195

networks. Regarding size I refer to the number of people belonging to a personal network, and regarding range I relate to different categories of social ties in degrees of spatial and emotional and social closeness. Overall, I conclude that the intensity of internet use and internet literacy interrelate with larger social networks, and also with the quality of mediated social ties. Firstly, because the maintaining of mediated social ties is interrelated with the use and knowledge of a wide range of media for inexpensive communication purposes – especially on the internet, and to a certain extent also opportunities for further social networking - on friendship sites, regarding online contacts – are adopted. Furthermore, the use of different media allows for a greater ease in order to maintain a social tie: such as the use of not only email and chat, but additionally profiting from other communicative features, such as exchanging links, e-cards, documents, pictures, and net based calls, which could add to or enhance the social tie's quality. Secondly, it concerns the bulk of communication "work", which can be effected by an experienced internet user within a particular time span, and the longer time spans frequent internet users spend online. Overall, internet literates can profit from less expensive and additional opportunities to maintain mediated social networks, in particular on a transnational level, partly substituting phone calls by internet means of communication[273]. Generally, the phone tends to be more important for local social networking practices. Here, as I will describe later in this subchapter, the mobile phone enhances the interconnections between those in mutual presence availability and enlarges the range and size of local social networks. Phone calls are also the favoured means of communication in order to maintain transnational social networks, due to their ascribed ability to enhance emotional closeness. However, the consideration here, ranging from domestic to international calls, are limited finances, a problem which concerns most media users in the setting. Even though internet literates have just as vast social networks maintained by phone as less internet literates, for less internet literates, regarding transnational social networks, financial conditions are more often perceived to be obstructive regarding the maintaining of social ties.

---

[273] An important part here is the substituting of phone calls on a national level. However, I have already related to that frequent internet users could also use internet on a local level with those who are present.

Interdependencies between the choice of media, categories of social ties and supportiveness

The choice of media depends on their user's skills and literacy, however it also depends on the communication partners, and how they are connected to media of communication. Hereby, internet favours the social connections to peers - since internet literacy is often correlated to age - concerning the local linking up with friends, or staying in contact with friends who for example study in another city in Cameroon. The internet also favours the connection to migrants abroad. Often people abroad are likely to be reachable by internet means of communication, and internet communicative features are more favourable in order to avoid high expenses. For non-migrants, the internet is seen as an apt means of communication regarding transnational social ties. The phone, however, remains indispensable to connecting to those who are themselves not internet literate. Moreover, the choice of media depends on considerations of the desired emotional closeness in communicative interaction, the importance of issues conveyed as well as reasons of politeness and moral conduct in communication[274]. Regarding transnational social networks, which are maintained by internet, relatives form one category of social tie, ranging from – in the few cases of eight interviewees – siblings, in most cases aunts or uncles, and in many cases more distant relatives, such as cousins. In four cases, interviewees claimed to have a boy or girlfriend, future marriage partner or husband abroad. Regarding this category of social ties abroad, phone also plays an important role. The other and considerably larger category of social ties abroad, where the internet also plays a more important role, are friends, differentiated in "close friends" and more commonly "school mates". Whereas strong social ties might not necessarily have increased in numbers with more intense internet use, weaker social ties are increasing proportionally. Thus, the larger the network, the larger the proportion of weak ties, whereas, the smaller the networks, a few strong ties are favoured. The evaluation of weaker social ties might shift on a transnational level: Since connections with migrants abroad are generally given a high importance, these could also be socially and emotionally less close social others, and the degree of emotional closeness could also shift. This could consist of enforcing a social tie in the sense of claiming

---

[274] I will relate in more detail to this in the next chapter.

197

relationship, and "work" put into the maintaining of a mediated social tie to a migrant abroad, and also, the more internet versed, the more likely youth themselves become a "player" in managing mediated relationships[275]. We could also say that the larger the network and the higher the proportion of weak social ties, the more heterogenic a social network, as it provides a range of social categories of relatedness and emotional closeness, such as kin, close friends, mates, up to online contacts. In general, the range and number of mediated social ties do not render information regarding their level of accurateness - activeness - and potential supportiveness. Some regular internet users have small networks similar to occasional internet users: a regularity of exchange however enhances social ties' qualities and potential supportiveness. It seems that in the category of occasional internet users, the likelihood that contacts break off is at its highest, or it is most acutely recognized[276]. Small strong networks of regular internet users could be as supportive and lively as those of frequent internet users, which include a great range of weak and not necessarily active social ties.

Mobile phones – exchanging social coordinates in presence availability

Whereas internet is – not exclusively, but to a large extent – in relation to reaching out to transnational social ties, the mobile phone – apart from maintaining translocal and transnational social ties[277] - is of great importance regarding practices of linking up with those in presence availability. Whereas the internet is more confined, social networking by the way of the mobile phone is a life reality for large groups of people. In Tufte's words (2002:247) mobile phones are flexibilizing the co-ordinates of people's everyday lives, which includes the renegotiation of time and space-bound appointments: structural coordinates of everyday life - when and where to meet – as well as relational aspects, who to meet or communicate with (compare Ling & Haddon 2003). It has also

---

[275] In some cases, youth might become the link between their families in Cameroon and relatives abroad.

[276] The communicative performance depends on work invested in a social tie – regularity, observing conduct, politeness and balancing of communication initiatives - in order to be ideally supportive. See next chapter.

[277] I will come back to the role of mobile phones in transnational social ties in detail in the coming chapters – 5 and 6.

enhanced youth's possibilities to interact with peers, and plays a crucial role in youth's dealing with sexual relationships. Mobile communication allows the pursuing of youth's interests out of adults' control in youth related spaces. The exchange of numbers has become a habit and normal part of social interaction[278], and is an important aspect of first contact to secure the opportunity of re-meeting. It is related to profiting from the potential presence availability of others provided by face-to-face interaction (Giddens 1984, Calhoun 1991) in an enhanced way by storing people's contact. In this sense, the mobile phone serves not only to maintain and coordinate old relationships, but also to explore new ones (Tufte 2002:248). Contact coordinates of others are in itself social capital: They are on the one hand taken and given out in order to profit or let others profit from social networking through mobile phones. On the other hand, other's social coordinates are also jealously guarded, as for example those of migrants are kept for oneself as an exclusive access to social capital. Furthermore, giving out somebody's number to somebody else underlies restrictions regarding considerations such as whether one is eligible to do so or not. Of course, giving out the number can also be restricted and denied by some people towards certain others[279].

That the mobile phone is a great tool in order to link-up and social networking, Horst and Miller (2005, 2006) have shown by relating to length and content of youth's mobile phone address books. Averagely, I have come across address-lists containing a hundred and more names[280]. People adapt strategies of how to name contacts in address lists in order to keep overview. Especially when a name is very common, it is necessary for example, to store the family name. Some people adopt a system with place, as the place of residence, work, or where one has encountered the person[281]. Or they store people's contacts giving hints to their profession or social connections, reflecting also one's own

---

[278] Licoppe (2009:81) relates to this as a ritualized exchange, where both exchange partners mutually engage in.

[279] It concerns for example females, who consider it as morally dubious to give out their numbers to men.

[280] For example also in messenger chat, and in email programmes, many people have a considerable list of addresses.

[281] Since in mobile phone address lists there is only space for a limited number of characters.

interest and specific attention towards them, as for example "divine comp", Divine repairing computers, or "Mirabelle Meta", Mirabelle staying at Meta quarters, or "Peter, Margrit S", Peter, son of Margrit Suh, or "Cletus German", for Cletus, a classmate in Germany, and so on. For the purpose of overview, it is common to create different folders in the address-book, similar to how it is done in messenger chat. I observed that many people used categories of social ties in their address lists, such as good friends, friends, mates, kin, cousins, family, girl or boyfriends, important people, business contacts, and so on. Furthermore, SIM cards could be switched for different purposes of networking to keep business and family networks apart, which is enhanced by most Chinese mobile phones disposing over double SIM functions. Accordingly, the importance of being connected through mobile phones, the loss or theft of a mobile phone is considered as catastrophic: "I had to start all over again (...)" (Dennis. Interviews 2009/F). However, only recently to the time of writing, Orange and MTN offer a service to retrieve lost numbers[282].

Numbers must be stored, in order that they inform directly as to the caller's identity by displaying the name on the mobile phone screen. Regarding important social ties, people usually store Orange and MTN or Camtel numbers alike, to be "fully connected". Having somebody's number stored is a sign of valuation – deleting could be seen as a sign of withdrawal, as a statement of an interviewee shows who described a disappointing experience with a close friend, who left from Bamenda to Buea in order to study, and who due to his absence forgot about her: "It came a time when he did not call again. (...). I was always the one calling him ... with time I got tired. What made me angry (then) was that one time when I called he picked and asked: Who is this? It means that he had already deleted my number in his phone. (...) Then I stopped communicating" (Andrea. Interviews 2009/O). However, there is also much confusion about phone numbers. People use different SIM cards, take them in and out of their phones, lose them, use other's SIM cards,

---

[282] It means that the number is still operational, and that the number could be retrieved unless they are stored on the SIM card. Regarding other numbers, which are stored in the phone's memory, it is necessary to wait until friends call, in order to gradually retrieve the contacts, which is facilitated through networks of friends. A few people told me that they were also noting down numbers on paper, sometimes due to a previous experience of loss of their mobile phone.

or call through a friend's phone or from call boxes[283], thus it is not always clear who is calling. Most people said that they would answer every call, also when the number was unknown, due to these reasons. This is expressed when somebody picks the phone saying: "Na who?" (Who is it?), and it turned out that the caller is well known to the person. A few said that they were reluctant to answer calls from unknown numbers – be they domestic or international. In such cases, an underlying fear of "obscure and dangerous numbers" alludes to narratives of numbers and text messages tied to occult forces, which would afflict a person when picking up such calls or opening text messages. Numbers represent people, storing and recognizing numbers thus serve to attain a certain control and overview over social networks and to identify valid and useful contacts.

Mobile phones – mediated communication influencing face-to-face social networking

Mediated contacting and communication ad in various ways to face-to-face social interaction, as I have described in the previous chapter. Whereas a cyber café is a "new social space" designed for mediated reaching out and linking up, the mobile phone enters spheres of face-to-face social networking and social interaction, where it is a new factor influencing such practices.

On a local level there are two prominent types of calls, whose contents are both related to social networking practices. One type concerns organizing daily endeavours and meetings with others, or following up a certain issue: "In an instance you can decide to call, because something is very urgent and important" (Barbara. Interviews 2009/R). The other type concerns social linking up (Horst & Miller 2005, 2006) in the sense of "pampering a relationship"[284]. "Calls are important, or text messages. The person realises: this person cares about

---

[283] Calling from call boxes is also seen as less prestigious than calling through a personal number. "It makes a call more personal, when calling from your own number. Also sometimes people do not pick up the phone when they do not recognize the number" (Kaspar. Interviews 2009/O). Calls to inform somebody about a new number are also important to stay connected, in order that the valid number is stored accordingly.

[284] Compare earlier in this chapter, regarding linking up, and see later in this chapter, regarding showing concern in calls.

me. (…) You can just ask them: How was your day, I love you, I am praying for you … it cheers up the person. (…) I can call or text when I realize that I have not seen or heard from somebody for a while" (Caroline. Interviews 2009/O)[285]. However calls are not solely communicative acts between two persons, but they can be collective happenings, and thus relate to face-to-face social networks. Often calls – in particular from close others such as family members, and in particular calls from abroad – can be collective calls in the sense that the caller talks to several people. "When my brother (in South Africa) buys this calling card, he can talk long. He says it's cheap, and he has to use it at once. He talks then to everybody in the house. Last time (when he called) we talked until I was tired" (Andrea. Interviews 2009/O). In other cases the callers can share calls by passing the phone around, and friends present to the caller talk to the person on the phone as well[286]. A friend's mobile phone can also be used to make a call, when somebody has to make a call immediately and has no credit on his phone: Caron and Caronia (2007:164) name such calls borrowed calls. This practice usually happens with the consent and in the presence of the mobile phone owner. Or face-to-face togetherness can motivate to access mutually known social others by giving them a call to talk, or in order to extend this togetherness by asking the respective person to join[287]. In this sense, having access to calls, to credit and connections, depends on others and thus on local social networking.

In this sense mobile phones have the potential to influence face-to-face social situations when people present talk to – or beep – absent others. Fortunati (2002:515,521) describes such calling in face-to-face situations by pointing to the caller's as being only "half present" or partly elsewhere, the absent person on the phone also contributing to constitute the local social space. To shift the focus of attention from a face-to-face to a mediated conversation is generally not considered as impolite or annoying by the physically present communication partners – the social expectation of an exclusive focus of attention when

---

[285] Compare to different types of calls in the next chapter.

[286] It can happen that the mobile phone owner feels the expense of the call slipping out of control and accuses his friends that they "finish the credit".

[287] Or, as I experienced myself a few times, people who were friends and all of us had spent time together, were beeping me at the same time in order to indicate that they were together somewhere, as an offer to join them.

socializing face-to-face is less rigid than in a Western societal context (compare Humphrey 2005)[288]. Even though some people apologize when their phone rings in the midst of a face-to-face conversation, answering the call is considered as a given, and there are hardly social spaces or situations where calls – especially those received – are not considered as appropriate[289]. This could be explained by the generally high evaluation of social networking and connecting: Calls are prestigious and can serve to demonstrate social importance, as an interviewee who was a student described: "That's what I discovered in the campus, with my mates (...) especially when they receive calls. They walk around and make sure everybody sees that they have a call. They feel important, people think about you..." (Caroline. Interviews 2009/O). As I have observed, international calls are sometimes indicated to others present, by showing excitement, standing up, asking to reduce noise of people talking or music, or talking loudly and sometimes in "good English" in order to make clear that this call is a "bush call". But the mobile phone also has an impact on face-to-face sociality by its mere presence. For example mobile phones are present during face-to-face conversations and could ring at any time (compare Krotz 2001:18). Caron & Caronia (2007:34,35) call this a "stand-by status" concerning the phone's impact on face-to-face sociality. Furthermore, using the phone – for example remembering that one has promised to call, or something urgent has come up - is always in the range of consideration of physically present communication partners.

I have tried to show in this subchapter, that the mobile phone and the internet have boosted social networking. New Media of communication transform the ways in which social networking occurs, and how mediated communication impact on face-to-face social situations[290]. Local and translocal, face-to-face and mediated social networking practices intersect. Contacts with future migrants are set up on a face-to-face level, being connected to migrants abroad could locally have an impact on people's chances to network. Calling habits add to

---

[288] Also a beep or a text message are enough to attract attention to one's phone and physically absent social others.

[289] In schools and lectures at university, at times in church service, and maybe in a very serious or intimate conversation, phone calls would instead be seen as disturbing.

[290] See figure F46, a young man manipulating his phone(s), inserting numbers.

face-to-face social networking practices, transnational social ties make people spend time in public cyber cafés[291], and so on.

New Media users prioritize social networking and New Media have likewise enhanced expectations for social networking, on a local as well as a transnational level. Limitations are therefore vigorously felt: in as much as the mobile phone enhances people's sense of potentially being connected to large social networks, it also evokes the continuous struggle to keep it in use. Likewise, in view that the internet could strengthen transnational social ties, it's use goes along with felt disappointments. Occasional internet users feel the high possibility that their internet mediated social ties break off, and internet literate individuals feel that the bulk of weak ties maintained in the internet are not as responsive and supportive as assumed. Networking practices are strongly tied to access to social capital. When taken on to a mediated level, this has an impact on social and moral considerations of practices of social networking, as I will show in the next subchapter.

F46: Manipulating mobile phones (Ndop)
F47: The entrance to a cyber café Bamenda
F48: Call card

Local and translocal social networking and accessing social capital

Social networking is closely related to the "securing of social capital" in the sense of having access to forms of capital attributed to the different social ties in one's social networks (Diefenbach & Nauck 1997:279, Massey et al. 1998:42, Sey 2011, Homfeldt, Schweppe &

---

[291]See figure F47, a local cyber café. Call cards are often used for international calls, as depicted in figure F48.

Schröer 2006). In Bourdieu's terms, such practices of social networking create social fields, which are determined by social action, and conditioned by enabling and constraining structures and resources related to different forms of capital, or rules and regulations that need to be observed within social fields. The "social field" is also related to Bourdieu's concept of "habitus" accordingly: Habitus consists of patterns of perception, thought and action, determined by societal and individual factors, and alterable through experience. Social fields are under continuous transformations, also regarding processes of accumulation of different forms of capital (Bourdieu 1993). In the local context, these would be related to lines of explanation for the need to link up with migrants abroad, moral expectations towards the ones who live abroad leading a "good life", and the moral conduct in social interaction, which needs to be observed and negotiated in mediated social interaction. An individual occupies a certain position within a social field or within the structure of distribution of different "forms of capital" (Bourdieu 1998:65). Bourdieu (1986) has developed the concept of capital by introducing dimensions of four different kinds of capital, consisting of social, economic, cultural and symbolic capital. They influence habitus and are a part of habitus structures themselves. One of the key characteristics of such forms of capital is their "convertibility" [292]: one form can be converted into another. In that sense, they are indispensable for collaboration, forming part of social structures and facilitate agents' activities (Espinosa & Massey 1997:142, Putman 1993:167).

Bourdieu (1993, 1998) stresses the coherence and inclusiveness of social groups - or „classes" - in specific social fields, and neglects discrepancies. In the field of social networking and accessing social capital, however, in my research, differences in perspectives and positions of social actors come to the fore, as I will show in this and also in the following chapters. In this subchapter, I will address ideal notions of solidarity and reciprocity, and how practices of social networking related to solidarity are pursued and negotiated on local and translocal levels. I led a set of additional interviews during my field stay in

---

[292] Bourdieu sees economic capital – prominently money - as a primary good, dominating over other forms of capital, and also necessary, even though not exclusively, to obtain other forms of capital (Diefenbach & Nauck 1997:279-281).

2010/11, in order to probe into notions of sociality and solidarity in more detail. Here, I was supported by Eric, my fieldwork assistant. Some of the quotes in this and the next subchapter derive from his interviews[293].

Social networking, social capital, and ideal notions of solidarity

Social networking is related to access to social capital, which could be transformed into economic capital – financial support – human capital – as access to another person's social capital and networks, or education - and symbolic capital – such as maintaining social ties with people abroad could render a certain prestige to an individual. As an interviewee expressed: "(…) If you cannot socialize, you cannot get real things" (Festus. Interviews 2009/F). Or as another said: "If I have people, I am already rich. We have this saying: ma money na ma people. I can always communicate with my people, even if I do not have money. This is a background security. (…) It is not just about money, but about people" (Divine. Interviews 2010/11). Wellman (1999:10) argues that social ties of different meaning and intensity provide access to respectively different kinds of support (compare Licoppe 2009:95). Weak social ties are apt at providing access to new forms of support and capital, whereas strong ties of solidarity allow mobilizing and maintaining existing resources (Wellman 1999:10). Solidarity could then be defined as a realization and another dimension of social capital, relating to a feeling of communality and togetherness (Faist 1997:77). It consists of mutual support in different dimensions, and relates to mutual expectations and feelings of duties, but also empathy. Norms of solidarity also serve social control, and enhance smooth collaboration within a group and diminish the possibility stepping out of line (Miztal 1996:211-214). I would adhere here to Coleman's (1990) view of solidarity as reflexively and consciously created and negotiated by expressing needs, expectations and interests[294].

Solidarity refers to the characteristics of social relations, but it also contains a general underlying aspect of morality: According to the

---

[293] I will indicate quotes from his interviews with his name abbreviation „EC". Compare chapter 7, where I will relate to our collaboration in more detail. See also the appendix at the end of this book.

[294] In comparison to a Parsonian solidarity, which is taken for granted (Miztal 1996:68,69, Parsons 1969).

explanations given, this moral basis could be justified in different ways, be it religious in the sense of Christian charity or rather humanitarian, or based on social relatedness. Some expressed an ideal of sociality and moral attitude: "Love, truth and honesty are the genuine substances on which solidarity can flow. No man is an island. When one has a good rapport with friends and family who assume responsibility, life is more worth living" (Kenneth. Interviews 2010/11/EC). Some emphasized an "African culture of solidarity" as typically African – or Cameroonian, or Anglophone – transcending different categories of social ties. "As an African, one should understand that solidarity is our bedrock. When I was sick, many people gave me food. When I had my baby, many people gave me things to take care of my baby. I won't want to have a friend who is not generous" (Minette. Interviews 2010/11/EC).

Solidarity depends on social relations of social and emotional closeness and relatedness. Above all, solidarity within families was emphasized: "You will always be there for a relative, even if you are not always close. But if my uncle for example is sick, or if I am sick, then we will run to see each other and help as far as we can. For a friend, solidarity is out of free will, for a relative, its out of obligation" (Anastasia. Interviews 2010/11). Family support was seen as a duty and a moral concern: "Family is forever, friends are always time bound" (Gladys. Interviews 2010/11/EC)[295]. The same interviewee related to the combination of obligation and free will regarding different social ties: solidarity and supporting others, she said, derived from factors such as "(…) love, social pressure and economic viability. Solidarity depends on the social relation, for example I am less solidaric with colleagues than within my family" (Gladys. Interviews 2010/11/EC). Norms of solidarity are not only confined to kinship and descent, but are also related to good social relationships, affection, policies and circumstances (Kaberry 1952:11,89). "The value of persons depends on what they have done for me. Not only in financial terms, but in terms of affection. I am not holding somebody closer at heart because somebody is a relative" (Divine. Interviews 2010/11) (Horst & Miller 2009:100). Solidarity among non-relatives, friends or those in the same social status group, is described as a moral feeling of communality and responsibility: "A

---

[295] In the sense that friendships can be temporarily limited, friendships fade out and people engage in new friendships.

friend in need is a friend in deed. Everyone is responsible. When someone is down we have to be there" (Gladys. Interviews 2010/11/EC). Or: "Others help to make your life, change your life, everybody is influencing each other, keeping you growing, upgrading yourself, mentally, spiritually" (Franklin. Interviews 2010/11/EC). These statements point to a sense of growing together and being responsible that others rise to the same level. Sharing is a central notion in order to build a good relationship: "You should always try to share with others, make them rise as well and bring them to your level" (Ivo. Interviews 2010/11). The notion of "sharing" runs to the core of the discourses regarding a moral base for accumulation and redistribution: "You should strive to see others succeed, you should help them always when you yourself are in the position to do so" (Divine. Interviews 2010/11). When a person takes part in activities of solidarity, he or she assumes a place in the community as a responsible member, in the sense contributing. "When you help people they will remember you for what you have done for them, especially when they made it themselves, they will always know who has helped them to reach where they have reached" (Ivo. Interviews 2010/11). In this sense, acts of solidarity also contribute to an individual's shift in status. Notions of an ideal solidarity oppose narratives of jealousy and individualism in the local context: I do not intend to raise the assumption that solidarity is seen as unproblematic in presence availability. There exist deep frictions across even the closest social ties. Solidarity and obligations are claimed upon situation and strategy, and could also be perceived as a burden[296].

Categories of social ties, obligations, and (limited) reciprocity

Faist (1997) relates all social transactions or social capital within social networks to the norm of reciprocity. For Simmel also, every social interaction can be seen as an exchange, "as a sacrifice in return for a gain" (Simmel 1971:51), thus he sees the element of reciprocity as a constitutive factor for all social relationships (Miztal 1996:50). Other authors, nevertheless, (Horst & Miller 2005) do not postulate such an applicability of the concept in all situations. For Horst and Miller (2005:763) linking up is promoted through a general asking and giving. "One should give when one is able to. The friend might return

---

[296] See above all chapter 6.

something to you another time, when he is better off" (Minette. Interviews 2010/11/EC). As Wellman states: "Although rarely symmetric, ties are usually reciprocated in a generalized way" (Wellman 1988:41). Wellman, Carrington and Hall (1988:170) relate to this issue as "network balancing": Reciprocity in this sense is going beyond exchanges and reciprocity between ties, but is related to whole networks. "(…) aid is sent to another member of the same social circle and not to the original donor" (Wellman, Carrington & Hall 1988:170). Reciprocity can also be temporarily postponed, as for example is the case with migrants who are sent abroad[297], people investing in them expect an "output" at a later moment in time.

Reciprocity depends on the nature of the social relationship in the sense of mutual obligations, and as well in the category of social capital, social ties offer. For example, as Wellman, Carrington & Hall (1988:168,169) also state, for networking and practices of reciprocal support in Canada, that a favour of a close kin is less likely to be returned than a favour regarding a social relation that consists on a "free will" base, as a friend (Wellman, Carrington & Hall 1988:175). As an interviewee expressed, relating to the duties of migrants abroad: "(…) You should send materials, build a house, or buy your brothers and sisters jobs and place them in good positions. (Then) they will respect that you have done something for the family. The family will love you. When you have a problem they will help. They know: thanks to our brother we are there where we are now…" (Valeria. Interviews 2010/11). Furthermore, the more specialized the support, the less likely it is returned in an equal sense: specific support depending on the positions of the involved individuals and their specific access to resources, cannot be equated. The uncle who can provide access to a government job, or a relative abroad who supports by remittances, are not considered as socially equal, and thus held responsible. According to Wellman, most social ties are asymmetric in nature, regarding amounts and kinds of resources and attention that flow from one to the other. In this sense sometimes quite unidirectional flows of favours, money and

---

[297] See chapter 2, and later in this chapter.

goods develop, related to status and attributed duties and claims (Wellman 1988:43-45)[298].

However, in another strain of normative discourses on reciprocity, interviewees relate to reciprocity as vital in order to stay on equal terms. "A relationship is not just something you throw or take for granted, you have to secure it and pamper it by balancing exchanges" (Festus. Interviews 2009/F). This is the case particularly with relationships "out of free will" and those who consider themselves as socially equal. As I have often come across in the Cameroonian context, when a social relation towards a mate or a friend stays unbalanced over a period of time without a clear explanation, the person who feels "exploited" would complain or react in an attempt to reverse the situation. "I am always calling her, she never calls. This is not polite. We are supposed to be friends... Now I stopped and I will only beep her" (Ivo. Interviews 2010/11). However, concerning migrants abroad, from the perspective of non-migrants, they have shifted in position, which renders claims on their supportiveness legitimate, even when they are peers.

Face-to-face and mediated social networking as strategies for economic survival

The importance of social capital accessed through social networks is related to the high insecurity of income and resources in general, as I have already highlighted in the previous chapters. Apart from a few people active in business, and some few who had a stable income from employment[299], most youth's economic situation is characterized by great instability. The issue of following different options, and relying on different sources of income, is a strategy widely pursued. Here, social ties are crucial: some people depend on financial support from local kin or even friends, and some on remittances from migrants abroad.

---

[298] Positions within social networks can themselves be seen as resources, since they shape the access to scarce resources and render power over their distribution, through their positions as gatekeepers or brokers (Wellman 1988:45).

[299] Which is rather the exception, because apart from the amount of money people might earn from salaries, which could be insufficient as well, employment is characterized by unreliable payments, often depending on the performance of the business, which means for some that they work for months without being paid. The only „secure" and stable incomes are perceived to be those provided by government jobs, which is a reason why most people strive to obtain such employment. In the private sector, international enterprises are seen as the most reliable.

210

Locality regarding social networks of solidarity and support within the local sphere of communities (compare Appadurai 1996) continues to be important, despite the importance that is given to some members who are residing out of the country. Wellman acknowledges that the more difficult the circumstances of scarcity of resources, the more important local social networks of support are (Wellman 1999:35). Everyday "survival" strategies also depend on a large extent on physically accessible help (Wellman & Gulia 1999:103,107), or in Gidden's (1984) terms, on presence availability of supportive ties (Wellman, Carrington & Hall 1988:148). I found it at times astonishing how youth "survive" by "tapping" other's resources here and there by temporally shifting access to resources through others within circuits of redistribution, at times being on the receiving, another time on the providing end[300]. Claims on other's duties of solidarity are more easily to pursue, when being physically co-present and when the social tie is integrated in a wider network of social ties, whereas somebody abroad could more easily slip out of control. Although for most people, much actual support derives from a local level, the idea that those abroad are supposedly the ones who could contribute more substantially to the betterment of one's economic situation, ranges in the background as a strong imaginary (compare Tazanu 2012:206ff).

Slater (2005), in his study on uses of the internet and mobile phones in Ghana, emphasizes the mobile phone's local ability to connect people to support, and that the internet is mainly adopted for online contacts in order to obtain possible support related to migration ventures. According to him, people's engagement with the internet is thus „consistent with a central livelihood strategy – emigration and escape – but opposed and unconnected to the equally central livelihood strategies effected through mobile phones." (2005:770). I

---

[300] As Warnier (1985:92) relates to Ardener, saving wealth is hardly possible due to the social pressure of redistribution. A common accepted way of saving and thus taking money "out of its circulation" – at least for a certain time – is saving in one of the various credit associations, called "Tontine" in Francophone, or "Njangi" in Anglophone Cameroon. Each time a member profits of the entire amount paid into the Njangi by other members. Warnier states that the fact that one needs a certain amount for the weekly or monthly contribution for the Njangi is widely accepted. A considerable part of my interviewees took part in one or several Njangi groups, on different levels of savings.

think Slater is right when he states that mobile phones serve to network and organize strategies of livelihood on a local level, but in my research the internet is also mainly used for mediated communication related to former face-to-face social ties, such as kin and friends abroad. These social ties also play ideally a crucial role in providing support for daily livelihood. I would thus tend rather to emphasize that social networking practices and interpersonal communication in relation to life chances are effected simultaneously on different levels and through different media of communication[301].

Strategic use of social ties: "claiming relationship" with migrants abroad

According to Simmel (1950,1971), in "modern societies" the individual is freed from the dependence on particular others related to a range of practical needs, thereby enhancing the chance of a valuation of social relationships as an expression of personality (Miztal 1996:53). I would strongly question this, also in regard to Western societies, and interpret Simmels' notion of social relationships rather in the sense of an ideal or model. I would claim that in "modern societies", social relatedness and strong emotional bonds of love and affectionate ties should not be seen as opposition to their containing social capital - especially related to social relations enhancing opportunities or providing support as part of an expression of valuation and affection. Friend's social capital can be used because they are friends, and they are not friends because they are useful (compare Miztal 1996). The symbolic valuation of a social tie can undergo an upheaval, when a social position is changed in a positive sense, and the tie could supposedly be emphasized as it could provide access to – a new kind of – social capital and related status: "I always make sure to associate with people who are more than me. I do not associate with people of whom I cannot benefit from ... regarding advice, morally, or financially. I love those kind of

---

[301] For example mobile phones connect people to opportunities for potential migration and its organization, and the internet is also used for communication on a local level of presence availability. Furthermore, these media could themselves also be used in order to make money, by serving as economic tools, such as business people organizing their activities and scammers using internet as a tool to generate their income.

212

people, not the ones who (…) would bring me on the wrong track. I don't like useless people" (Moussa 2010/11).

I have already highlighted the importance rendered to social ties abroad. Even though ties abroad concern only a small part of people's overall social networks, they contain a considerable part of the imagined potential of social capital. Thus, in the case of migrants, social ties are often strongly influenced by the symbolic difference that is attributed to place. The change in position through migration could even transform or reverse flows of support in respect to age, hierarchies and responsibilities, in the sense that "being abroad" has greater value, as for example a younger brother or sister could be expected to send remittances to his or her elder siblings or relatives. "They know you are in white man's kontri where there is money. We can never compare white man's kontri and Africa. So they expect a lot from you…. It only depends on somebody is making money. (…), even when the mother is abroad. Here the children would take care of her, if she is abroad, everybody would depend on her and complain" (Valeria. Interviews 2010/11).

Among my interviewees, as a tendency, the existence of transnational social ties was particularly highlighted. Most of my interviewees who did not have kin abroad, claimed to have at least a friend or former class mate abroad[302]. Depending on the individual's local access to social capital, a desire and need to connect and engage with the migrant – even when not particularly socially or emotionally close – could arise, expressed in attempts to "claim relationship", which could have different bases: At first, it consists of the expectation that a migrant is reachable and attainable through communication media. It starts from making sure that one disposes over a person's contact coordinates, attained before the person left for abroad, or reactivating social ties with former school-mates, acquaintances, and so on: very often, it is now possible to find them on Facebook. Then it includes regular contacting and communication. When somebody migrates it is used as an opportunity in order give consideration to the contact, also

---

[302] When I asked more in detail about the frequency and quality of the contact, these descriptions were not always pointing out to a long term friendship and a close social affiliation, or a frequent interaction: "I have a friend in Holland, but we do not communicate regularly" (Herbert. Interviews 2009/R).

since contacts abroad are less and more manageable than the bulk of local acquaintances. In addition, there is the assumption that due to New Media of communication, and migrant's potential to make sure that they are well connected to those technologies, smooth contact should be possible. Here it is also highlights differences regarding social and emotional closeness. With kin the base for claiming relationship is in this sense "naturalized". Friends are more ambivalent and prone to disappoint. Especially here, the notion of a high possibility of the breaking off of social ties is widespread. Ultimately, solidarity is viewed at as a socially confined right and respective duty on the side of those being in a better position. Contacts to migrants are inflated in their being meaningful in the view of the local conditions, which reflects a less promising social reality (Nyamnjoh (2005:252).

Diverging strategies and evaluations of social networking set by claims on obligations

I have already shown that practices of social networking depended on position, location and perspective. With regards to people in Cameroon, new technologies of communication are used in accordance their ideals of social networking, which are the maintaining of wide networks of close and more distant contacts, of emotionally and economically important strong and weak social ties (Miller & Slater 2001, Horst & Miller 2005, Wellmann 1999). Here, the underlying attitude is "link-up", thereby negotiating opportunities, connections, mutual reciprocation, obligations, potentiality of "being of help" related to finances, gifts, social connections and advise, and importantly "being abroad or not". Some – in particular in Cameroon – related to this claiming of relationship in a rather non-normative way, to explain the implicitness of the inclination of turning towards those who have shifted in position. "When you become a big man, they will come to you. Even those you hardly knew before. Then your responsibility towards others will grow. It is normal that they will turn to you then, and it is also an honour for you" (Ernest. Interviews 2010/11).

In this sense, migrants living in diaspora often become the "target" of expectations, and "the ones with whom people want to stay in contact with". In order to deal with diverse connections, which most often include expectations and claims, migrants adopt strategies in order to control social networking practices. The habit of maintaining wide and far flung social networks and to invest in social capital for potential support and profit, shifts in the situation of diaspora. It could be called a shift regarding "zones of closeness"[303]: In order to deal with expectations of too many people, networks are reduced to closer social ties of family and intimate friends, and herewith, to morally more acceptable claims. A non-migrant expressed understanding here: "Once I am out there, I have to reduce contacts to only a few, I cannot be of use for so many people. Also relationships can be strained when you are abroad. People only look at you as a resource" (Divine. Interviews 2010/11). The situation in diaspora enhances new priorities regarding social networks, which are more likely to be examined regarding qualities of an emotional closeness and intimacy[304], to the disadvantage of weaker ties in general. At times social closeness with kin can also be re-evaluated by migrants, when certain social ties turn out to be burdensome and deceitful (compare Tazanu 2012:151), and others prove to be trustworthy and morally supportive. As a migrant stated: "When you are abroad, you find out to whom you are really important, and who only sees you as a source of money. Your estimation of people can change with the experience, how they relate to you when you are abroad" (Blaise. Interviews Diaspora).

In this subchapter I have described how claims on solidarity are taken on to a mediated level of social exchange and networking practices, where they are prone to negotiation, which is particularly expressed in oppositional views and practices of migrants and non-migrants.

Being part of local and translocal social networks of support

Youth – even when not financially well off – are integrated into social networks of support giving and receiving, and they are not always merely on the "receiving end"[305]. In particular, they are the ones who are

---

[303] Tazanu (2012:176) speaks of "a family within the family", relating to core familial ties.

[304] Compare to the next chapter, as well as the conclusion chapter.

[305] It is however obvious that people want to portray themselves in the best possible way by calling upon an idealized version of their activities related to supporting and assisting others.

supposed to be future benefactors. Young people express their feeling of responsibility to contribute to their family member's wellbeing. Meanwhile, before achieving such a position, they depend on others, but often also for the time being, others depend on them, for example concerning support towards younger siblings, or small acts of solidarity among friends. In absence of formal social security systems support is obtained through social networks. Since New Media of communication are means of handling support giving, the instantaneous getting into contact is also assigned to expectations that problems could be quickly solved by linking up with somebody, who could render help in one way or the other (compare Tazanu 2012:238ff).

Support rendered or received mainly contains the gross categories of material and non-material support. Both categories also involve moral implications. In the array of material support, financial support is considered as most important. I also relate here to the pooling of resources for migration ventures. Material support could range from durable investments and contributions up to small courtesies. The category of non-material support figures very prominently in the range of support giving and receiving – I subsume it under the notion of "showing concern" – a notion, which is often used in the local context. Such kinds of support could be rendering advice and information, or providing somebody access to one's social network, as for example in relation to job seeking or migration opportunities. The range of support giving and receiving is then differentiated between social relations in presence availability and those physically absent, where mediation is solely based on communicative exchange, which has to substitute physical presence[306].

The pooling of resources for migration ventures and (moral) indebtedness

A migration venture is an investment in social networks, and it is in return enhanced by social networks and ties. Stark and Lucas (1988:466) relate to an intra-familial "contractual arrangement", and Fleischer (2006:7) states, that migration could be seen as "a strategy that shifts the

---

[306] This subchapter relates in particular to the 1st sub-question of guiding question 3 – how sociality and solidarity transform from a face-to-face to a mediated level, concerning here physical presence or absence of people especially.

focus from individual independence to mutual interdependence". Already decisions to migrate are hardly taken in isolation (compare Nyamnjoh 2005): Adepoju (1995: 329) emphasizes the role of senior members in the household who often decide who should migrate and who should not do so. Usually the person with the greatest potential is chosen, depending on the individual's character and level of education (compare to Fleischer 2006). Fleischer speaks of the support rendered to an aspiring migrant by his extended family, under the duty of rewarding them with support in return, for example by sending remittances back home. Investments in education could also be considered in this way: "When they supported you so that you could make a good education, they also think that you should at least try to migrate, so that there is a valuable outcome" (Christian. Interviews 2009/R). A major difficulty for migration ventures, which are financial resources, could be overcome by money being pooled together in a family for the migration venture of one of their members. An individual young person would not easily be in the position to gather money for such an endeavour. Already at this point of the venture, the financial support of family members is involved. It could thus lead to a great deal of pressure for aspiring migrants, when they have already gone through several attempts to obtain a visa, to finally make it. Investing in aspiring migrants is seen to be profitable for their sponsors, as a certain outcome could be anticipated. "My mother sold a part of our land in the village, in order to pay for the fee for the university (in Greece)" (Christian. Interviews 2009/R). Migrants abroad play an important role. They might contribute financially to a migration venture of, for example, a relative, but often they render help in the organizational and informational field. Often sources of financing are widely varied, stemming from different people. "Everybody in my family was putting all their money together, so that I could pay for the visa expenditures, so I really pray now that I can travel soon" (Peter. Interviews 2009/FS). This secures the base for later indebtedness and obligations towards wide social networks of kin and friends at home and abroad, for the migrant (compare to Nyamnjoh 2005, Fleischer 2006:21).

Financial and material support on a local and translocal level

When relating to migrant's financial involvements on a local level, it is very often designated with the term of remittances (Espinosa & Massey 1997, Fleischer 2006, Levitt 1998, Walton-Roberts 2004, Atekmangoh 2011, Stark & Lucas 1988, Tazanu 2012). As in many comparable countries, a considerable part of the GNP in Cameroon stems from remittances. According to the World Bank, there were about 167 million Dollars transferred to Cameroon from abroad in 2008, and probably the transfers were probably considerably higher, due to more and increasingly informal channels for sending remittances (Evina 2009:17, Adepoju 1995). Money transfer institutes[307] such as Money Gram or Western Union are highly present in Bamenda. The mobile phone companies are also starting to become involved in this market: the MTN service called "mobile money" was introduced in 2010, and it was then already widely used and advertised[308]. I do not want to delve deeper into the topic of remittances, but I would instead like to look at this topic from the perspective of personal social networks.

Financial support is emphasized when migrants abroad are concerned, because migrants are seen as more economically viable and thus also expected to act supportively. Furthermore, sending money is often the only way and for sure the easiest way for migrants to support friends and relatives in Cameroon when being physically absent. However, the potential for being cheated is highest when financial transactions are involved, as in examples when provided finances are not used as intended by the benefactor. In the local context, it is common knowledge that migrants generally prefer to effect financial transactions for foreseeable and important investment, which also involves, in this sense, also considerable amounts, and in turn support could be less regular. Such financial support is often bound to a specific investment, such as school fees for example, treatment in case of sickness, repairing a house or contributing to a social event. Non-migrants said that they do usually not render accounts to their benefactors, emphasizing that they

---

[307] See figure F49, which depicts again another money transfer institute, called Express Exchange.

[308] These services are advertised as being quicker and simpler than the usual money transfer institutes. See figure F50. The procedures of sending money consist of paying and receiving an amount in cash in an MTN branch, then the transfer is effected by a confirmation through MTN and the sender provides a code to the person benefitting.

are in continuous close contact with them through New Media of communication, and enjoy their confidence[309].

Over the range of the interviewees (2009 and 2010/11, altogether 67 individuals), there was only a minority of people who counted on regular allocations from kin abroad, eighteen people said that they had already received financial support from abroad several times – even though not always regularly, but they could turn to these people when they have a problem or shortage, and they would often be in the position to render help[310]. Another few people stated that they had received financial support a few times, which was not benefiting them directly, but was instead family related. Some indicated, that they could from time to time also rely on friends abroad, when smaller amounts – below FCFA 20,000 – are concerned, and a few other people stated that they had received credit for their phones[311]. Material support – including gifts sent or brought by a visiting migrant, have an important symbolic value. One interviewee got a suit, another a pair of shoes from an uncle and a friend abroad, respectively. Apart from that, the importance of durable investments was mentioned in order to support relatives in Cameroon, such as building a house, which could be rented out, or investments in farming activities or in a business, which could sustain the family.

F49: Money transfer office, Bamenda
F50: Publicity for sending money, by MTN

---

[309] Compare to chapter 6, regarding giving financial support from the perspective of the migrants.

[310] Generally, most people do not dispose over a financially viable contact abroad. It here also depends on the social and emotional relatedness of people, since it is usual instead that a closer relative could be counted on as substantially supportive.

[311] I did not go into this in a sense that I could render account about annually transferred average amounts for example.

In general however, since having an economically viable and as well supportive contact abroad is the exception rather than the rule, most people rely more on local contacts for their "coping in everyday life", when financial or material support is concerned. Apart from those who were in one way or the other supported by parents, seven interviewees claimed to have other generous benefactors on a local level, be they kin, or, in two cases, non-related others. Usually, local financial support instead concerns smaller amounts of money, including money for emergencies. If money for "emergencies" is asked from migrants abroad or from local social ties, it depends very much on the quick responsiveness, and of course availability of such ties, on either level. Most people told me that "small money" was instead asked for from somebody in presence availability, apart from in urgent cases, it was also in order not to bother a migrant abroad, in the view of another greater contribution that could come from this side[312]. On a local level, the money is often bound in the sense that the investment is direct, the materials needed are purchased by the benefactor, for example. Material support is rendered particularly among family members, as for example support towards siblings regarding school materials, food provisions, drugs, clothing, and any other vital contributions. Exchange of small favours, such as inviting somebody for a drink or giving a gift and also the sending of mobile phone credit, is viewed as an apt means of consolidating a social relationship, in a wide array of social ties, such as family, friends and lovers. Furthermore, support consists of providing food - or rendering help in cooking, which is a very common practice of people living in neighbourhoods, an example of solidarity between people who are not related but living in spatial propinquity. Non-material support is not always clearly differentiated from material support, as such examples show, since material support such as a gift, sending credit, or any small attention, have the component of cheering somebody up, an expression of one's appreciation for the person. "Gifts bind together. When my grandmother sees the gifts I have sent, she thinks of me and that there is somebody out there sharing the fruits of her sweat with her" (Gladys. Interviews 2010/11/EC).

---

[312] There could have been also a sense of dignity involved, that they do not like to always be dependent on migrants.

Spreading information through face-to-face and mediated social networks

In the array of non-material support I want to select such kinds of support, which seem to be important in particular in relation to New Media use, in which rendering information is important. I have highlighted that in migration ventures it seems to be crucial whether people dispose of "social capital" abroad. Migrants abroad are seen as not only financially viable, but also knowledgeable, in particular where information and advice for migration ventures are concerned[313]. However, information is also spread on a local level among groups of friends for example. Even though information in many cases seems to be a „by-product" of mediated communication by internet and mobile phone – and face-to-face – it is considered as a vital part of rendered support. Information is thereby appraised on the evaluation of the assumed competence and trustworthiness of a social tie.

One aspect of rendering information is providing links regarding various topics of interest. Many people prefer to obtain a link by a reliable friend than through a search engine. Some occasional internet users, but also regular internet users, almost entirely depend on such sites, not just out of an inability to search for information, but also because there is no need for further information – the sites known are sufficient for their purposes. Such sites are related to different fields: Prominent is the field of migration opportunities, above all regarding educational purposes. Widely known sites such as "braintrack.com", as well as "4icu.org", give an overview of universities in Europe and North America. Furthermore, regarding opportunities for financing migration and education, the sites "tuitionfreecolleges.mtnhome.org", and "scholarship-positions.com" are popular. Of course the official site of the US DV lottery "dvlottery.state.gov" and the current application site, as "dvlottery2010.com" are wide-spread, and for other visa information the site "migrationexpert.com". The site "jobsindubai.com" reflects the current popularity of Dubai as a "haven" for jobs in different sectors. In the field of "youth culture", the most prominent sites are related to football, such as "goal.com", or the official websites of mostly European football clubs. Such sites are popular also among people with a low level of internet literacy: They too have one or two addresses of sites, which

---

[313] I have related to this in chapter 2.

they check from time to time. The yahoo site "yahoo.com" is the most popular among the sites, since almost everybody has a yahoo email address. When opening the yahoo website in order to log in, many people regularly check out the news on the yahoo site, using it as an easy opportunity to "browse" by just clicking on links given regarding certain topics which they are interested in. Among scammers, many sites are in circulation. Most often these are sites with the latest proxy lists[314] for example, or new free ads sites, which are not yet blocked for Cameroonian IP addresses. They are spread with a connotation of secrecy among inclusive groups of "friends". Such information and links regarding popular sites are spread on an online as well as an offline and face-to-face level of social interaction, by accordingly sending links but also by mouth-to-mouth advertisement.

Showing concern as moral support in local and translocal spheres of social interaction

The kind of support which I would subsume under "showing concern", figured mainly in realms of intangible support, but it can also contain material support in the sense of rendering attention, for example through the giving of a gift. I would differentiate "showing concern" regarding two aspects of support giving – or receiving – which is a rather mental dimension of moral and emotional support, and a physical dimension of assisting. Regarding "showing concern" physical presence is important – however, such support can also be mediated and rendered through media of communication, at least to a certain extent. Furthermore, "showing concern" through media of communication is effected on a local level of presence availability as well as on a translocal level. In any case, this dimension of support giving and receiving is felt as being most affected by physical absence.

In the dimension of moral and emotional support I relate to support, which refers to moral attributes of "good relationships", in an ideal sense. These are clearly mental, intangible and non-material: "As a child I grew up to know that the best things in life are not visible. Like love, kindness, goodness. One only feels them: having a shoulder to cry

---

[314] Lists to find transparent IP addresses, in order to use them to place ads on free ads sites, which are blocked for Cameroonian IP addresses, in order to pretend to be located elsewhere than in Cameroon.

on, a friend to share your joys with" (Minette. Interviews 2010/11/EC). Or: "It is not a matter of amounts of money but happiness, cheering somebody up or sharing something together" (George. Interviews 2010/11/EC). Moral and emotional support can also contain a sense of a positive influence and interaction. Regarding communication it involves listening to each other, giving the person a heartfelt advice, cheering each other up, encourage each other, trying to keep up trust in a relationship, and "being in accord with others" (Pride. Interviews 2010/11/EC). As a further aspect of mental support prayers are mentioned. Moral and emotional support is seen to be possibly fulfilled also in physical absence, but with difficulties related to the substitution of physical presence through communication: It contains a continuous staying in contact. "Devote your time, energy and money for the person, be there for her. Phone calls, sharing of pictures and videos, getting information about the person's life. If you cannot meet through media then the relationship may weaken. You have to meet continuously" (Kingsley. Interviews 2010/11/EC). By keeping the contact lively, moral and emotional support can be effected. "When communication is good, time and space does not limit. Past time shared impacts much on the degree of trust and closeness" (Gladys. Interviews 2010/11/EC).

The other dimension relates to physical co-presence and thus a local level. It concerns assistance, which involves a clearly material component, but the sense of "showing concern" is very important thereby, and it also has a strong moral and emotional connotation. It is not the mere material assistance in the sense of sending money or goods to support somebody, but it is taking time and at times making a physical move in order to "be there" physically. "Doing something" for somebody, in the sense of devoting time, is also a substitution of rendering support of money or material support and vice versa. However, relating to "social events" physical presence and "showing one's face" is evaluated as most important. Not to be present can only be excused when a person is absent in the sense that the geographical distance is too far[315]. Especially around festive periods, such as Christmas and New Year, when many events take place, such as

---

[315] Very importantly this concerns migrants abroad, who are then expected to make a considerable contribution in monetary terms, but it can also relate to somebody in Yaoundé or Douala, and somebody who is busy job-wise.

marriages and cry days – death celebrations to remember the deceased – people are very busy spending time at such events. The closer to the concerned person, the more the duty to contribute rises. "Let's say, my cousin is having an occasion. I am bound to go there and do something to help (she is not so close to her cousin). It is really important. I cannot stay back and just come as a visitor. We are belonging to the same family. With a friend it is different, it's more out of free will. It also depends on what type of friend it is" (Anastasia. Interviews 2010/11). However, "being there" is an expression of concern in the sense that consoling somebody who has lost a close person, expressing condolences, offering assistance in difficult times, giving moral and emotional support, or cheering with somebody, such as in the event of a marriage or born house – when a baby is born in a family. Showing concern through physical presence is also related to visits, spending time together, such as taking a friend out for a beer, or dropping in at a friend's house for a visit. "I like to hang out together (with friends). It concretizes your friendship when you meet physically" (Isidore. Interviews 2010/11/EC). Or: "I am impressed when having a friend who is showing concern for me, maybe just beeping, (…). Just that somebody thinks of me, including me into his programme, not forgetting about me" (Felix. Interviews 2010/11). The sense of being together is emphasized[316] and the issue of spontaneity plays an important role. It is lamented that this kind of casual closeness and spontaneity is lost when somebody is physically absent, and it is seen as a quality of socializing, which is difficult to substitute by mediated contacting. "It is very sweet having a friend who is very close, especially on a local level. Time in a relationship counts so much. (Instead) when you have not communicated for some time, you lose touch" (George. Interviews 2010/11/EC).

Giving and receiving support under conditions of physical dislocation

Physical, and along with it social and emotional distance can be considered as a problem for support giving and receiving. Thereby, it is

---

[316] I had the impression that it is not always communicative exchange that is given the highest priority. Attending for example a social event is valued highly. Less important are elaborate conversations. Spontaneous being together and social visibility – who is socializing with whom - is given great importance among youth in an urban context.

deplored that migrant's concern decreases when they are abroad. Showing concern is due to physical absence first of all based on communication. Feelings of emotional and social closeness are then considered as a base to legitimize, but also to a claim on solidarity and support. Thus, tied up in the term "concern" are also expectations regarding support, typically expected from migrants – remittances or material investment.

In order to enforce claims, spatial distance is seen as potentially obstructive, since migrants abroad tend to slip out of control. In this sense, distanciation of people abroad is strongly related to migrant's "disposition to change": non-responsiveness and lacking a sense of solidarity, in view of the idea that it should be possible to stay connected and support each other despite physical absence and spatial distance: "People going abroad cut down their contacts to only the closest ones. Then also they are not to control, just easily disconnect and no responding. (…) the explanation could be that back there it is so hard for them. They will only take care of the closest family, sending money (to them), but they will not care about the extended ones. But they should think of everybody, because one is related! It is not only about money, but not staying away from people, keeping contacts, showing concern" (Simon. Interviews 2010/11). In this sense, showing concern is mentioned as a main feature of evaluating the responsiveness of a social tie abroad, on which supportiveness is based. It then relates to the responsiveness and willingness of people to maintain a tie, less than on the geographical distance. As an interviewee expressed clearly: "If you move out and do not keep communication, it is not distance (that is the problem), but not showing concern any longer" (Pride. Interviews 2010/11/EC). Another interviewee related to the felt distance as an emotional distance, which does not depend on geographical distance, but on physical dislocation: "The distance cannot disappear, even through media. Although with phone and internet, there are means. But even the ones who are just settling in Yaoundé - the distance is already much too much for spontaneous meeting and sharing. People do not even have to travel abroad to make the relationship distanciated" (Anastasia. Interviews 2010/11). Co-presence was seen as favourable – feelings of closeness are easier to maintain when somebody is physically around: "You are close to those who are always around you, you know

225

what is happening to them. When you lose this connection then you will be less close automatically" (Simon. Interviews 2010/11).

According to others, however, distance is related to in the sense of geographical distance, which thus makes a difference between social ties abroad and in the country: certain advantages were mentioned regarding a social tie maintained in Cameroon. Regarding valuation of support giving, an interviewee expressed: "I give more importance to the relations in Cameroon, they are there, not too far... but those over there, the distance is there, they can only give you moral or financial support. Some people give more importance to international relations because they think these are more viable financially. (But) it is easy with your friend here, to discuss things at length, handling the situation even at the moment (...)" (Terence. Interviews 2010/11/EC). Included here is the assumption that people in Cameroon are in presence availability, an idea, which contradicts the life reality of most people, who might only see their friends and relatives who live in another city in Cameroon at Christmas or a family event – but they are "felt" to be closer. Tazanu (2012:156) relates to the interconnection of a feeling of closeness and geographical distance as the illusion of closeness within nation state borders, and that Cameroon is seen as the authentic ground of sociality. One crucial issue to support such assumptions are statements, which relate to an apparent easiness for mediated sociality in the country: For example, when it concerns the prices for communication. Calling "to just say hi, and ask how the person is doing", is considered to be less expensive: "The fact that one can communicate more easily in Cameroon makes relationships at home more cordial than those abroad. When I have a problem, all I do is put some credit in my phone and call at night"[317] (George. Interviews 2010/11/EC). Calling – or being called by – migrants abroad is said to be difficult and less spontaneous since they are considered to be busy, cannot answer the call, and have to follow a lot of protocol in their daily lives. This uncertainty of maintaining connections with migrants relates to their tendency to withdraw and not show concern.

New Media of communication play an important role for people in the sense of being potentially reachable or able to reach out also for

---

[317] Because of cheap offers for nocturnal local calls, which means that communication partners could take time to chat.

moral support, expressed in notions, which indicate a sense of co-presence, such as „rendering a helping hand", "doing something for somebody", or „giving a shout" related to the act of calling. In this sense "showing concern" also has a mediated dimension, which is considered important on a local and a translocal level, under circumstances of presence availability and physical absence alike. This mediated "showing concern" figures in the dimension of moral and emotional support: it can consist of talking, trying to understand and sharing things mentally. By taking sensory and emotional limitations into consideration in social situations of durable physical absence, it is generally attributed to the internet and mobile phones that they can be apt means of conveying concern, by regular and also by spontaneous acts of communication. Therefore, showing concern from the part of migrants by spontaneously calling, or taking part in the lives of those "at home" a mediated sense, such as calling in when an event takes place, is highly appreciated. It opposes to the idea that calls from abroad always need to have a specific purpose, and it means, that those abroad have an idea about the life courses of those "at home".

In this subchapter I have tried to show how support giving and receiving can transform in relation to location, distance and their symbolic evaluation. Geographical distance and physical absence is seen to have an impact on feelings of closeness and supportiveness regarding different categories of social ties. Being connected through New Media of communication is seen to offer a great potential for connectedness, which is however sometimes obstructed by the effect of different life-styles and –worlds, as I have shown regarding the issue of presence availability as important in acts of support giving and receiving. Overall, the dimensions of support of giving and receiving are closely related to an evaluation of the quality of social relations. I will look into the qualities of social ties when being mediated through New Media in the following chapter.

Conclusion - Creating liveness by adding face-to-face and mediated social networks and opportunities for support

In this chapter I have tried to show how local or transnational social networks and practices of networking inform each other, concerning the interrelations of migration, profiting from remittances, status obtained through transnational social ties on the local level, and mediated communication practices influencing local social interaction: I have examined networking practices of youth, which range from local to transnational and from face-to-face to mediated social interaction. In Hannerz' words (1996:73), both, immediate and mediated social ties become part of people's "structures of attention". I have tried to show by introducing categories of social ties based on presence availability, that networking on a local level can contribute to – future – translocal social networking. Regarding notions of solidarity and support, face-to-face and mediated social ties are subject to moral considerations. On the one hand, evaluations of being connected and solidaric – adopted vis-à-vis specific categories of social ties – are taken on to a mediated level. On the other hand, practices of social networking transform under conditions of opportunities of New Media to initiate, maintain, and claim relationships. In view of imaginaries of migrant's affluent life-worlds increased expectations arise from being connected and having access to migrant's social capital.

I have emphasized that social networking is considered as a priority in New Media technology's uses – Horst and Miller (2005) have related to link-up as the priority of social networking over content of communication. Even though networking motivations might be not much different than before introduction and access to New Media, what has been transformed in the era of the internet and mobile phones is the greater scope and range of social networking opportunities. Furthermore, the agency of the users is enhanced as New Media of communication offer opportunities to network along notions of solidarity regarding socially close kin but also beyond, in the field of peers, and related to youth's specific interests. New Media thus create opportunities for monitoring social capital and networking practices, and following up individual life projects.

The aspect of liveness becomes very apparent here. Regarding liveness as an imagined potential, the creating of elaborate and stable social networks relates strongly to the imaginations of the potentiality of "the West", which here relates to the valuation of the assumed social capital of social ties abroad. Or the sense of liveness as a potential can be produced in practices of linking up on a local level, by exchanging and storing phone numbers. It also concerns the expectations vis-à-vis those who are physically absent to be potentially always in reach, including the possibilities of disappointment in the face of an expected liveness. Liveness as "work" would then be the investment into maintaining these social networks. This sense of liveness is created through continuous and regular communication – face-to-face or mediated – in order to keep social ties "alive". Media habits also point to such effort, such as frequenting the cyber café, the money spent on credit for phone calls, and the continuous evaluation of how much to invest related to the "output" needed, expected and hoped for. Support rendered can then be seen as fruition of the effort to maintain apt social networks. These dimensions of liveness intersect with the aspects influencing social transformations under condition of different nuances of mediation, related to physical presence or absence, and the qualities of social ties as a perceived social and emotional closeness.

In the next chapter I will try to show how social ties are cultivated through diverse technological features in New Media, in a range of social and emotional closeness. Qualities of social ties, and the ways mutual trust and evaluations of closeness are built and maintained, might undergo transformations when social interaction occurs without physical cues, and transformed references regarding current life-world conditions of the communication partners.

5

Physically distant but emotionally close – transforming qualities of social ties

Practices of socializing and social networking, which I have described in the previous chapter, are interrelated with people's dealing with space. In the current chapter I will look at how social relations are transforming under conditions of physical absence, as well as how feelings of social and emotional closeness are negotiated in mediated communication. "(…) Increasingly, people who are near emotionally may be geographically far apart; yet they are only a journey, email or a phone call away. Thus developments in transport and communication technologies not merely service or connect people but appear to reconfigure social networks by both disconnecting and reconnecting them in complex ways." (Larsen, Urry, Axhausen 2006:1)[318]. New Media intensify and multiply the maintaining of interpersonal social relationships and networking. Hereby, New Media are adopted in various ways in order to bridge physical and social distance most effectively, in attempts to maintain an ideal connectedness in a habitual dimension of agency. Furthermore, negotiations of sociality, as well as of moral legitimacies and solidarity, can be construed in differing ways related to life situation – and location – of communication partners. Different strategies of communication are adopted, related to different expectations New Media users have in regard to different communication partners or social categories, which I have looked at in the previous chapter. Sociality is evaluated according to the quality of contacting and maintaining a sense of closeness: I will relate to such qualities of social ties by addressing notions of social and emotional closeness, trust and intimacy. Such qualities could be supported – or limited – by communication media: They are influenced by their specific mediality, technological features, and media conditions. In this sense, New Media do not only transform sociality in their potential to enhance mediated

---

[318] It points to how social mobility is tied to physical mobility, and how in the case of New Media users in Bamenda, people engage in practices of virtual mobility in order to bridge distance.

connecting, but they similarly enhance processes of distanciation within social relationships[319]. Liveness could be seen as shaping mediated interpersonal social relationships, when communication features are used as tools to "control space and time" by monitoring closeness and intimacy, here mostly related to New Media user's judgements, or a practical-evaluative dimension of agency. In this chapter I will relate to guiding question 3, how New Media users negotiate liveness in mediated social ties[320], hence, to liveness as a sensory experience and as work. Furthermore, I suppose that qualities attributed to social ties, and feelings of closeness in social interaction are transferred to media realms, negotiated under conditions of mediation. The data I am relating to here mainly derives from my field stay in 2009.

Reflections on "closeness", "trust", and "intimacy" as qualities in social interaction

As I have discussed in the previous chapter and will look at in the course of this chapter, New Media of communication do not induce completely new practices of socializing: Appropriations of New Media are deployed along the lines of social networking practices and moral assumptions in the sense of a continuity of habitual practices. These are however also coined by projective orientations, as expectations and desires for transforming one's life though connecting – and thus an evaluative judgement of New Media practices for these purposes (Emirbayer & Mische 1998, Wilding 2006). I will thus need to look at preconditions of sociality, enhancing social interaction and communication, including in a mediated situation, where physical co-presence needs to be bridged by – at first – communicative and other modes of interaction. Love – or empathy – and trust are such media, which serve as a basis for human sociality (Rasmussen 2002:101, Gross and Simmons 2002:533,534, Barber 1983, Hjarvard 2002). These notions are attributed to an ascribed quality of social ties, and a

---

[319] In the same sense as the perception of being involved in a globalized world of increasing mobility and interaction, could also lead to a feeling of immobility and disconnection, as I have discussed in the previous chapters.

[320] Especially I will address the 1st and 2nd sub-questions, how sociality and solidarity, as well as how qualities attributed to social ties transform in social relations when being mediated.

qualitative evaluation of mediated social interaction. The meanings of these notions are in the context of mediated relationships related to the question of whether and how the use of New Media of communication can produce and maintain a sense of closeness. This subchapter is mainly based on interviews from the fieldwork period of 2010/11.

Trust as a basis and media for social interaction and everyday life

Trust has been seen as a basis for social and political collaboration in different societal realms. According to Giddens (1991), trust coordinates social interaction across space and time. He here differentiates trust here into abstract systems – faceless commitments – which are media such as money, expert systems and institutions, and personal trust – facework commitments. For example Govier (1994) differentiates personal trust and social trust. Personal trust is trust in known others with whom we have a certain personal relationship and personal experience. Social trust is trust in people, who take over certain roles in a society, with whom we most often have a limited personal experience (compare Govier 1994:31), such as doctors, lawyers, teachers, and so on. For my work, I will relate to personal trust, however, and also the expectations of roles[321], which ought to be performed by social others in certain positions are important here. According to Giddens, personal trust plays an increasing role in the sense that individuals actively pursue personal social ties, since "trust in abstract systems is not so psychologically rewarding (…)" (Miztal 1996:90). Thus, trust – as well as mistrust – in others is produced on the base of social interaction and social obligations that are related to social relations and social roles. "When we trust other people, we expect them to act decently towards us, and we are inclined to interpret what they say and do according to those expectations." (Govier 1994:31). Personal trust is usually – but not necessarily, as I will try to show – decreasing from the closest and dearest person to more distant ones, from those sharing a common background, such as nationality, ethnicity, or religion, up to "strangers" (Miztal 1996:99). Trust is also situational – somebody might be able to trust in a certain instance under certain conditions and in others not – and it is relational – somebody trusts in certain persons according to "markers of trustworthiness" (compare Miztal 1996). Trust building is

---

[321] Compare to the introductory chapter on social norms and roles.

also taken to a mediated level, challenging the "traditional" mechanism of building trust to a large extent. These mechanisms could be cues of self-representation given in face-to-face encounters, depending on expectations of culturally and socially "correct" comportment, and the attitude or inclination to trust or not to trust (Govier 1994:33)[322]. In particular in mediated interaction trust can be at risk and sources of estimation are limited[323]. As Giddens (1994) and others state, trust is indispensable for collaborative action, and mistrust is more burdensome (compare Tufte & Stald 2002:78). We will see that trust might be favourable in the coordination of social ties, but it is not necessarily intrinsic to social relationships. In regard to everyday interactions of people, social trust has also been called "social glue", basic trust, or confidence (Tazanu 2012:128, Govier 1994:31,32, Giddens 1994). Confidence is a trust in the stability of the social world and conditions, related to a certain predictability of social life and routine interaction (compare Miztal 1996:105, Berger 1978). Trust is needed in everyday interaction with social others[324], relating to the lack of full knowledge about other's motivations and intentions, the inability to monitor others' behaviour and the "contingency of social reality" in general, and insecurity regarding the outcome of actions and interactions[325]: Trust is dealing with one another's freedom to fulfil and reward trust, or to disappoint (Miztal 1996:83). In this sense, trust always involves taking a risk (Beck 1992, Giddens 1994) and continuous "work on trust".

Qualities of social ties: Sweetness, intimacy and closeness

The notions of "sweetness", intimacy, and closeness have different relations to physical presence or absence and they are media, which reduce distance (compare Silverstone 2007:123). Concerning mediated social ties, the notion's use or non-use or transformed meaning can

---

[322] In the case of strangers and lacking knowledge about the person, cues of self-representation could be a social role, comportment, dress, voice, and a general estimation of "human nature" (Govier 1994:33).

[323] Sources to estimate trust relate to cognitive sources of information attained through observation and judgement of others' comportment, others' reputation related to third parties' evaluations, and a personal emotional estimation.

[324] See figure F51, here related to trust in business relations.

[325] This is what Giddens (1994) names "ontological security", a feeling built in processes of socialization and the security concerning immediate social ties and habits of everyday life.

render insights into evaluations of transformations regarding qualities of social ties. The ideal of striving for harmony on a negotiated basis of mutual confinement seems to be a human basic need (Bourdieu 1977, Luhmann 1973, Durkheim 1964, Miztal 1996). Kaberry (1952:33) states that the Nsaw look on quarrelling as destructive for social communities, since it creates disharmony, in the sense that it destroys the basis of livelihood and wellbeing. Social harmony is also described as "keeping relations sweet" (Kaberry 1952:41) in the sense of keeping favourable, gentle and friendly relationships, also "profitable" related to access to social capital and opportunities. However, regarding this notion I obtained differing responses from interviewees: for some a "sweet social tie" is related to intimate social relations or romantic relationships between the sexes. Furthermore, "sweet" is described as "(...) something that is good, nice, perfect" (Franklin. Interviews 2010/11) in the sense of an idealized sense of sociality. "Sharing ideas, having drinks together, going out, meeting girls, having a good time with friends - stealing the father's car together with a friend, this is sweet. Thinking about good times together with somebody with whom one had shared them: these are sweet memories" (Kingsley. Interviews 2010/11/EC). Here "sweetness" seems to contain a spatial connotation, in a sense of co-presence. However, on a mediated level, "a sweet conversation" can be related to spending time together online or on the phone. "Sweetness" in any case captures a strong feeling of emotional closeness as a quality of a social tie, as well as a sense of togetherness. "If the relationship flows well, it is sweet. Quarrelling however, is normal in a relationship, if you never had a quarrel with a friend, this is not a real friend" (Terence. Interviews 2010/11/EC). The attribution of "good" then instead describes and points to a less strongly emphasized emotional quality, to "up and downs" in social relationships over time. A "good" social relation would however be seen as morally rewarding and containing a "correct performance" in social interaction.

Intimacy is another quality of a social tie. Intimacy is a notion, which indicates processes of inclusion and exclusion, defining a chosen circle of close persons in regard to others who are considered as more distanciated. Intimate social relationships are exclusive relations, and intimacy – the same as trust – is based on reciprocity. Intimacy has a sense of face-to-face, a temporal and a spatial connotation: there are specific times and places for intimacy (Hantel-Quitmann 2002:25,26, Auhagen 2002:95). These times and places for intimate relations are characterized by a focus on the specific person or persons one shares an intimate relation with, thus intimacy is also situational. A quality of intimacy could also be created through mediated interaction by spending quality time with a strong emotional emphasis (Hantel-Quitmann 2002:26). Intimate social relations also had sexual connotations for some of the interviewees. In this sense intimacy as a quality of a social tie or interaction seems to be very close to the notion of "sweetness". Personal trust and intimacy can interrelate with notions of social and emotional closeness: Being close to a person means "knowing each other". In this sense, closeness is related to sharing a common background, cultural and social values and frame of reference. This knowing is then almost exclusively related to a previous face-to-face involvement with each other in physical co-presence, and it is related to social relations of great attention, as kin, friends, lovers, husbands and wives.

In the following subchapters I would like to relate to different strands of "closeness", social and emotional. Social closeness, in the sense of a biological – or naturalized – relatedness is used to describe a gradual sense of closeness, from closest family to a distant cousin. The notions of social and emotional closeness are easily mixed and often not differentiated: "Close is what I use to describe as my inner circle, such as family and real friends. In this case my colleagues do not count" (Gladys. Interviews 2010/11/EC). Here, social and emotional closeness is decreasing towards more distant social ties. However, this must not always be the case: a good friend might be perceived as closer than an aunt who is socially close, but distant in an emotional sense[326].

---

[326] See figure F52, in an MTN publicity, relating to ties of friendship, conveying an emotional way of being together, implying that mobile phones help to "glue friendship".

F51, 52, 53: Telecommunication publicity, and a notice in a cyber café, which allude to trust, closeness, emotions.

Social and emotional closeness – claims on relatedness and a sense of intimacy

Close social ties are based on mutual trust, but they are also the precondition of an individual's ability to develop or to learn to trust: The family is seen as the node of trust, where elementary trust (Giddens 1991) is built up in processes of socialization and within bonds of affection. In social ties of relatedness such as family and kin, the social connection is based on a sense of intimacy, which has been acquired in a situation of co-presence (Larsen, Urry, Axhausen 2004:3). As one young woman expressed: "Trust is where there is no fear. Where there is love. Where there is no competition. Not to be afraid to make mistakes. For instance, with my father, there cannot be competition" (Gladys. Interviews 2010/11/EC). Nevertheless, trust is not automatically related to family ties, even though the basis of trust tends to be rendered within such intimate social ties, however, exactly here, trust is sensitive on any incident of disappointment. I would argue that the practical meaning and the moral ground of relatedness of families in terms of solidarity have a strong inclination in view of difficult life circumstances: The basis of social closeness is closely linked to ties of obligations and solidarity[327]. The basis of claims to solidarity is stronger in ties of social relatedness, than in those lacking it. In this sense, social closeness is defined by kinship, and not dependent on the emotional quality of the tie, at first: "A brother is always a brother".

---

[327] See figure F53: a notice in a cyber café, that relatives should not be given free services, despite relatedness.

In communicative interactions through media, emotional closeness can play an important role regarding ties of social relatedness, but not necessarily always. Thus, trust does not seem to be a prerequisite for social closeness: "There can be a close relation, but no trust. Maybe somebody has another purpose in mind. Along the way somebody gets hurt (...)" (Kenneth. Interviews 2010/11/EC). However, it does not mean that the emotional aspect of closeness is not evaluated as important from the perspective of interviewees in Cameroon. The desire to achieve "closeness" in an emotional sense is also very much a given, especially related to socially close and intimate relationships such as close kin and friends. The notion of intimacy serves here as a good description of how emotional closeness is built and maintained in social relationships, and tied to personal trust. A shared life-world is seen as a prerequisite for building an intimate relationship. In mediated interaction intimacy then relies on readopting intimate social relationships in a situation of spatial separation (Auhagen 2002:100, Giddens 1984). "(...) I look forward to forgive and earn the trust of my intimate friends. I have casual relationships - if you do something wrong I do no longer care about you. With intimate friends, I forgive them. That is the difference coming with trust and intimacy" (Isidore. Interviews 2010/11/EC). A sense of understanding is related to a sense of emotional closeness (Auhagen 2002:94). Communication – verbal and non-verbal – and intimacy are closely related, also a mutual opening up and disclosure of the self demands trust. In a situation of physical absence, intimacy relies decisively on communication. Trust and intimacy are both qualities, which are tied up in feelings of a perceived emotional closeness. "Closeness means two individuals, who are in good terms and trust each other" (Franklin. Interviews 2010/11/EC).

The notions of trust, intimacy and sweetness, which I have described in this subchapter, are to a lesser extent used as emic expressions, but they are subsumed by "closeness": I will therefore use the notion of social and emotional closeness, in order to describe qualities of social ties.

New Media and mediated communication

In this subchapter I intend to look at the internet and mobile phones in relation to technologies of communication, how they are adopted by media users, and how they contribute to creating a sense of social and emotional closeness – or in other words: their ability and constraints in transcending physical dislocation and differentiating life-worlds of the communication partners. I will concentrate here on mediated social relations, which have been built in previous physical co-presence. The internet and mobile phone related technologies I will be looking at here, are adopted and evaluated by media users related to their abilities to enable liveness in communicative exchanges, such as a sensory experience. Media of communication differentiate regarding their mediality and thus, how they are adopted by New Media users for communicative interaction according to specific local conditions.

Beeping – a media to link up

Among means to stay in contact adapted in a most peculiar way is "beeping", which means to call somebody and let it ring once or twice, without the intention that the person called should answer the call. The called person can then see the missed call on the mobile phone screen and knows who has been "beeping", if the contact is stored in the phone. In this sense beeping only serves its purpose concerning known others, whose phone numbers are stored in the mobile phone's address list (compare Donner 2007:9). Beeping is free of charge, but a minimum credit is needed to effect "a beep". Beeping must be seen as a peculiar mode of link-up, thereby, it is not an uncompleted call, but a mode of communication of its own. In beeping, as Horst and Miller (2005) express, the emphasis is laid on being in contact, and the connecting or link up to each other. The beeping phenomena could be seen as an effective means of coming back to a social tie, reminding each other of the relationship as well as of mutual positions and duties. Beeping is a "sign" of the other person, expressed without words, integrating a range of interpretations.

In general, "beeps" can have different meanings, including in relation to different people and situations. The general categories of beeping are greetings, coordination, and "call-me-back" beeps. When beeped only once within a certain period of time, it could be seen as a greeting: It conveys a "I think of you, I have not forgotten you". Such beeping is integrated and part of other modes of communicative interaction. Greetings could be sent locally and as well to migrants abroad. In the latter case, the topic of beeping might be even more important, since contacting in general might be less frequent, and calling is not always affordable for people in Cameroon. On a local level, beeping often also serves to coordinate space and time relations, in the sense that it organizes encounters and actions (compare Donner 2007:10). Such beeping needs to be agreed on before, or it occurs in the course of social interaction. As an interviewee expressed: "I only beep with some good friends of mine, and we agreed for beeping, in certain situations. For example, if we have made an appointment, when I am beeping it means: I am about to arrive" (Miller. Interviews 2009/O). On a translocal level this kind of beeping could be used in regard to issues, which have been discussed previously by other means of communication, such as having an appointment in online chat platforms is organized by a text message, and a beep serves to confirm that one is now online. This shows how such beeping only makes sense when a contact is on-going, and a mutual knowledge of other's activities is provided. Beeping several times within a short temporal interval means "call-me-back". Apart from beeping, call-me-back could also be sent as text message[328]. Beeping could be effected just out of a caprice, when wanting to link up without spending money on a call. Due to this particular kind of beeping, practices of beeping are seen as doubtable.

Beeping underlies norms of conduct regarding the nature of the social tie within certain constellations. By beeping people position themselves in hierarchical social relations. Most of my interviewees said that beeping was a means of communication between age mates of the same social position. Beeping towards a higher-ranking individual is perceived as a sensitive issue. However, the closer and better established a social tie, the more likely beeps can be used in kinds of codes and in unconventional ways, mutually agreed on. "My uncle told me I can beep

---

[328] If the credit has finished, an automatically generated call-me-back text message can be sent for free. Telecommunication companies do not earn from beeping, see figures F54 and F55, regarding publicity for voice calls, which I have hardly come across, and SMS, which are not as wide spread as beeping practices.

him when I am short in credit. If he can, he would then call me back. There is no problem with that, we arranged like this" (Clothilde. Interviews 2009/F). Contrarily, when a social contact is not well established and the communication partner's reaction could be doubted, beeping is handled with care. Another aspect is the expectation that the more economically viable communication partner ought to take over the expenditure for the interaction – notably for calls. This estimation is the reason for beeping higher-ranking persons in the sense of call-me-back in certain situations. However, depending on the expectations towards the person in question, and the issue of communication, beeping would also be restricted: When being interested in accessing the social capital of a person, one would call that person to show respect, in a sense of investing in order to obtain potential support (compare Donner 2007). In the case of customers for example, beeping would not correspond to rules of politeness: "I am the one who wants something from the person, so I always have to call the person myself. You should treat your customers well" (Ernest. Interviews 2009/O). In this sense, reflections regarding beeping or not beeping, or beeping whom and in what situations, are adopted along lines of rules, roles, positions, and duties. "With elderly people it's not so polite (...). But with people who are close... In Africa we hardly call our parents, we beep and they call. You do not just beep anyhow" (Barbara. Interviews 2009/R). In the example of a gendered use, men are likely to call women as a part of courtship, whereas women are likely to beep men in order to maintain access to potential support. "Girls always like to beep. Each time you ask a girl why she was beeping she will say she has just beeping credit. (...) If a boy beeps you, you will just start to insult the boy. We believe that a boy has to call and a girl has to beep. (...) From 250 FCFA downwards we call it beeping credit. It is enough to call, but you better use it for beeping, when not ready to spend for calls" (Andrea. Interviews 2009/O).

A general underlying "rule" is that "one should not misuse beeping", as it could be easily perceived as intrusive and impolite. Apart from unproblematic ways of beeping, call-me-back-beeps underlie at times the suspicion of the "beeper" wishes to profit. Many people expressed their disaffirmation in this sense: "I usually ignore beeping, I just take it as greeting, even though it is sometimes clear that people would like me to call back" (Kaspar. Interviews 2009/O). Some would like to state that they do not beep, in order to emphasize that they were viable enough to call. Beeping was criticized as having become a veritable "culture of beeping": "When everybody beeps, you cannot take beeping serious anymore" (Ernest. Interviews 2009/O). Some related to beeping depending on frequency: "If one person is always beeping, I do not like it. If a person beeps me who rarely beeps, this is something else" (Patience. Interviews 2009/O). It was then related to a specific social tie, which rendered beeping legitimate: "I do not react every time to beeping, it depends who is beeping me" (Richard. Interviews 2009/O). Some stated that they would call back, when the person was not habitually beeping, in case it was an emergency. In this sense, beeping is confined to certain persons and situations as an accepted mode of communication.

Concerning migrants, estimations of politeness could weaken vis-à-vis exaggerated imaginations about migrant's ability – and duty – to assume the costs for communication. The possibility for migrants to also beep is largely denied. Migrants beeping non-migrants in the sense of call-me-back would be regarded as non-accurate[329]. They are expected to call, partly suspending usual considerations of age and gender. In this sense, beeping could lead to disenchantment: When migrants become the target of too many people beeping them, it could be perceived as "really disturbing". Interpreting beeps is closely related to previous communicative exchange and mutual arrangements, out of such a framework - in particular in a transnational contexts – beeps could raise anxiety[330] and the urge to call. This is especially so when beeps come from emotionally close persons. In regard to more distant persons migrants are inclined to rather interpret these beeps as demands for support, and not pay any attention to them[331]. Furthermore, beeping

[329] However, a beep as a greeting from a migrant could be acceptable in such cases where the relationship is close.

[330] Migrants take, for example, into consideration, that international calls might not be affordable for some people in Cameroon, and thus might feel the urge to call back. Most however tend with time to reject beeps in view of their experiences that they are habitually rendered the duty to bear the cost for communication. Compare to Tazanu (2012:69) regarding creating suspension through beeps on a transnational level, where their interpretation becomes more difficult, because the communication partners are not familiar with each other's daily activities and situations.

[331] See in chapter 6, how migrants deal with beeping practices of non-migrants.

needs to be balanced, to be seen as apt modes of communication in transnational social ties.

The habits of beeping have also lead to certain practices related to receiving calls or beeps. As a result of beeping, people wait a moment[332] before they answer the phone, to ascertain whether it is "only a beep" or a call. Most often, when receiving a beep, remarks are made regarding the person beeping, relating to the quality of the social tie, or the inequality of expenditures for communication. The beeping phenomena thus gives the call – also the short call – a new meaning, such as expressing a heightened importance of the social tie.

F54, 55: Telecommunication companies do not earn from beeping: Publicity promoting voice and text messaging

Mobile phone: simultaneous communication and the drawback of high costs

Simultaneity of interaction creates the strongest impression of "being close to each other": speaking to one another by phone is seen as the best way to create and maintain social and emotional closeness and immediacy. "You get that person closer. Hearing the voice is good... It makes it alive. You can know that this person is ok. Online (…) you do not know what I am going through, but from my voice you could detect…" (Caroline. Interviews 2009/O). Hearing the voice creates a feeling of emotional closeness, as the voice is something related to the

---

[332] Another way how to deal with beeping is to try answering the call as quickly as possible in order to purposefully make the "beeper" lose credit when the call is accepted, which is not intended by the caller, especially in cases where beeping and calling – according to the perspective of the person – are seen as unbalanced. Or, as Tazanu (2012:86ff) describes, calls from abroad are answered instantly because they are expected to be calls and not beeps.

communication partner, his way of expressing, intonation and choice of words, as something unique and close to a face-to-face interaction with the respective person. In this sense, as interviewees showed, it was easier to "grasp the other person", which related to a familiarity built in face-to-face interaction previous to physical separation. It could however, also be considered as a means of evaluation when not clear about the person's motivation, in the sense of establishing authenticity. In regard to business contacts, one interviewee stated, that "One can better estimate the seriousness of the person, when hearing the voice and having the opportunity for inquiring directly from the person" (Ernest. Interviews 2009/O). Another important aspect concerning authenticity, which is achieved by calling a person, is that by recognizing the voice, a certainty is established regarding the intended recipient of the message. Simultaneity in calls or chat is also considered to be more effective when a specific message needs to be conveyed. "When telling somebody something important on the phone, misunderstandings are less likely to arise" (Anastasia. Interviews 2009/R). A need for immediate "results" was emphasized. "Cameroonians are impatient: they want to have an immediate answer, that is why they like to call instead of waiting for an email reply" (Linda. Interviews 2009/F). All these evaluations could be related to calls being preferable in order to convey important issues[333].

A phone call is valued highly as an act of granting attention to the person called. Calling somebody regularly includes an effort to keep a good and valuable relationship. Calling also usually denominates something about the communication partner's economic and social position. A higher-ranking person should be called[334], when economic expectations are involved. When a social relationship is however stable and close, the higher-ranking person is often expected to call. Furthermore, as I have already related to in the section on beeping, men are likely to call women to express their interest in them. Concerning migrants abroad, expectations that they are supposed to call non-migrants, have become wide spread. Apparently, regarding strong ties, the phone plays the most important role, as a media, which is seen as

---

[333] Compare to the sending of text messages, where the sender cannot be sure whether the message is exclusively received by the intended recipient.

[334] Not to forget that many people – in particular when elderly – are reachable by phone only.

244

accurate in sustaining social closeness and to convey appreciation of the social tie, and in order to enhance emotional closeness, as one interviewee expressed, that calling each other "keeps the connection vibrant". All these estimations highly depend on the fact that phone calls are expensive: thus, calling or not calling is apt to be a constant issue of negotiation. As I have already related to when discussing the beeping phenomena, unbalanced calling could lead to tensions in relationships.

Most people are inclined to ponder the advantages of the perceived closeness and quality of communication to the disadvantage of phone calls being expensive. Calls[335] on a local level of potential presence availability are thus rather held short - as in the example of the below 59-second calls, but they are also more frequent (compare Licoppe 2009:77). They could serve to link up, to greet or to convey an important message, or to coordinate meetings and undertakings. Such calls can also concern demands for support, or sending credit for one's phone, and so on, towards supposedly economically more viable people. Calls just for the purpose to "chat" are said to be rare[336], many express that a phone call needs to have a purpose, in the sense that "something important" needs to be discussed. In this sense, chat-calls on a local level are not even highly regarded: It is generally expressed that women would tend to chat more than men (compare Boneva & Kraut 2002), and to the general notion, that some women have the tendency to profit from others, as two female interviewees expressed: "Women like to chat. With boys they don't really chat. They believe their time is precious" (Andrea. Interviews 2009/O). Or: "Some women just sit like this and beep, expect call backs (...). That's how they use their phones, not for important issues, just greet and chat, gossip, blablabla. Men are not like this and use the phone for important things. A man can make good use of it. Some women spend credit carelessly, (...). Also they are not the

---

[335] I will also relate to different categories of calls in chapter 4, compare also later in this chapter.

[336] Regarding domestic calls, if longer chats are desired, they could be made at night, when the rates are lower: MTN Yello Night, from 10pm to 5am in all networks for reduced fees, or Orange free favourite numbers at night.

ones who struggled to put credit in their phones"[337] (Patricia. Interviews 2009/R).

The financial constraint is also seen to have an impact on the quality of communication, in the sense that the duration of a phone call is most often determined by severe time constraints. I have related to the short calls below 59 seconds: Here communication is held very brief. When related to specific information it is necessary to come directly to "the point", or to just "say hi", or convey a "call-me-back" message. Such a communication is framed in the sense that the calls stops or people hang up consciously at 59 seconds at the latest in order not to pay more – or they have only limited credit on their phones. Callers have to control the length of the call, be it, as some said, that they have "a feeling for the time", or "when you call you get a sound, it signals that you have reached a minute. If you want to stop you quit" (Andrea. Interviews 2009/O). An expression of this "time consciousness" when calling is that many people keep their finger on the stop bottom while talking (compare Tazanu 2012:87). "(…) you just push the stop bottom, some end it (the call) on the ear, they do not even put the phone down (…). (Barbara. Interviews 2009). In the option of "billing per second", people make calls as short as possible or as long as necessary to convey the information: "I can make like ten short calls with 300 FCFA. You can say a lot in a few seconds. It depends on what you have discussed earlier. You are not saying hello, (…). (Barbara. Interviews 2009/R). The conversation has thus often no ending and there is no time to use proper leave-taking formalities. It is expressed that when calling one has "to rush over things".

For most young New Media users, expenditure on phone credit is considered as a continuous limitation. As I referred to in the previous chapter, when introducing different categories of occasional, regular, and frequent internet users, the intensity of internet use could also give insights in people's use of the mobile phone. Occasional internet users maintain most of their contacts by phone, their social networks maintained by phone are however often not larger in size, than social networks maintained by phone of frequent internet users. The main

---

[337] This interviewee was herself a business-woman and often liked to lament the "laziness of our women here", and their dependency on men, probably to differentiate herself as a hardworking woman - she was married with children.

difference here is that frequent internet users are likely to have larger networks overall including a greater number of transnational and weak social ties, such as friends and mates, often of their age. These social ties are also most likely to be substituted by internet based features of communication, whereas certain social ties are likely to be maintained by phone, because they are of great importance, or because these others are not connected to internet. Thus, regarding people's expenditures for mobile phone credit in these different categories, occasional internet users do not necessarily spend more for phone credit than frequent internet users. However, in the category of frequent internet users, the limitation of financial means is less felt, since they seem to be more flexible at adapting their communication habits to their financial viability: For example internet literates could decide to rather chat than call, they could skilfully combine phone and internet means of communication, and they could call through the internet. It rather depends on people's financial situation, and their activities for livelihood, whether they are likely to spend more on phone credit. Among my interviewees, people who had a job, were in business, and scammers, or those who seriously tried to pursue activities to better their life situation, generally spent more on phone credit than those who did not have a source of income. In all of these categories, high priority is given to calls according to their evaluation as an apt means of maintaining smooth communication – a prioritization, which is limited by the economic realities. "We like to get the voice, but the credit is too expensive. If not, we would call all the time" (Patricia. Interviews 2009/R). Calls need to be balanced with other means of communication, which seem to make economically more sense regarding certain situations and communication partners.

For many people it is not easy to give an exact estimation about their expenditures for mobile phone credit[338], because they only buy credit in small bits – such as 100 or 200 FCFA, just sufficient for a few domestic short calls. These on-going small expenditures for prepaid credit in order to "keep the phone in use" are favourable for people who do not have greater amounts of money at their disposal at once (compare Horst & Miller 2009:77). Furthermore, some calls are done on a call box, others

---

[338] I could realize that when asking for their expenditures in interviews and also informal conversations.

from the personal phone. Thus expenditures are highly variable, and range from three to five thousand up to twenty thousand FCFA per month[339]. Considering that most of my interviewees had no regular income, their expenditures for phone credit are amazing. Furthermore, they are very aware of calling opportunities and offers, in case they are the ones taking over the costs for the call. It involves a continuous balancing of expenditures with the need for calling.

Interviewees expressed that longer calls were necessary to get an update from the communication partner, when "somebody is far" for a longer period of time. However, few people call abroad, due to limited finances, but also due to the expectation that migrants ought to call them. However, some stated that they regularly called migrants in shorter calls. "I am in frequent contact with my cousin in the US, we are very close. I call her about once a month, but then we only talk ten minutes, but she calls me regularly every couple of weeks and then we talk up to two hours, about life, about her work, and my work, about people we know, and many topics"[340] (Divine. Interviews 2010/11). It is expected to be possible to chat extensively when being called by a person who can afford making long calls, such as migrants, in particular because such calls are scarce. Equally, however, complaints are heard that migrants have the tendency to "chat for too long", including their presumptions that non-migrants are always available to chat on the phone.

Text messages – inexpensive, but not widely adopted

Apart from beeping and hoping for a call back, an inexpensive means of mobile phone related communication is text messaging. In the context of Cameroon, however – at least amongst my interviewees – writing text messages is not too wide spread. This has something to do with some youth's reluctance to spend time writing text messages, as they would not be used to expressing themselves in written communication. Internet literate people are then more apt to texting via internet platforms: it is easy to type messages on the computer's

---

[339] In some cases even more so, when people's businesses are strongly based on phone calls.

[340] About 9 people out of all categories of internet users said that they also called abroad for lengthier calls, using special offers for calling, depending on the country – the US is inexpensive to call – and if they had somebody abroad.

keyboard and sending is free[341]. Furthermore, text messages are not seen as adequate when a message is important or urgent: it is said that many people do not pay attention when they receive a text message, and only see it later on, and also it is known that text messages are sometimes received with some temporal delay. Another important aspect is thereby the insecurity about not only the reception of the message, but also whether the right person receives it. In the Cameroonian context, as I have described in chapter 3, mobile phones are handed around, and many people have a habit of checking on their friend's phones for messages or answer calls on their behalf, when the owner of the phone is not present[342]. In the sense of linking up, as Caron and Caronia (2007) describe the use of text messaging for youth in the US, in Cameroon it is to a large extent replaced by beeping. The same is true for texting for coordination purposes, which more often involves short calls and beeps. Another issue is the evaluation of the costs: a domestic text message can be sent for 50 FCFA, which is the same as for a short call, which could explain the preference for the latter.

Nevertheless there are two genres of text messages, which are regularly adopted by youth in the Cameroonian context. These are "romantic" and "friendship" text messages", including sending out text messages for the purpose of well-wishes on a special occasion such as Christmas, New Year, Birthday, and so on[343]. These are also adopted on a transnational level. Both could be lengthy, and consist of poems, or descriptions about love and friendship, and personal feelings. Horst & Miller (2009) also found that in Jamaica, that the use of text messages is limited, but mention text messaging in courtship. Several people showed me romantic text messages, in particular young men are said to write such text messages to their girlfriends or women they intend to "convince"[344]. Text messages are subject of similar evaluations of

---

[341] There are several pages where text messages are written. Most often people visit the MTN site, when they use an MTN SIM card in their phones. About 5 of my interviewees wrote SMS in this way on a daily basis.

[342] Important are thereby scams effected among groups of friends, such as for example by getting hold of the respective message before the intended recipient, and to pick up money from Western Union on their behalf, when a code has been sent.

[343] Which has the advantage that the writer can easily consider many people by writing one message.

[344] It is said that through such messages, men convey "their romantic side" to women.

conduct as beeping: "It depends to whom you write. You should not write a text message to your employer, but it depends on rules and regulations - with elderly people it's not so polite" (Barbara. Interviews 2009/R). Text messaging is rather practiced among age mates and close ties: "In text messages you can say a lot, but short hand type writing – you can only write to a friend who understands, it's the same you say in a call of 30 seconds, I use abbreviations, but somebody has to understand it, be used to it" (Barbara. Interviews 2009/R)[345]. Gender differences are also noted: "Guys usually hate texting, it's just not their way to communicate (…). They only send text messages to their girlfriends. With girls its different, they text to male and female friends" (Barbara. Interviews 2009/R). Explanations are for example that men communicate for "important issues" needing an immediate answer – and women rather for chat and maintaining relationships, and furthermore that men's generally higher economic viability and status makes more likely to call than text. It is also mentioned that a text message could be used as a means of communication when there are tensions in a relationship and a confrontation with the person on the phone or face-to-face is expected to be difficult. "A text message you cannot hear, you just see it. The message cannot shout, it cannot slap you… Sometimes it's easier if you have to communicate something in particular" (Caroline. Interviews 2009/O).

Text messaging could be used to link up with migrants abroad, but most of my interviewees were also more inclined to beep or send emails on a transnational level. This evaluation includes high expectations for call-backs on the migrant's part. Texting to migrants is instead seen as a means of reactivating a contact when one has lost connection for a while[346]. Text messages are not considered as provoking an immediate response, when an urgent issue needs to be conveyed to a migrant. Furthermore, depending on the country, a short call from a call booth for international calls below one minute could be even less expensive than a text message from a personal phone.

---

[345] Text messages contain interesting uses of words, abbreviations and codes (Horst & Miller 2009, Caron & Caronia 2007). I do not intend to go into more detail on writing styles of text messaging here.

[346] Thereby, migrants stated that they at times receive text messages, which do not indicate a name, and they could not identify the sender. It highlights people's assumption that migrants have their numbers stored in their phones.

Internet chat: simultaneous communication and the constraint of internet literacy

The other medium creating closeness due to its simultaneity is internet chat. Chat conversation is of hybrid nature, it is neither written nor oral, as Schröder & Voell (2002) express: writing has become simultaneous and interactive. Most often messenger chat, which is the chat section of yahoo, is used, since most people have a yahoo email address[347]. I relate here to face-to-face social relations maintained through chat, and not to online relations at first hand. Using chat relates to calculating costs for phone calls – domestic and international – in comparison to chat as an inexpensive opportunity to stay in contact. Chat seems not to correspond to the same level of esteem when calling somebody or being called, because of less investment and because users are more likely to use chat with social ties of same social status and position – and migrants abroad. Regarding constraints for chat, it is related to time constraints in the sense of devoting time to online conversations, rather than to financial constraints. People would add additional airtime if they chat and do not want to interrupt the conversation, or chat is adapted to the prepaid airtime. Chat is conditioned by a certain level of internet literacy and "mastering the keyboard". Chat is thus related to people's ability to type relatively fast in order to have the impression of an on-going conversation.

For people belonging to the category of regular and frequent internet users, chat is sometimes the preferred means of communication. The sense of immediacy and simultaneous conversation is also related to in discourse about chat, or how people expressed: "talking to each other in chat", "see you soon" when saying goodbye, or to "spend time together online". Chat is considered to be "like a conversation", "you have the feeling of talking to each other" (Divine. Interviews 2010/11). This shows the transforming meaning of written communication: Chat is to a large degree informal, similar to oral conversation. In this sense, chat users "write as they talk", what Crystal (2001:25) has called "wired style". Conversation is built up mutually by reacting to the answer of the communication partner, including utterances such as "emotional pointers" – these could be short colloquial expressions or "sounds" used

---

[347] There are additional ways to chat, in issue related chat rooms, and on various friendship and dating sites, to briefly mention the most common ways to chat.

in face-to-face communication, playing with writings, exclamations or internet specific abbreviations, such as "hhmmm", "wow", "great", "ya massa", byyyyeeeeee ooooooh!!!", "ah ic", "this na 4u", or adding emoticons. According to Turkle (1995:183), emoticons replace physical gestures and facial expressions: Many people who use chat said that they added emoticons, since with their help they could express subtle nuances of emotional feelings[348]. The majority of the interviewees used Pidgin in chat, since they felt it closer to their language use in oral conversation. "It is very normal to use Pidgin when talking to a friend, like we would meet face-to-face, we would also talk in Pidgin" (Clement. Interviews 2009/F). Regarding understanding in chat or phone conversation, it was expressed that "you can clear things immediately". "Its chat… immediately you say how? Communication is moving. The person says fine. Directly from your questions the reply comes. Then the voice is not so important, because you know from the way the person talks, it's the right person. You type quickly and it flows" (Patricia. Interviews 2009/R). This statement again highlights the issue of authenticity, whether it is really the designated communication partner: Here, simultaneity renders certain cues for verification.

In chat people can devote longer time spans to communication without having the urge to convey information quickly, but can take time to chat "about what happened to us, about this or that person we know, about all sorts of stuff, sometimes nothing important, but it is nice just to chat" (Emil. Interviews 2009/R). It transforms the conversation, compared to a temporally restricted phone call. Conversation is often casual, spontaneous, relating to issues, which need an immediate response and exchange. It of course also relates not only to the conveying of specific messages or the discussion of certain issues, but it is also unconstrained regarding changing the topic or integrating small talk, due to less time constraints. Some interviewees used web-cams[349], however, in most public cyber cafés they are not available due to being prone to theft, and if available, extra amounts are charged. Thus, a few interviewees had their own, taking them along when they

---

[348] Most often smileys serve as emoticons, - expressing happiness or sadness, or the smiley waving in order to say goodbye, kissing, blushing, laughing out loud (lol), angry and impatient emoticons are also part of conversations.

[349] Using webcams points out to both, a desire for emotional closeness, but also for authenticity.

intended to chat. If loudspeakers are available in a cyber café, some chat users make net calls.

Even though I suppose that moral and social conventions in chat communication are not as elaborate as in phone calls, I would not agree with Crystal (2001) who highlights the absence of agreed conventions in what he calls "netspeak". It however might be true that such conventions are more pliable according to the individual, the person's adoption of chat technology and importantly conditions such as language and typing skills, time and financial constraints[350]. Some contested issues are for example marking presence or hiding, the exclusiveness of interaction, and framing in chat communication. People could log in as "invisible" when they want to control and limit arising chat, by appearing not to be online. Some people said that they would use it as an opportunity to just talk to specific others, or if they had no time to chat, they would not offend others who wanted to chat, when trying to downturn an arising conversation. However, some people also related negatively to "being invisible": "I don't like logging in invisible, when I log in I am ready to chat with people and take my time for them" (Clement. Interviews 2009/F). Regarding an exclusiveness of interaction, versed chat users often chat simultaneously with several people, fast typing skills thereby balance the slowness of the communication partners[351]. In this sense, an undivided attention towards one communication partner is sometimes not given in chat, however, the communication partners are often not aware of it. The more balanced the typing skills between communication partners, the higher is the possibility of an exclusive attentiveness towards the communication partner[352]. Regarding framing of chat conversations, they can be

---

[350] Furthermore, chat has not been involving a greater part of population in order to build out confined social norms of interactive communication practices for a broader public, as with phone.

[351] However, it is a matter of degree: people who consider themselves to type fast, said that it was not convenient to chat with somebody whose pace of typing was much differing, since the conversation was not fluent and confusion could arise with time gaps between posing questions and rendering answers (Crystal 2001:152). For example, threads of conversation can mix up when a chat user receives an answer for a question he has posed some moments ago, while the conversation has been proceeding.

[352] People also make appointments "to meet in chat" by phone or text message, coordinating time when to meet online, in such a situation emphasizing the possibility of an exclusiveness of the chat interaction.

253

explicitly framed, in the sense of integrating a clear end of the chat and exclusive attention of chatting with a specific person, but they can also be less framed. A chat conversation can sometimes just stop, be it that airtime is finished, or a power failure occurs. Similarly, the conversation can be slightly delayed due to flickering on and off of the connection, related to the common low voltage problem in Bamenda. Sometimes people put the "I am busy" sign on their ID[353] to indicate their status, or a chat is going on and "fading out". At times people interrupt chat conversations because they have to work on something else, or receive a call, and so on. In this sense, chat conversations often relate to offline situations, related to the current situations of the chat users while communicating. Chat as well as phone calls create feelings of a shared presence[354].

Generally, chat depends less on spatial distance of the communication partners. Among frequent internet users, chat is often used on a local level. However, as a tendency, for most internet users the focus of chat is translocal. Phone calls are also used on different levels of spatial distance, with the difference for chat, that expenditures do not differ for national or international levels. Migrants are said to be often online, which favours being in contact with them through internet related means of communication, and often, chat. Chat is thus not a

---

[353] The messenger ID is the chat address. In messenger chat, users can chose their "status" of communication, if wanting to state that they are "busy" or "idle", or logging in invisible or visible. The ID contains a picture – it can be a personal picture or any other, and it is up to the user also to add a "slogan" compatible to the current "mood". Some of my interviewees with whom I was involved in chat communication often added slogans. They for example announced a current important issue, such as "Paying homage to my grandfather", when a user lost his grandfather, or expressing "disenchantment": "I am a hustler", "real hard where I live", „man was born free but everywhere he goes he is in chains", or their attitude in life: "get or die hustling", "from"grass to grace". Or scammers at times announced their success, for example by: "Come and have a share", "everybody l@@king for money: I got it", also expressed by putting a corresponding picture in their ID, such as a bundle of money, a sports car or a dollar sign, and so on.

[354] In chat, people do not necessarily need to constantly communicate. Often communication partners convey to each other their potential availability online – especially when they are online for a longer time span – by just saying "hi", greeting each other and exchange a few words. They can come back to each other later on, but meanwhile they have the "feel" – by seeing the other online – that the communication partner is potentially "around".

means of communication for the whole range of social ties, such as the phone. However, apart from the condition of internet literacy, technological conditions often hamper deliberate chatting in Bamenda, due to power failures, slow lines and low voltage, or other disturbances in public cyber cafés[355].

Email as interpersonal communication and adding to conversation

Email is the main means of communication on the internet for many people, especially in the category of occasional and also regular internet users. Email is non-simultaneous and thus – contrary to chat – it does not involve an interactive experience of exchange. Messages arrive time delayed and are designed to stand for themselves. In this sense, emails could also include a certain level of formality, however, depending on the style of writing and intention of the composer. In such cases, apart from formal writing style and polite expression, people adopt formality regarding the design of the email, in the sense that they start with a kind of letter head, used indentation for the starting of a sentence or capital letters, titles and greeting formulas, similar to those used in letter writing. Adopting such formality in an email could in some cases be deliberately done to enhance the importance of the communication act. Formality refers to decreasing closeness in social ties, to communication with strangers or officials. However, people who are used to email or chat conversation, are more apt at writing in a more colloquial style, and most often they write in Pidgin.

Often email is used to "just stay in contact", in the sense of linking up, greeting, asking how the other person is doing, and to convey a short update. Most email conversations I had come across, were kept rather short, from one or two up to a few sentences. However, there are exceptions: some people said that they wrote lengthy emails, valuated as a means for very personal communication. One interviewee expressed another advantage of email: "You can take your time to think before you write and because of this, you convey a good and well-reasoned message" (Festus. Interviews 2009/F). In this sense, email renders a certain control over the conveyance of messages, whereas in chat or in a phone call it is more up to spontaneous interaction. "You take more time to reflect about an issue" or, as two male Cameroonian stated

---

[355] Compare to chapter 3.

concerning relationships to women, or other "delicate" relationships and issues, it was useful, because "You are less excited when writing emails". As Wellman and Gulia (1999:352) state, written communication gives people more control over timing and content of their self-disclosure, and thus "closeness". The same interviewee related to the durability and fixation on written messages, and thus a different handling of it: "There are situations where I think mailing someone is better than calling (...) because reading has another impact than hearing. For example, there is information that people will like to print out and read after. If it is talking, and the person was not recording, that one will be gone for good. With writing, you can always go and check on the details again" (Festus. Interviews 2009/F).

It depends greatly on the use of email, how the communicative interaction is perceived, intense, mutual or singular and independent. Since email is not simultaneous it depends on the frequency and regularity of email messages, whether the exchange can be perceived as on-going communication. The less somebody is internet literate somebody is, the more likely email is the main means of conversation on the internet, and the more likely regularity in communicative exchange needs to be maintained through other means of communication, such as the phone. Amongst occasional internet users, messages tend to take some time until they are answered, and they are thus less apt in following up a continuous thread of conversation. Email depends to a large extent on the responsiveness of the communication partners, in order to create a communication performance and mutual interaction. In chat or on the phone, the communication partner needs to be "present" through technology, but email involves a high level of insecurity related to the acknowledgement of the message: the risk is involved that people would not receive a message – or they could easily pretend not to have received or seen it. Email is therefore not seen as accurate when an urgent or important message needs to be conveyed. There could be a difference regarding the accurateness of email in the case of people abroad, who are often online, as one interviewee expressed: "My uncle in UK is always online at work, so it is not a problem, when writing an email to him, I will have an answer the same day or a day later only"

(Precious. Interviews 2009/R)[356]. When used regularly, email can be an apt means of exchange, as part of an on-going conversation, since people most often use the reply button to respond to an email. Some people preferred email to chat under certain conditions: For example email was considered to include a higher degree of privacy: "When you open your mailbox, you are aware that there could be information which you do not want others to see. When you open messenger chat, the conversation just pops up, so that everybody could see it"[357] (Francis. Field notes 2009). In this sense, many people preferred to send an email instead of an offline chat message, when somebody was not online, or they preferred non-simultaneous communication, when they did not have time to chat.

For many people, belonging to the category of more frequent internet users, however, email as a means of communication had less importance, since they preferred simultaneous communication, used chat, or communicated on friendship sites, or by net-calls. The more social ties are maintained through other communication channels, the more likely email is to serve as a complementing media. Email thus often relates to a previous conversation through other media. Email is also used in order to attach and send pictures to them. Nevertheless, since Facebook has become wide spread, pictures are often shared here instead of being sent by email. Text messages or chat can be used to announce that one has sent a message: "I am sending text messages to tell that I have sent something to their box, if it is important" (Precious. Interviews 2009/R). Email for these people has an importance regarding getting an information update or in general as complementing direct communicative exchange, in the sense of sending information, links, jokes, or bible slogans (compare Licoppe 2009:119). "When I browse in the internet, and when I come across something interesting, then I think of some friends who would like it and send the link to them" (Manfred. Interviews 2009/F). This kind of additional interaction was related to by some people as "fun": "When I send a joke to my friend I already know

---

[356] Furthermore, on a transnational level, email is taken as a reliable means for exclusive communication, whereas on a local level, writing an email often involves an insecurity whether the message has been read by the intended recipient or not. I have mentioned in chapter 3, that it often happens that private email boxes are hacked and entered.

[357] See F56. Privacy is often very limited in public cyber cafés.

that when he will read it, he will be laughing, and this thought alone gives me pleasure" (Clement. Interviews 2009/F). Some people also regularly send and receive "forwarding emails"[358]. In this context, so called "family letters" (compare Wilding 2006:132) are worth mentioning, although I only came across them twice amongst the interviewees in Cameroon, who stated that some relative abroad wrote emails informing the relatives in Cameroon about the latest news. In Switzerland, three migrants told me that family letters were posted among their kin, and one said that he himself would do so occasionally.

Frequent internet users use email not only for direct communicative interaction, but also for impersonal communication: some people told me that when they are on a page which they are interested in for example, they would subscribe to mailing lists in order to receive regular information. However "one can feel bored after some time, but you can easily stop it again" (Matthias. Interviews 2009/R). Clothilde (Interviews 2009/F), who worked in a cyber café, had developed a ritual of logging into her mailbox every morning to read bible slogans and commentaries, which were sent to her daily. "This is my start in the day. I really love this, it gives me peace before the first customers are coming". Other people receive church news on a regular basis by mailing lists[359], and some obtain daily horoscopes or are updated regarding the latest technological developments when interested in computer and mobile phone technologies[360].

---

[358] There are different genre of such forwarding emails: the religious genre, with bible slogans or little "moral stories". Another "genre" is the "friendship email", including moral stories or poems of "real friendship" or "true love". A further "genre" is about "incredible stories", often related to bodily anomalies, such as sicknesses, mutations, but also religious and spiritual "miracles". Especially the latter genre most often contains pictures in the sense of proof. Another, widespread "genre" are the messages containing a "forwarding duty", for example forwarding it to twenty people within 24 hours. It often already highlights the outcome, in a positive sense, what has happened to people who forwarded the email as demanded: they won in a lottery, became rich overnight, they were healed, became happy and successful. In these messages, it warns of the consequences, when the message was not forwarded.

[359] See F57: this friend working in a cyber café was always updated about the activities of his Nigerian based church.

[360] Overall out of all interviewees there are about 19 people who regularly received such news lists emails.

F56, 57: Youth in the cyber café, old town, Bamenda

In regard to the migrants, on the one hand, email is used as an inexpensive way of linking up, or as a reminder, in order to make sure that the exchange continues. It is expected that migrants ought to acknowledge and quickly respond to emails, since they are thought to have easy access to the internet. Often such emails contain a suggestion or demand that migrants should call, as well as complaints why the migrants have not taken initiative for communication. On the other hand, some migrants prefer to be contacted by email, and decide on the reception of the message, whether they would call. In these ways, email ads in interesting ways to mediated communication.

Interaction on friendships sites, or pursuing other's activities by commenting on pictures

A peculiar way to communicate is the communication and interaction on the various friendship sites, such as most prominently, Facebook, but also Hi5, Tagged, Quepasa, Linkedin, Friendfinder, and others. Most of the earlier popular genres of networked publics, chatrooms and forums, are topic related. Here, most of the people "meet" in such publics are strangers who share interests. In comparison Facebook and other friendship sites are friend- and peer-oriented, generally denominated as "social media" (compare Danah 2008, Zhao 2008). Most of the frequent, and many of the regular internet users engage on such sites. Whereas some only log into a friendship site, when they receive an email indicating that they have a message, other's activities on these sites ad to their daily habits of internet use. In particular, Facebook has become very popular recently, especially since 2009 more and more of my acquaintances had become active on Facebook[361]. In the case of my interviewees, most of their contacts on friendship sites are people previously known face-to-face, and many also are in presence availability. Friends range from a few relatives and close friends up to a large range of weak social ties and online ties. In their denomination, there is just one category called "friends", making no differentiation in degrees of closeness (Danah 2008). Users have different opportunities to interact on Facebook. There is a personal messaging system to exchange private email messages, and there is the opportunity to use chat on Facebook: When logging in, a user is visible to others and can be addressed. The public section is a peculiar part of the communication on Facebook, it consists prominently of following one's friends activities[362].

Most common, also for those who are not intensively active on Facebook, is looking at each other's pictures and commenting on them accordingly in the public section. For many, Facebook and other sites are the main locations where they keep their pictures, apart from their mobile phones. Whereas pictures can be shared on the mobile phone with those in presence availability, with those absent pictures can be shared on Facebook. Many of the users regularly update their albums by adding new photographs. It is also common also to add a description to the pictures explaining the situation, locations and persons depicted in them. Long comment threads can occur when people then comment on their friends' pictures, the people concerned answering, and threads of communication between two but also more individuals can occur by commenting on each other's comments. The comments are usually short, often joking remarks related to the picture but also beyond. Apart from referring to visual features, comments are also connected to greetings, they could be less but also more personal and intimate, or they relate to the picture and situation depicted in it, revealing background information. For example, somebody had uploaded a picture, depicting himself in traditional dress and the regalia of a "big man" – the photograph was apparently taken in the Cameroonian countryside.

---

[361] For many youth it is important to have a Facebook account, to be "up to date". Sometimes more internet literate youth create accounts for their less internet literate peers, similarly as happens with email addresses.

[362] On facebook users can become a member of groups, express their interests by conveying "I like", or they can indicate their current status and activities or leave a short message in the public section by "posting on their friends' wall".

Some of his friends added joking statements, as "Hey! U bi chop chair?" (will you inherit the position of the family head?). Upon this comment, various comments were added by the person concerned, about family issues and explanations about the depicted situation, whereof a range of friends in return commented on these contributions.

The authenticity of the pictures is also an issue here. It concerns personal pictures and any picture obtained on the internet, which is put on Facebook. In the domain of personal pictures, it depends on the closeness of friends, whether they could accurately interpret them, for example in the sense of locating the person. Amongst more distantiated "friends", the possibility of people's mobility is always taken into consideration. As expressed in a statement, which I came across more than once: "U de look good oh! Long time bro. U don lef pays? (did you leave Cameroon?)". Or, as in the case of a young man, who posted a picture of himself in front of a nice and modern building, which provoked comments from his friends asking whether he had "fallen bush". He first jokingly confirmed before later revealing that the picture had been taken in Douala. Regarding personal pictures, the reference to appearance and physical looks is common, in the sense of adding to an interpretation of a person's wellbeing, especially when the person is located abroad[363]. It could then also contain a more or less direct utterance of claiming relationship or expressing one's desire to profit from the connection. My migrant friend's picture, showing him at a party in the US, provoked such comments as: "Chop bro! (eat, enjoy). How u throwe me so (how can you neglect me like this)! A dey wait ya invitation (when are you inviting me (to your place)?".

Pictures render the opportunity to connect to each other's lives, providing an accurate update, serve as a departing point to involve in a conversation, and render additional social cues. Thereby pictures can provoke a wide range of possible interpretations. The combinations of pictures and written statements of a possibly wide range of Facebook friends gives rise to a peculiar form of mediated social interaction, interrelated with other means of communication[364].

---

[363] Compare to the sense of outer physical features of migrants, in chapter 2, and Tazanu 2012:147.

[364] This part of my observations is interrelated with my own experiences of being linked up with friends on Facebook.

Sensory experience in New Media of communication connected presence

Technological features on the internet – also concerning mobile phones – require a choice of modes of communication, contributing to meaning and consistence of communication acts (Rasmussen 2002:101). Wilding (2006:132-134) speaks of "layers of communication" when relating to how various modes of communication intersect with each other. According to Wilding, families who maintain mediated contact over distance, add various layers of communication by adding new technologies to already established ones, in order that overall, "Each new layer of technology was arguably being used to communicate more efficiently with existing modes, (…)" (Wilding 2006:131), adding to the frequency and regularity of communicative exchange[365]. It leads to specific combinations of communication technologies in everyday communicative interactions, depending on habits, skills, financial means, social position, purpose of communication, and the communication partner. Intermediality is related to such combinations of technological means and their communicative features: As I have pointed out in this subchapter, in the sense that different modes of communication ad to each other, also different medialities add to each other. Hearing the communication partner's voice on the phone, involving in a chat conversation, or looking at pictures on Facebook, enhances the sense of a "grasp of the communication partner". Furthermore, in communication, borders between the oral and the written blur – for example a webcam could add to a net-call or chat, links are attached to email messages, or pictures commented on Facebook.

This points out to the integration of different senses into mediated communication practices. Even though technology must reduce the experience of the other to certain senses, by combinations of different media, a sensory liveness could be enhanced in mediated communication. Thereby some senses are preferred to others, such as hearing the voice in comparison to written communication, or seeing the

---

[365] For the communicative exchange between migrants in Perth and their families in Ireland, Wilding (2006) says that former modes of communication, such as letters for example – and of course the land line phone – continue to play an important role, but co-exist and are combined with more recent modes of communication, such as the internet and mobile phone. In Cameroon, we have to consider that most people hardly had access to phone previously.

person and his environment in a picture[366]. What is considered important is a sense of authenticity (Berger & Luckmann 1980), if "you are talking to the right person" (Patricia. Interviews 2009/R), and "how the person is doing": cues being conveyed through hearing the voice or seeing a recent picture. However, as an interviewee expressed: "Generally it does not matter so much what media you use, if chat, email, with or without webcam, if phone… what matters is what comes from the person's mind, and how you can convey it within communication, it can be written or spoken. (If) it comes directly from the source, genuine, you can feel it. It conveys something of the person, you can feel that in words" (Festus. Interviews 2009/F). Communicative interaction is intentionally used in order to add cues, which serves to estimate the communication partner's state of mind and motivation, to follow up his life, or to put him into context.

Continuous communicative exchanges via communication technologies contribute to the communication partners' imagination of their social relationship, in the sense of maintaining and (re-)creating a sense of liveness. A sense of liveness in mediated social interaction depends on the ways in which both communication partners use technologies of communication in order to continuously relate to each other and thereby maintain a feeling of connectedness and closeness – instantaneously, when communicating, and in the sense of maintaining and working on the social relationship. Communicative interaction is thereby not seen as compensation, but as an attempt to continue a relationship even under conditions of physical dislocation. "At least, (because) one has tried not to lose the contact, and one has tried to follow up the story of the other person, one has not lost each other" (Festus. Interviews 2009/F). Wilding (2006) and Licoppe (2009) call this a connected relationship or connected presence: "More significant than what is said in these exchanges is the moment of exchange itself, which reinforces a sense of the relationship between sender and receiver", and also the fact that one communicates[367] (Wilding 2006:132). Simultaneity is thereby central, as effected by phone or chat. Meyrowitz (1985) speaks

---

[366] However, I have mentioned, that a re-evaluation of media can occur considering the constraints of their use.

[367] Note that Wilding (2006) does not speak of necessarily simultaneous communication, she relates to email as instantaneous communication due to the fact that its exchange is so much quicker than postal letters.

of an instantaneous mediated encounter, relating to Goffman's (1983) work about face-to-face encounters (compare Moores 2003). I suppose that simultaneity in communication is the most important cue, which conveys closeness and certainty about the communication partner's authenticity[368]. As an interviewee stated: "Face-to-face (interaction) minimizes all other constraints in communication. (But with) a friend, when talking to her on the phone, it can be very powerful. It depends who is talking to whom. It can be very effective. We used to hold our hands, now we are holding our hearts" (Gladys. Interviews 2010/11/EC). It is about creating a virtual space and time of togetherness (Goffman 1983), mutually set up in communication practice[369].

As I have tried to show in this subchapter, technologies and their specific mediality, conventions, and conditions influence mediated communication. I have been looking at how New Media technological features are used to continue social relationships under conditions of physical absence. In this sense, they are consciously adopted in order to create and achieve a certain desired level of liveness in social interaction, in the sense of liveness as a sensory experience and as work.

Negotiating a sense of closeness: observing conduct in mediated social interaction

Positive conduct relates to what is seen as morally and socially acceptable in social interaction in a certain context. Mediated communicative interaction is not always smooth. Disagreement and disappointment over closeness and – interrelated with it – over conduct can arise. Mediated communication occurs along social and moral norms of social interaction, whereby these can also become negotiable under specific conditions: such as – most importantly – a transformed mediality in communication, as well as changes in position – geographical and social, and related imaginations of others' abilities and duties regarding sociality – for example reachability – and solidarity. In

---

[368] Compare to Tazanu, who speaks about the „feel" of New Media, and of directness – simultaneity - as the preferred modus of social interaction across borders (2012:61ff).

[369] Compare later in this chapter. Also I will further relate to this issue of a space of common social interaction in media in the conclusion chapter in this book.

particular migrants become "targets" of link-up communication, and are expected to be the ones to call and take over the main expenses and responsibility for communication. Thus, conduct can be transforming induced by media technologies, within mediated communication, and by a lacking frame of reference and diverging frameworks of social and cultural worldviews given by differing structural systems of the environment. What Giddens (1984) calls "tact", can be related to the handling of social ties related to the overall communicative performance, layers of communication and combinations of media, in specific social situations regarding specific social others in morally confined and accepted ways. When tensions and frictions occur in mediated social relationships, it is most often related to a non-observance of tact in mediated social interaction. As I intend to show in the following subchapter, conduct can be seen as transforming from face-to-face to mediated communication, and in particular, as I have tried to show in the previous subchapter, in regard to transnational social ties. It includes evaluations of mediated communicative performance, such as the regularity of social interaction, responsiveness of the communication partner, politeness, and expectations expressed in communication[370].

From face-to-face to mediated social interaction – transforming mediality and evaluations

According to Raab (2008:235) face-to-face interaction is still the most successful mode of a mutual review of experience, built up in face-to-face modes of communication and interaction of "successful communication". In a face-to-face encounter, as Berger and Luckmann (1980:35) say, a personalization and authentification of the social other takes place. In such direct communicative interaction, understanding – or "grasping" as Calhoun (1992) calls it – of the other becomes possible by including as much information as possible. In face-to-face interaction, in bodily clues, from appearance, ways of talking, behaviour and mimic, agents make assumptions regarding their communication partner's character and inner motivations. Here, Goffman distinguishes between expression, which is non-verbal, such as bodily communication, and communication as verbal social interaction (Rafaeli 2009:71). In

---

[370] This subchapter relates to the 3rd sub-question of guiding question 3, regarding frictions and tension, which can occur in social relationships due to mediation.

mediated communication there is in this sense a limitation of expression and of regulated feedback. Furthermore, written mediated communication lacks dramaturgical expression, communicative clues such as loudness, gesture, pausing, emotions, and so on (compare Döring 1999). In direct social interaction the context is continuously integrated into the encounter in the sense of a continuous verification of the common base of interaction and environment in which the interactions are embedded (compare Endress & Srubar 2003). Thus, undoubtedly, communicative exchange in mediated social interaction lacks certain social and emotional cues, induced by a lack of physical presence: the social relation has thus to be represented by communication acts and is reflected in communicative terms (Rasmussen 2002:100). "Before the presence was there, we could always meet, seeing each other, now you can just talk. All is talking. They can send a picture, but it is just a part of communication" (Anastasia. Interview 2010/11). As I have already indicated, social and emotional cues could to a certain extent be provided in mediated communication, enhanced by a variety of communicative technological features. However, it is mostly seen as an approximation, since New Media can only provide a partial sensory experience of the other. Furthermore, mediated social interaction alludes to a shared cultural background, which corresponds to a common home, people commonly known, and cultural values and norms, which communication partners can relate to in a mediated situation. Furthermore, when relating to social ties, which have pre-existed under conditions of co-presence, people tend to link media experiences to former experiences with the person in co-presence, but also to previous media social interaction (Howells 2003:225-227, Giddens 1984). In this sense, face-to-face and mediated social interaction cannot be seen as antagonists, but they closely intersect.

While mediated communication and the appropriation of New Media follow to a certain extent moral predicaments about modes of being connected, transformed conditions contribute to partly suspend them (Krotz 2001:68, Hahn & Ludovic 2008). In such ways, non-migrant's underlying expectations of being instantly and seamlessly connected and the communication practices of migrants, create confusion about conduct in a framework of contesting interests.

Regular communicative interaction – or cultivating social ties through New Media

Regularity in communicative interaction is one of the most important concerns[371]. Social relations need to be cultivated, and thereby communicative exchange and other interaction needs to be more or less regular, integrating different means of communication. Regularity depends on the circumstances of communication and the specific social tie. Regular communicative exchange can include different degrees of frequency of exchanges. It could be, for example, a monthly phone-call from an aunt in the US, or an almost daily interaction by either phone or chat with a beloved person. The first example can be considered as a means of maintaining the connection towards a strong social tie, securing social closeness. The latter example then relates to a desire to keep the connection as lively as possible in the sense of an emotional closeness. It also depends on the overall quality of the social tie, to what extent regularity of communicative exchanges is needed[372] and strived for. All interviewees had their "core" of most valuable and strong social ties, which were maintained regularly. In general social relations – from less close through to close ones – have their "peaks " of interaction. For example Ivo (compare interview 2010/11) informed himself about universities in Holland – his friend studying in Holland had sent him a few links. Around this time Ivo often wrote to his friend, who helped him to fill in the forms. Or Mike's uncle died and he contacted his brother in the US for financial support in order to organize the burial, and they phoned and chatted intensively over a period of about three weeks. The issues involved are often related to a social tie, which is maintained in a regular manner, as an example that continuous work on a social relationship enhanced access to social capital. However, not necessarily so, as other examples show: as in the case of Peter (compare interview 2009/FS), who planned to go to India for a computer networking course found out that a classmate had studied in India, and they ventured into intense communication, though they had hardly communicated in the last years.

---

[371] Compare to the previous chapter related to the intensity - regularity and frequency - of internet use.

[372] As friendship ties, which do not need to be activated all the time to remain apt, see previous chapter.

Regularity is related to the frequency of communicative exchange, in the sense of the time gaps in between mutual communicative messages. When the time gaps are short, a real follow up of conversation can occur, in the sense that people pursue a thread of conversation, by making reference to the previous communicative interaction. Often the regularity is also taken care of by the communication partner, such as relatives or friends abroad, regularity and frequency of exchanged communication also depend on their responsiveness. It is expected that they come back to their kin or friends in a regular manner, if not it could be evaluated negatively. "When my brother does not call me at least twice a month, I start feeling worried" (Jack. Interviews 2009/FS). In verbal – oral or written – expressions, a long time gap is related to as "long time", or "longest time", in order to announce a new attempt at getting in contact, or as a kind of reproach or excuse: regularity is also a subjective perception. Regularity is dependent on habits of media use, such as checking one's mailbox or opening messenger chat when logging in. Regularity is most often achieved in different ways such as a combination and balancing of different media of communication, and complementing it by linking up practices such as short phone calls and beeping.

Whereas in potential presence availability, seeing or calling each other is not always emphasized so much, and can be continuously delayed, concerning migrants abroad, regular contact is expected. When the mediated interaction does not correspond to this expectation, migrants are generally seen as withdrawing (compare Tazanu 2012:60,61).

Negotiating conduct: Language, framing and attentiveness

Giddens (1984) relates to qualities of direct social interaction as a reflexive monitoring of conduct in and through co-presence. The rules of social interaction or encounters are according to Giddens language, framing[373] and conduct, in the sense of respecting socially and morally accepted modes of interacting. This can also be transcended to conduct

---

[373] The opening and closing of situations, or their bracketing (Einklammerung) (Goffman 1983).

in mediated social ties according to media's specific mediality[374]. I have several further points to add here which could be subsumed under conduct: Apart from choosing the appropriate media and regularity and frequency of communication, I will relate here to attentiveness, responsiveness, politeness, and handling expectations.

At first I will refer to language, framing, and interrelated with it, attentiveness. The described means of communication first of all differ regarding oral or written language. I have highlighted the importance of oral conversation in building a sense of closeness and a grasp of the communication partner. Furthermore, oral conversation on phone circumvents obstacles regarding computer and internet literacy, and more generally the ability to read and write, which is also a reason why it can include a wider range of communication partners. In chat, even though it depends on written conversations, the feeling of chat being "like an oral conversation" is strong among chat users, borders between the spoken and written blur. Whereas in emails messages stand for themselves, in chat, communication is interactively set up, similar to colloquial oral conversation. In chat Pidgin is very often used, which is as a tendency less often the case in email where messages can be more formal, and thus sometimes conveyed in "good English". In chat, but also in text messaging, a kind of youth language can emerge, consisting of abbreviations and short forms. In calls, since they most often have time constraints, a specific way of oral expression occurs. It relates to conveying messages as quickly as possible. This contains avoiding extra communicative features, setting priorities, using keywords, and limiting speech of the communication partner, when it does not serve the purpose of the communication. In short phone calls, expressions of politeness could be left out, such as greeting and farewells. Such types of calls depend largely on previous communication. Beeps are related to convey a message without speech. Their handling and interpretation depend on context and person and the thread of communicative interaction.

The framing of social interaction has media specific characteristics as well: the situation of interaction is usually framed – had a beginning and

---

[374] See figures F58 and F59, relating to moral conduct such as an indication to bring back a stolen mobile phone, and a note in a cyber café concerning the observation of rules of conduct offline and online.

an end and specific mode – but the framing could become peculiar in mediated communication. I have mentioned short phone calls, where communication can be cut off without the use of leave-taking formalities, be it that the caller consciously "presses the stop button", or has limited credit on the phone. Furthermore, in chat, framing is sometimes lacking: chat conversations can – depending on conditions such as airtime, power failures, or other issues influencing the communication environment – just interrupt or fade out.

Framing in chat conversations is related to attentiveness. I understand attentiveness as attention given to the conversation and communication partner, in the sense of a mutual adjustment, which depends on simultaneity as a prerequisite. It indicates the authenticity and identification of the communication partner, the level of intimacy in the communication, and thus to the nature of a social tie. In chat, focussing on one communication partner alone or chatting with several people simultaneously also depends on the speed of typing of both communication partners. The more balanced such skills, the more likely they are to focus exclusively on each other.

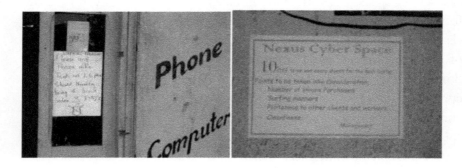

F58, 59: Allusions of communication technologies with social and moral conduct, Bamenda.

With migrants abroad, there are certain specifics related to language, framing and attentiveness. First, attentiveness of migrants towards non-migrants is related to as a general claim. More specifically, regarding language, I have related to a tendency of domestic calls being rather short and international calls being an opportunity to chat without predominant time pressure[375]. Some migrants told me that when they called people in Cameroon, they had the experience that they started to talk elaborately, above all about their problems and needs, which raised ambiguous feelings from the migrant's part. Regarding framing, some migrants, conscious that they could not lead endless international phone calls, conveyed to me that they might be tempted to push the stop button[376] in order to interrupt a call. In such a case, technological problems can be used as an excuse, most possibly the network in Cameroon, which is usually given credence. However, such failures in migrant's calls - as it is reversely often the case when non-migrants call - could raise ambiguous feelings among non-migrants. "Why is he not calling again?" I witnessed a non-migrant in Bamenda asking, when he had received a call from abroad, and he then suddenly realized that his communication partner "had disappeared".

Negotiating conduct: Responsiveness, and expectations of being reachable

What I want to call responsiveness of communication partners' one of the most crucial issues – a sense of a lack of responsiveness, which provokes most tensions and frictions in mediated social relationships. Such discourse is also prevalent on a local level: it is based on the expectation of being connected. Regarding migrants abroad it is linked to the belief in migrant's economic superiority and duty to actively contribute to the maintenance of connections. Responsiveness interrelates also with regularity and frequency of communicative exchange between two communication partners, which are important characteristics, on which responsiveness is valuated. It includes the balancing of communication initiatives, as well as promises to call back. It is often lamented by non-migrants that migrants would promise to call, but never keep their promise. It also includes the expectations vis-à-vis migrants to take over the costs of communication, being responsive to demands for support, and reacting to attempts by non-migrants to get in contact with them. New Media offer various means for linking up communication, such as beeping, short calls, text messages, or also

---

[375] See Tazanu's description of receiving a „bush call" (2012:83ff).
[376] In a few cases, I also pushed the stop button myself when calling to Cameroon, when it became too lengthy.

email, asking for contacting on the part of migrants abroad. In this sense migrants often lament that they are accused of being non-responsive, or not taking the initiative for contacting, whereas they are seeing themselves as a "target" of being contacted by so many people[377]. It is often regarded as one of the great advantages that New Media of communication provide transnational contacting coordinates: "In earlier times before internet and mobile phone, it was normal that people went and they were only reachable for people in Cameroon, when they called them and gave a fix phone number. In this way many just disappeared. But nowadays with internet and mobile phone, it is easier to stay in contact, and you expect that they leave their email and inform you when they change it" (Peter. Interviews 2009/FS). In view of such opportunities of staying connected despite physical dislocation, it is used as an allegation, when these contact coordinates are not given: "People go abroad and do not leave an email address behind, or they change it. So it depends on their goodwill if they ever contact you again" (Herbert. Interviews 2009/R).

The expectation of the communication partner's being reachable is crucial, which is seen as a given on the basis of being connected through communication media, as the internet and especially mobile phone. Furthermore, because the mobile phone is a mobile technology and a personal device, and mobile phones ought to be switched on, constant availability is expected. The same is also expected from migrants abroad. When people abroad switch off their phone – or put it on silent – it could be perceived as a refusal to communicate, often reacted to by allegations such as "Your phone is not going through", "You do not pick", and so on, as an expression of the frustration and futility felt when trying to make contact[378]. "I cannot understand when they do not pick their phone. I always have my phone on to receive calls at any time" (Lizette. Interviews 2010/11), or "They always have excuses when not wanting to communicate, such as they are busy and stressed. They become so anti-social when they are abroad, I don't like it" (Manfred.

---

[377] See the previous chapter, where I have related to "claiming relationship" with migrants abroad, and chapter 6, how migrants deal with the expectations of reachability.

[378] An issue here is "falling into the answering machine". It was often mentioned by non-migrants that when they tried to contact migrants abroad, as an allegation, they did not only try to contact in vain, but they even lost credit.

Interviews 2009/F). Regarding other media of communication, it is expressed in a similar sense that the person in question would not respond to emails, or never be online. I have already shown that inconsistencies in communication are sometimes attributed to technological inconsistencies or failures. In this sense, statements such as that "the number was not going through", or one has not received an email, serve as excuses, which could be accepted in some instances, but could also contribute to doubts, if they are only used as excuse in order to camouflage a non-interest in communicating.

Negotiating conduct: Politeness and handling expectations

Politeness is an important factor in communicative interaction, as well as in mediated communication acts. Some face-to-face cues are missing, and misunderstandings can easily arise. "What I send through the internet has an impact (…). Since the communication is not face-to-face, your eyes are fixed to the letters, your mind is judging, and so on. (…) The words you use to get in touch with others through the internet are very important" (Festus. Interviews 2009/F). Apart from the words, the style, such as the degree of formality for example, the language used, in a phone call also the intonation of the voice, and the framing of the message needs to be adjusted to a specific social tie, and to the issue of communication. Lacking politeness is especially felt in narratives about call back beeps or short phone calls: "With us here, if you want to pass a message, you just go direct. If I want to tell my friend something, she sees my number, so she knows, I just start talking. But with other persons you have to greet and you cannot just stop. You (can) call and say you want to talk just a minute, (but) it ends that you spend more. (With elderly people) you need politeness … if you want to ask something, you have to be polite" (Andrea. Interviews 2009/O). As the interviewee expressed, what could be accurate for an age mate, needs not necessarily be so vis-à-vis an elderly person, and also migrants, in particular when a demand is involved. Not only the style and framing of communication, but regarding positive conduct the choice of the media is also important. Somebody could "make a mistake" in etiquette, but also regarding the use of the "appropriate media", such as a call back beep for example directed at a higher ranking person. However, politeness and choice of media are embedded in a specific communicative performance, which goes on over time with a specific communication partner. This relates to what Giddens (1984) calls "tact". It means that communicative interactions are valuated as an overall performance. For example a call-me-back message could due to financial constraints be accepted in certain instances, but needs to be balanced with other means of communication. The closer and better established a social tie, the more unbalanced communicative performance can be, and it is still seen as morally acceptable, but it also depends on an agreement between communication partners.

The non-observance of rules of politeness however often provokes discussions and disenchantment, which are related to expectations which might not be fulfilled or seem to be exaggerated from the perspective of migrants. Crucial issues are the expectations regarding taking over the costs for communication and rendering support: For example many migrants felt that they were only contacted for reasons of support giving, without showing interest in their well-being and feelings. "When the phone rings I always expect somebody calling me, asking for money" (Alice. Interviews Diaspora). Migrants perceived this mode of contacting as misconduct, when demands were not embedded in a "good" communicative exchange. "You cannot just be quiet and after months you send me a call me back message, expecting I have immediately to call back. When I do it then, the person is asking me for money" (Blaise. Interviews Diaspora)[379].

A change in position can lead to a re-evaluation of the social tie and attributed responsibilities and claims. This inequality was emphasized, claimed and performed by both, migrants and non-migrants, but often in particular by non-migrants, as a base to profit from the relationships with migrants. Even though claims are seen as legitimate, when they are experienced as exaggerated and impossible to fulfil, and when they arise regarding a wider circle of people "claiming relationship", and thus become a burden on migrants, they are also contested (compare Tazanu 2012:199ff). Similarly, under conditions of constraints regarding communication, such as the high costs of calls or lacking internet literacy, short calls and beeping are accepted by migrants, but a positive

---

[379] However, non-migrants also said that migrants seemed to only call when they needed somebody to do something for them on the local level. The reproach of opportunism is an issue on both sides of social ties (Tazanu 2012:191ff).

conduct in media is also evaluated as a base to continue a social relationship, since all social interaction between migrants and non-migrants are mediated. Thus, communication acts could become sensitive issues, and could influence qualities of trust and perceived closeness related to a contingent base on which a social tie can be evaluated and examined.

Evaluations of closeness under conditions of differentiating life-worlds

When observing moral notions of positive conduct in mediated communication and social interaction, New Media are seen by many as ideal means to maintain qualities such as closeness, trust and intimacy within mediated social ties: "Communication has come to stay. Be it phones, internet, (…), and so on, they have facilitated the linking of loved ones. Within a twinkle of the eye they give you a ring, and so many wounds have been healed. These media have reduced the walls. (….)." (Kenneth. Interviews 2010/11/EC)[380]. In this sense, a non-fulfilling of such ideals would be attributed to a non-fulfilling of moral ideals of connectedness, since "When the communication partner is good, time and space cannot not limit (...)" (Gladys. Interviews 2010/11/EC). Nevertheless, the ability of New Media to maintain closeness in social relations is also questioned, because the immediate life-world and conditions of daily lives would involve the greatest part of people's attention and energy. "Thanks god for the internet, it is great, but distance is still there. It is different with the internet, (…) chat can continue, but not too often. People are involved in their own lives, you will not forget each other, but it distanciates" (Francis. Interviews 2010/11). Different environments and lifestyles were mentioned by interviewees as possible obstructions for emotional closeness: "(...) We live and operate in different geographical, economic and socio-cultural settings, which have varied constructs. While friends who now live

---

[380] Interestingly, the tendency to affirmative statements regarding media's ability to bridge distance in a positive way, was more strongly emphasized in my fieldwork assistant's interviews. It is difficult to explain this tendency, I assume that it has something to do with the level of formality in the interviews, and the interviewees' awareness that they were contributing to a research conducted on behalf of a European researcher. My own interviewees, interviewed on the same topics, featured a great deal of complaints regarding media's deficiencies in transnational social ties.

abroad may consider our local constructs as archaic, ... we live in the different societies. They may have outgrown from our values and constructs. This creates a barrier for effective communication" (Terence. Interviews 2010/11/EC). Furthermore, the communication partner's sincerity could be doubtful: "After my friend went to the US, he did not communicate regularly anymore. He hesitated to answer my questions, and I am not sure if he told me the truth" (Herbert. Interviews 2009/R). It seems that in particular in mediated transnational social ties, the maintaining of a feeling of being close to each other could become a difficult task. Many people lament social ties breaking off or weakening. Not because they would not have tried to maintain them, but: "With time, we had not too many common issues to discuss anymore. (...) He is still my brother, but we have been close before, and now it has become difficult to understand each other" (Francis. Interviews 2010/11). In this sense, stable frameworks of social relatedness and a feeling of closeness can become fragile, when common frames of reference lose their validity and a face-to-face level of exchange is missing.

Nevertheless, a mutual feeling of closeness was not always highlighted as a particular priority. Some non-migrant's narratives about closeness are not most importantly oriented to an emotional quality of the social tie, but rather, how interaction can generally continue "smoothly" in absence of the respective person, including making claims on duties. Some emphasized these mediation processes as a normality to a certain point, when somebody has to leave his or her immediate family and loved ones. The issue of "becoming a man", and "standing on one's feet" in the sense that a person could contribute now to sustain others and play a viable role, is seen as a favourable condition. In this sense, it is related rather to a continuity of the social tie itself as a matter of "relatedness" to be claimed. However, claiming social relatedness alone is not always seen as sufficient anymore. Depending on the social tie, in situations of mediating a social tie over distance and common frameworks partly lose their validity, a "good" communication performance gains more importance, in order to continue a "good relationship". This is especially so from a migrant's point of view, but also to an extent from a non-migrant's perspective. By the notion of "good communication performance" I understand that positive conduct is observed, as I have described it in this subchapter. It has to do with a sense of investing work in a social tie, whereby the specific mediality of media of communication plays a role. The evaluation of emotional closeness to be necessary in mediated social ties in order to keep them positive – and also potentially supportive – is related to bridging the differing perspectives of migrants and non-migrants in particular. These evaluations are also tied to an understanding of the situation and conditions of the life-world of migrants abroad, from the perspective of non-migrants. As one interviewee said: "I need to exchange with him (the brother abroad) and understand a little how he is doing and how he is living over there, so that we do not become strangers to each other" (Francis. Interviews 2010/11). Thereby, closeness and trust maintained in mediated social relationships in turn serve to generate understanding.

More wide spread narratives on transnational social ties tell a less successful story of connectedness. I have shown that migrants tend to limit their social networks to those most close and that they become selective towards "emotionally rewarding" social ties. Many non-migrants feel deprived from the promising opportunity and intention to continue – or re-activate – relationships with migrants abroad. This is especially the case when they are not positioned within the exclusive social network given priority to by the migrant. When ties break off, it is related to migrant's tendency to change when they were abroad[381].

Mediated closeness: "good communication" and communication contents

So far, I have looked at technological features and practices of media users in order to obtain an impression of their attempt to substitute physical absence with mediated communication acts, and to bridge differentials of life-worlds and – conditions. I want to look at details concerning communication contents here, which are particularly related to a negotiating of a sense of closeness, and its difficulties, through media of communication under condition of physical dislocation[382].

---

[381] I will come back to the arising of frictions in relationships between migrants and non-migrants, in chapter 6.

[382] I have not concentrated specifically on contents of communication by mobile phone and internet based communication technologies in my research. I found it difficult to obtain access to them, even though I often had the possibility to listen to

277

These are communicative practices, which serve to "locate" the communication partner, "keeping each other up to date", as well as "what remains unsaid", or issues which are not communicated. Auhagen (2002:104) speaks of "good communication" in the sense of personal disclosure and sharing of deep-rooted meanings[383]. However, I doubt whether this is the main intention of the communication we are looking at here. There is not always "profound" communication necessary in order to create a feeling of common understanding and closeness. Especially in regard to New Media there are various opportunities to express emotional closeness, also in the sense of linking up, as an expression of mutual attention and affection or "showing concern". In this sense, by "good communication" I do not only relate to the content of communication, but to what I have called observing positive conduct in mediated social interaction and communication.

Influences on communication contents: choice of media, constraints, and user's habits

Contents of New Media conversations are influenced by various factors. As I have already stated, financial constraints and computer or internet literacy have a considerable impact on contents of communication, by favouring or discouraging forms of communication: For example short phone calls or emails, or not using chat. Following up an issue also influences contents of conversations, which in turn relate to regularity and frequency of communicative exchange. In addition, social and emotional closeness have an impact on communication, such as social closeness which can enhance conversation about an issue, for example concerning support[384], or discussing "family matters", and

---

phone conversations accidentally, or when browsing with people, some also opened their emails or I witnessed chat conversations. Nevertheless, I cannot claim that I had a systematic insight in these conversation contents. Asking people about contents often proved equally lacking in insight - I obtained answers such as: "Just what you usually write to a friend" (Simon. Interviews 2009/F).

[383] According to Giddens (1994) trust in "postmodern societies" is based on a "revealing oneself to others" because within circumstances of interrelated frames of reference the knowledge of others might be limited, whereas in "traditional societies", knowledge of others seems to be pre-given by a shared framework and location.

[384] I see communication dedicated to support giving and receiving as a peculiar content of conversation.

emotional closeness can include conversations estimated as being intimate.

Communicative acts can be differentiated as link-up, communication acts, where the communication content is subsidiary to the linking up effect, and communication acts where the content of communication gain more importance. I have pointed out the importance to link up, which can be the purpose of a large part of communicative exchanges. It contains beeping as a means of non-verbal communication, used for greeting or call-me-back messages. Greetings – link-up – are also important in other means of communication, such as those, which can also be used for more elaborate conversation. It concerns phone calls especially, which are kept short due to financial constraints, text messages, or short emails. Regarding phone calls, Horst and Miller (2009) relate to "expressive calls" as a means of expressing emotions towards a person, rather than conveying a message in an instrumental sense, employed in link-up practices. Such messages form part of a range of mutual communicative acts related to a specific social tie.

Phone or internet based media of communication also play a role when the content of conversation is given more importance. Here we can differentiate instead between untargeted – open – and targeted communication (Licoppe 2009:78). I prefer to speak of focussed and unfocussed communication (compare Goffman 1973), because every phone call can be seen as targeted. Communication is focussed when it is related to conveying information or to discussing and following up a certain topic, whereas unfocussed is communication which serves above all to chat, in the sense that the content of communication is subordinated to the link-up effect[385]. In this sense, link-up calls can be seen as unfocussed, because they are not topic related, but serve to greet and a short exchange of amenities. Furthermore, calls from Cameroon to abroad are most commonly effected in the sense of such link-up by short calls. Short calls in the local context often serve to convey information, and are rather focussed. One type of focussed short call is related to the coordination of activities, for example about where and

---

[385] Of course, many acts of mediated communication may contain both, this differentiation rather serves as a typology.

whom to meet, or organizing meetings and events[386]. Short and focussed exchange most often relate to previous communication. Such information can also be conveyed by text message, or email – the latter in particular on a translocal level. Email is non-simultaneous and in this sense not a means "to chat", but to convey information, or to greet, thus it can be used for both, focussed or unfocussed communication. On a local level, most people expressed that they do not usually call just to chat, calls would have a purpose, something important needed to be discussed, or a solution for a problem to be found. Horst & Miller (2009:67,68) relate to such calls in Jamaica as "counselling calls". Regarding calls, a few interviewees said that they occasionally talked at length with a person abroad, provided it was the communication partner, who called. These calls can be focussed to discuss "an important issue": some migrants expressed that they could not afford and were not willing to spend money on unfocussed calls, they would always have a reason for calling. However, some people - also migrants - emphasized that they liked calling to chat, especially with somebody out of presence availability, when they had not talked to the person for a long time. The less one can meet face-to-face, the higher the probability for unfocussed calls and the longer the phone calls, and these calls are, as a tendency, less frequent than short focussed calls. In chat, unfocussed social interaction also occurs. Especially on a translocal and transnational level chat is used as an opportunity to "just chat".

Putting the communication partner in context in mediated communication

In mediated communication it is not only the physical presence which is missing in the social interaction, but what seems to be even more felt and leading to inconsistencies, is the missing link to the life-world of the communication partners. Knowing the environment of the person helps to stay up to date regarding the person's life and contributes to an understanding and putting the person's actions and messages into context. If the environment of the conversation partner is known, the work of bridging consists of staying up to date through information about events and third persons. If the environment is not

---

[386] Compare to linking up and coordination calls in the previous chapter, as tools for social networking.

280

known, the daily lives of communication partners – such as migrants abroad - are difficult to construe. Cues are few, based on descriptions by the conversation partner, which are usually not sufficient to really locate, or grasp the person abroad in her daily life and social networks. As one interviewee expressed relating to his siblings: "I am very close to my sister in Douala. I cannot see her often, but I know her environment. I know that (takes a look at his watch) right now she must be in her car driving back from work going home. It is very different when you know how the person lives … with my brother, I have not seen him for eight years now. We are communicating, but I do not know much about how he lives over there (in Belgium). Distance has distanciated us from each other (…)" (Francis. Interviews 2010/11). The mutual reference to the life-worlds of the communication partners seems to be an important constitutional element in conversation, in order to maintain or create emotional closeness.

Ideally a certain conjunction of mutual references can be achieved in communication. It contains the integration of the not shared points of reference regarding the spatial aspects – the new environment and life world of the migrant – and temporal aspects – the migrant's daily routine, and his staying up to date what happens "at home" – into the relationship in order to stay close to each other. There are common sets of questions that are asked in order to bridge missing parts of information, questions such as "Where are you (right now)?" or asking about the communication partner's current activity, or for known and unknown social others in their environment. There are general interests in a range of topics, representing a "feeling for the everyday life over there", such as climate, food, recreation, and people, as an interviewee asked me: "How do you spend your weekends over there? How is social life over there? How are you socializing with your friends and family?" (Richard. Interviews 2009/O). Questions and interests are related to a feeling of place, regarding the migrant's environment and daily activities, supplemented by information, descriptions, narratives, and visual images. "When they (the migrants) talk about their experiences, or when they send you pictures, you can much better imagine. So I always like to ask them questions" (Julius. Interviews 2009/R), or: "I learn very much from him (the friend) this way (by asking questions), how life over there is going, and I can get first-hand information through him" (Felix.

Interviews 2009/O). In this sense, on the one hand, locating the communication partner by trying to get a sense of place, and a sense of closeness, involves continuous work through communicative acts.

On the other hand, regarding the imaginations of the environments of migrants, some reference points take force as mere presumptions and are often not closely examined or asked for by non-migrants. It is presupposed that migrants work abroad – even when they are students, since work is the primary reason for migrating. Many of my interviewees did not know, how their migrant friends or relatives abroad earned their income. Among non-migrants, there is a saying that "one should not ask migrants what they are doing abroad", in the sense of avoiding to embarrass them. Regarding social ties, there can be a mutually known social network, for example family members in the migrant's country of residence. Apart from that, the family in Cameroon ought to be the primary concern of the migrant. The fact that migrants have their new social networks is apparent, but not considered too relevant, it could however have a negative connotation from the perspective of non-migrants: migrants could become too much involved with their new friends to the expense of their families. The situation of communication is often not questioned, what is important was that migrants call and are reachable.

We could thus speak of an interpersonal and placeless world, or space, which is imaginatively co-constructed in communicative interaction between communication partners. In Appadurai's (1996:30) words, it is an attempt to make a lost or not achieved world come to life. This space is set up by a common history and memories of each other, shared points of reference concerning a formerly shared life-world and imaginaries of the communication partner's life-world, mutually built up in the continuation of communicative exchange. Since non-migrant's imaginations are strongly superimposed by local narratives and presumptions, and also the migrants only convey specific details about their lives and activities, non-migrants' points of reference regarding migrants' life-world can only be fragmentary. However, also on the part of the migrant, they may doubt whether they are being accurately kept up to date and informed, regarding what is happening "at home". This gap creates uncertainties concerning an accurate contextualization and evaluation of the communication partner's communication practices, words, deeds, and motivations, inconsistencies, which often occurs in communication acts, performance and contents.

Updating each other in order to take part in each other's lives

In lengthier calls and chat communication, conversation could more elaborately go into a sense of keeping each other up to date, including small talk, discussing problems and specific issues – in this sense these conversations range from unfocussed to focussed communicative interaction.

Apart from descriptions of life abroad and asking questions, the content of conversation between conversation partners in Cameroon and abroad, is keeping each other up to date with what has happened related to anything important in the family, friends, neighbourhood, events, work, and so on. The update has a sense of a mutual taking part in each other's lives. Thereby, it is considered easier to talk about Cameroon, as a location with shared knowledge, than about migrant's environments. People in Cameroon thereby feel that the issue of update is often unbalanced, and they feel "the gap", as I have called it before. As one interviewee put it relating to her friends, who were studying in different countries in the West: "Often I realize, that our conversation is mainly about Bamenda, their families here, about local events and people we both know. They do not often just say something about their lives, without me asking and insisting" (Anastasia. Interviews 2009/R).

Updating each other also relates to a conveying of emotional states, positive and negative, talking about daily worries, problems and insecurities. On the part of the migrants, many stated that in mediated communication with kin and friends back home, conversation was – from their perspective – too often confined to complaining, to "problems" – and thus monetary issues. "People are always complaining on the phone... I have been sick and have to pay the bill for the hospital. This and that event is coming up and I have no money... and so on. Sometimes I think it is always about money" (Godlove. Interviews Diaspora). However, migrants also are said to complain, which at times provokes a sense of bitterness and disbelief in their sincerity. Talking about problems is then often tied up in talking about issues regarding support giving and receiving, often emotional topics, which create tensions. Another aspect of an update is an emotional sense of communicative exchange, regarding emotionally close people in the sense of "moving together", as an interviewee called it: "With intimate friends you have a common history and you are up to date with what the other is going through. In this sense you share a development together, your relation has been developing through time. In a close social relationship you need to share a presence, in order to have a future" (Festus. Interviews 2009/F). To follow up the story of the other person under conditions of physical absence, is seen as vital to stay emotionally close to each other. "You need to communicate, even when somebody is far, the person feels the closeness. You must be aware of the person's challenges, you can help, pray... just being quiet, he or she might take the wrong path. You should know their challenges. You should share things with them, (then) maybe somebody is far, but close to you" (Caroline. Interviews 2009/O). Such communication also circulates around the social tie and its qualities itself, talking about a relationship or friendship and its base. Communication among emotionally close people is complemented with communicative features expressing closeness, such as "I miss you, I love you, wish you were there" or "We are together" – a very common saying in face-to-face as well as mediated social relations. Conversations between lovers might also include erotic aspects conveying the longing and lack of seeing and being physically close to each other: This is often related to as a "sweet conversation". Often – between friends and mates – conversation is about "old times" and memories. In communicative exchange with less emotionally close social others and less regular conversations, the positioning of the communication partners is an important aspect, in the sense of clearing social status, regarding location and occupation, such as "I am teaching now in a school in Yaoundé", or "I have been travelling abroad", in the sense of locating oneself spatially and socially towards each other.

Updating each other relates to work invested in mediated social ties, in order to enhance and maintain feelings of closeness. Updating can have a sense of a mutual taking part in each other's lives, but it can also be adopted in a strategic sense. It serves to selectively convey information and hold back other issues, to position oneself, trying to convey expectations, and shaping imageries. The update can consist in merely informational aspects, as well as emotional and intimate issues.

Updating thus ranges across social ties of different emotional and social closeness.

What remains unsaid between migrants and non-migrants

Relating to mediated social interaction, silence is mostly adopted in the sense of non-communication, and avoiding certain issues[387]. According to Christensen, Hockey & James (2001:70), there are different forms of silences: First, about that of which there is no need to speak, the taken for granted. Secondly, about that which is known but should not be spoken of, and third, the silence of that which is not conceivable to articulate in language. According to Schütz (1967) speaking is relating to what needs and what is worth speaking about, what needs to be cleared. Only what is not self-evident is brought into speech in everyday conversation with known social others. This relates to a shared background and mutual knowledge, to which communication partners take reference in conversation, or on presumptions, which are usually not questioned.

In transnational social ties mediated though the internet and phone based media of communication, communication partners depend heavily on verbal expression. A social relationship needs to be continued on the basis of talk. As an interviewee expressed: "You need to work on both, the communication and the relationship. You talk, that's the relationship, and its alive, that's communication" (Caroline. Interviews 2009/O). This work needs to be effected under conditions of blurring references. In a situation of physical absence and the communication partner's lack of knowledge of his environment, migrants are in the situation that they have to describe verbally or in imagery what they are going through. Some migrants said they had tried to explain issues concerning their everyday life to people close to them in Cameroon, trying to create understanding. However, structural differences, different organization of society and daily life, work and sociality, or policies and regulations related to their migrant's status, are not easy to describe. Furthermore, certain issues are taken for granted by non-migrants: thus, for example for an asylum seeker to explain that he is not allowed to

---

[387] Silence within communicative exchange, in the sense of a break or a pause, only occurs in a limited way in mediated - oral and written simultaneous – communicative performance.

work is not easy. As well related emotions are not easily to convey to others far from having an understanding about their situation. Furthermore, migrants would avoid putting themselves in an unfavourable situation and losing face, when conveying their real situation, and they would not like to lose their position of high status attribution, by telling the truth, or, in the opposite sense, to prevent others from claiming to take part in their "success". Conversely, often migrants lamented that their communication partners would not show interest in their life situations. These factors can lead to a reluctance to exchange information[388].

Thereby, migrants adopt strategies to avoid certain topics. Apart from invention, it is mostly silencing certain issues, and trying to orient communication in other directions, by, for example, only talking about Cameroon. Often unsaid issues are perceived on both sides, by migrants and non-migrants, which could then lead to a decreasing sense of closeness and a sense of social and emotional distance. Unsaid issues – such as underlying tensions – come to the fore and could be expressed in communication performance, in the sense of avoiding practices: The most wide-spread and most lamented form of silence is avoiding communication, or non-responsiveness. Not calling can say as much as calling, especially when somebody is expected to call. Physical absence and differing life-worlds imply migrants' slipping out of reach, despite media of communication, of which their silence was a proof: "Distance has been the cause of so many misunderstandings in relationships. For instance, I don't know what my brothers abroad are going through. They may be in some difficulties that I do not know. But I am angry with them for I am writing and they do not write back. I don't know what is happening to them. It is because of distance, they make new friends, live in other circumstances" (Pride. Interviews 2010/11/EC).

Conclusion: Negotiating, creating or avoiding liveness in mediated social interaction

In this chapter I have examined qualities attributed to social ties, their evaluation, feelings of social and emotional closeness in social

---

[388] Compare the reasons of silences in social interaction of migrants and non-migrants, in the next chapter.

interaction, and how these are transferred to media realms, where they are prone to being negotiated under conditions of mediation: here notably, physical absence and differing life-worlds. When relating to liveness here, the first strain of considerations concerns a sensory liveness, which contributes to experiences of mediated communication. In social ties where emotional closeness plays a role, media of communication as the internet and mobile phones offer opportunities of simultaneity and additional cues to enhance emotional closeness through multi-medial and interactive features, providing a basis for a sense of closeness, building trust and intimacy to the extent that it is wished for related to a specific social tie. The claim for sensory liveness is closely tied to a habitual dimension of agency, relating to an ideal sociality. Accordingly, as I have described, liveness can be pursued – by choosing specific media related to specific social ties – and liveness can be neglected[389] or escaped from, by not seeking or avoiding such cues, as in the examples of non-responsiveness of communication partners – such as migrants who are not reachable, not taking the initiative for communication, or avoiding and silencing certain issues. The second strain of considerations concerns moral and normative evaluations of accurate conduct in the sense of liveness in the dimension of work. Especially in mediated situations, work needs to be invested in social ties in order to create and maintain emotional closeness, and to secure their accurateness and validity, which relates to a practical-evaluative dimension of agency. Such liveness is pursued by adding different communication layers, combinations of media, and by regularity of communicative interaction. Furthermore, it contains following a sense of positive conduct in mediated communication, related to responsiveness, balancing of contact initiation and taking over expenditures for communication, and notions of politeness in conversation.

I have tried to show that in mediated communication regarding transnational social ties, such conduct in New Media use can transform, and are subject to specific situational conditions and contextual imaginations. Here, it is expressed most strongly, that an ideal of face-to-face social interaction serves as a template to evaluate mediated communication. Likewise, as it seems that this ideal sociality cannot be

---

[389] Also compare to the next chapter.

287

achieved in mediated sociality, New Media of communication are adopted by users in order to best pursue individual interests and to sound out different modi of communication and sociality. In particular migrants' evaluations of conduct and of liveness shift accordingly. It can lead to tensions in social relationships, when such shifts lead to ambiguous and contradictory interpretations and evaluations of the communicative performance across borders. Such tensions come to the fore in New Media use for communication, and issues concerning such differentials and inconsistencies of closeness can be addressed – or kept silent – in contents of mediated conversations. I have related here to a gap in the ability to locate and contextualize communication partners in differentiating life-worlds. As it shows, New Media of communication cannot sufficiently bridge these inconsistencies. Here, it depends on the willingness of both communication partners, to work on this gap, in the sense of dealing with unfolding situations and moments in these transnational contexts and transnational social ties.

Moral predicaments of being connected can become an issue of intense negotiation depending on the position and location of the communication partners within differing contexts, as I intend to examine in further detail – including the view of migrants – in the following chapter.

6

Practices of Connecting and Disconnecting
Negotiations of Social Relations between Migrants and Non-Migrants

Prevalent imaginaries and imageries of "the West" and "a life abroad" and a certain opacity related to experiences and life realities of migrants can lead to conflicting views and even tensions between communication partners, between those abroad and those "at home" in Cameroon. Here I intend to make a link back to chapter 2. All the dimensions of agency are concerned here, however, in particular from the view of migrants, their judgement of the situation often leads to a re-evaluation of social ties and practices in the transnational context. As New Media offer enhanced opportunities to stay in contact, they also enhance expectations, and similarly, they may support feelings of deception and deprivation, when expectations are not fulfilled, as I have described previously. I have related to New Media as tools for social networking as well as to the experience of closeness, or its opposite, dwindling connections when social ties are mediated over distance – geographical, social, and emotional distance. In this sense, practices of New Media use not only serves to connect, but at the same time people – and in particular migrants - adopt practices of New Media use in a sense of "disconnecting" – or avoiding liveness, as I have described it in the previous chapter, in order to deal with demands and claims from their kin and friends in Cameroon. However, I want to explore disconnecting in particular as another dimension of mediated interpersonal social relationships – practices, which concern differing notions of what it means to be connected over distance. Underlying here are notions of legitimacies of social mobility, success and respective claims, related to from the different perspectives of migrants and non-migrants. In this chapter, I will relate particularly to guiding question 3, how sociality is negotiated in regard to the opportunities for liveness offered by New Media technologies. Specifically these examinations are then related to the 3rd sub-question regarding frictions and tensions which can occur in social relationships due to their mediation. I will specifically integrate the view of migrants here, and the interviews I have conducted among Cameroonian migrants in Switzerland.

Migrants, non-migrants, and generations – conflicting views and colliding interests

In the following subchapter I want to relate to colliding interests of different generations, as well as migrants and non-migrants. Frictions and tensions have always existed, but they have gained importance under conditions of being in close contact, also in situations of physical absence. Hardin (1993) relates to Gidden's (1984) conflicting models of orientation, concerning the deliberately taking over of social roles, positions, and actions "(…) that individuals determine to be in their own best interest." (Hardin 1993:39). Social roles orientate themselves to respective social norms, which ought to be performed and reproduced through them: importantly norms of sociality and solidarity. Under transformed conditions and alternative pathways and notions of success, such norms might be newly interpreted and renegotiated.

Youth, tensions between generations and the struggle for influence

In regard to the relations between generations, historical issues and developments could play a role, influencing today's relations between generations. Argenti relates to collective experiences and memories in the Grassfields going back to the times of upheaval. According to him, prevalent distrust and suspicion between the generations might partly be based on the deep rooted antagonism that was risen at the time when chiefs and elders traded youth as slaves or later on as enforced labour forces to the coast (Argenti 2007:5,23). Also Warnier attests to a "deep feeling of hostility" prevalent in Grassfields societies, especially in regard to relationships between young men and their fathers, involved in conflicts related to succession or marriage – and access to resources in general (Warnier 1996:120, Pradelles de Latour 1994). The struggle with colonial powers was used by youth as an opportunity to gain emancipation from elders. "In order to achieve autonomy, young men similarly used Christianity, salaried work, military service, skilled work, even mass migration. Their acceptance onto the colonial scene led to conflicts between elders and minors". (Bayart 1993:113). However, this emancipation should not be overemphasized, since, as I have mentioned in chapter 2, monetary remittances were important topics during colonial times, and youth who migrated for salaried work were tied into social networks of duties to support their extended families in their villages of origin, trying to keep migrants in their service. However, youth also try to delineate themselves from previous generations and negotiate their own position.

In more recent times, youth's attempt to reform society have been crushed by government forces, as in the unrests in the beginning of the 1990s and again in February 2008 in the protests against prices for staple food and fuel. Additionally, youth make statements regarding a renewal of society, for example by their appearance as moral actors, as described by Fokwang (2008) for Bamenda's youth associations[390]. As Fokwang (2008:264) expresses: "(...) through their crusades, they highlight the elders' failure in restoring social harmony and willingness to supplant the elders who have allegedly reneged on their moral functions". Or by engaging in scamming activities, by profiting from compliance of the state's representatives, such as the police, youth blame and unmask the corruption of the political system and elders as the real base of ills in Cameroon[391]. Youth are in this sense often suspected of engaging in subversive activities, in one way or another.

Conflicting views - claims on solidarity on the basis of strong imaginations

Having a responsible position in one's family and society is very much assigned with power, and power in society is most often attributed to certain groups, gender, age, and the ability to accumulate and redistribute wealth. In the Cameroonian context, as I have tried to show, it is difficult for young people to accumulate wealth through hard work and having a job, or to be in a position to meet expectations of redistributing and taking over responsibilities, and thus, becoming adults. Under such circumstances, it is vital to have good connections with influential others, and being able to count on solidarity networks

---

[390] As cleaning up campaigns, taking a stand in moral discourses on HIV Aids, punishment acts towards youth who are seen as not comporting in a morally confined way, and so on (Fokwang 2008:173-175, Fuh 2010).

[391] However, in stark contrast with the youth associations described by Fokwang (2008), which base their actions on a strong sense of their moral superiority.

of extended families. Under conditions of the great valuation of migration ventures and attributed imaginations of the West as a realm of abundant riches and opportunities, it is not only elders who serve as viable contacts, but also migrants, many of them age mates, fellow youth who have shifted in social position. In an environment of scarce opportunities for income, competition is related to achieving success in migration ventures, but above all, it is related to a link-up with those, who have "made it". Thereby, discourses on a legitimized position of youth in society are followed up by youth themselves, related to migrants abroad, in order to keep them under control and enforce their solidarity (Tazanu 2012).

Whereas Bourdieu points to positive effects of social capital, Portes (1997) also stresses the possible negative aspects of social capital, and consequent practices of exploitation, in the sense of what he calls "enforceable trust". "Reciprocity and bounded solidarity, for example, may be used by group members to demand excessive favours from successful friends and relatives, producing levelling pressures (...) may place onerous constraints on individual freedom (...)". (Massey & Espinosa 1997:142, Faist 1997:76). Baros (2001:23-25,300) speaks of a rise of conflict potential and tensions between members of communities and societies, the difficulties of comprehending the other party's situation and motives for their actions, and the problem of diverging perspectives in a situation of migration. It could enforce prevalent tensions between generations – but also migrants and non-migrants – and struggles of power and influence within family relations and beyond. Warnier (1993a:71) relates to the high expectations of parents towards their migrant children outside of the communities of origin. The latters' success is evaluated and judged in terms to what extent they are able to support their kin back home. In the sense of sanctioning the non-fulfilment of social norms of solidarity, Warnier (1993a:71) states for Bamiléké society after independence, that a "préfet" who failed to fulfil these duties, was buried "dans les champs là où l'on enterre les mauvais morts". Regarding migrants abroad, they are attributed high status, but at the same time, they are continuously apt for being deprived of status in view of exorbitant demands, which forces them to withdraw from – or at least to adopt strategies to deal with – such notions of solidarity.

Differentiating views on migrants on the base of a good relationship

Even though I emphasize tensions and frictions between migrants and non-migrants: related to perspectives of New Media users in Bamenda, it shows that the views and perspectives are not monolithic. Often, the interviewees also stated understanding for the situation of their relatives and friends abroad, revealing that – despite strong imaginations of the West as a location for great opportunities, they had a certain awareness of migrant's life situations and difficulties. This is strongly related to the "working on the gap" arising through differentiating life-worlds[392], explanations given by migrants relate to their practices of communication and rendering support, and constraints they are encountering thereby. This understanding is based on the responsiveness of migrants, feelings of closeness and trust, and what I have called "good communication"[393].

It seems that many interviewees are aware of the balancing of investments in a "good relationship" in the sense of observing positive conduct in communication, and a probable outcome in the sense of obtaining support. In this sense, some interviewees emphasized that they watched their own conduct carefully in communicative exchange - for example that demands for support would be well integrated in a flow of positive social interaction and communication. Some said – on the base of an evaluation of a specific social tie abroad – that they would ask for hardly any support[394]. Some interviewees acknowledged difficult life circumstances of migrants in diaspora as explanations for their reactions and responses. Here, it was also admitted that migrants might be short in financial means: "My uncle is jobless at the moment, so I will not ask him for money, and I can also not expect him to call all the time…" (Ivo. Interviews 2009/R). Regarding asking for support, interviewees seemed to be conscious of not overstraining the respective social tie: "I do not want to put him (the uncle in the US) under pressure, so I try to ask only if it is really necessary, and only when I have a good reason for asking" (Elvis. Interviews 2009/FS). Sometimes, the idea of migrants being busy with work, as a reason why they are not always responsive,

---

[392] Compare to the previous chapter.

[393] Compare to the previous chapter.

[394] This could however also be seen as a strategy in order to hope for the migrant's sympathy when telling them about their daily struggles, and their appraisals of their life circumstances.

was mentioned. "Life is hard over there, people are very busy with work" (Peter. Interviews 2009/FS), or: "In the evenings people are tired, and sometimes they are too tired to call" (Olaf. Interviews 2009/R). Additionally, migrant's being under stress is an issue, as an interviewee said: "The mentality of people changes when they are abroad. They are under psychological stress." (Clovis. Interviews 2009/R)[395]. Some also related to the reactions of migrants in such situations, which would not be consistent to bring them relief from their difficult situations: "It is strange that migrants in such situations withdraw from their people at home in Cameroon. Why do they not seek close contact with them, to make them feel better?" (Clovis. Interviews 2009/R). Some interviewees then sought for explanations for such withdrawal: one of the factors for stress, which is mentioned, is that migrants would feel under pressure to support their families: "It must be sometimes hard for migrants, they can never ask for help, as it is them who are expected rendering help all the time" (Titus. Interviews 2009/F). And: "People have wrong ideas about life abroad. They think, that people abroad have buckets of money they will just pour on them. This makes migrants feeling under pressure. But everybody who has travelled or is in contact with people abroad, knows that it is hard over there" (Titus. Interviews 2009/F). Such statements are extended to general appraisals and critique of Cameroonians' attitude of relying too much on the contributions and support from their relatives abroad: "Young people think they can just sit around and eat the money their relatives abroad send to them" (Godlove. Field notes 2009. 28.07.09). Or: "This waiting for people from abroad helping them is like the mentality of Africans to blame colonial times and Western countries for their situation, and expecting help from them. One is passing responsibility on to others and has less energy to solve one's problems" (Immaculate. Interviews 2009/O).

In this sense, many interviewees expressed understanding for reactions and situations of their relatives and friends abroad and were self-critical in regard to a perceived dependency on migrants. This might have also an aspect of pride and non-migrants' desire to emphasize, that

---

[395] Here it is interesting to compare to Tazanu's (2012) findings. Altogether his interviewees empasized even more decisively migrants' responsibilities at all cost. I suppose my position as a „white man" – compare chapter 7 – and Tazanu's position as a migrant himself is an explanation for this difference.

they are not always dependant on migrant's help, but try their best to "make it" themselves.

The migrants – between the country of residence and dreams of return

I have interviewed migrants in diaspora in Switzerland, as well as people with migration experience in Bamenda. Often migrants deal with new life experiences and sometimes hardships, and they feel the lack of understanding from people back home who do not have such experience. Nevertheless, many migrants see the realization of their success in an ultimate return to Cameroon.

A sample of interviewees with migration experience

I have started developing a network of Cameroonians in my hometown Basel in the year 2007. I encountered them mainly through friends and then gradually within networks, which were mainly related to Francophone Bassa and Ewondo ethnic groups – two further Cameroonians I met in Germany. I started to investigate contacting Anglophone Cameroonians specifically in 2009[396]. Regarding the samples of interviewees, I have interviewed 17 Cameroonians altogether, hereby 5 of them Anglophones, and another 2 originating from the Western Province, the neighbouring province of the Northwest Province, across the language border. 5 of the 17 were female[397], 9 were below the age of 35 years[398]. Another 9 had lived in Switzerland for more than 5 years[399].

---

[396] Maybe due to my small sample of interviewees, I did not note differences in attitudes and practices of maintaining social ties, and notions of solidarity, between Francophone and Anglophone Cameroonians. Possibly there are none. Differences rather arose in relation to age. See migrants' uses of New Media, later in this chapter.

[397] I can only explain the low share of the females in my sample by making speculations. I followed social networks of known migrants, and I did not explicitly follow female migrant's networks, even though I am aware that some exist. However, since I have a small sample and laid focus on interviews in Cameroon, I decided to leave the sample like this. I have also had many informal conversations (see Field notes 2008/09/Migration, as well as Field notes/diaries 08/09/10/11) with a number of returned female migrants in the Cameroonian context, in order to have a female perspective on migration. See later in this chapter.

[398] In this sense, on the one hand, my sample is only partly representative, in view that I concentrated on youth in Cameroon. On the other hand it is also interesting to see differences between New Media use of young and elderly migrants, and

Additionally, I had many conversations with people who had had migration experience in Bamenda, some were in Cameroon for holidays, others regularly spent extended time in Cameroon but were based abroad, and others had returned after having spent a certain time abroad. Among those who had permanently come back, or for a longer period of time, was for example an elderly lady, who had been married and lived for decades in Austria, but was now divorced and returned to Cameroon permanently in her retirement. Regina, a lady in her early forties had studied and worked in UK, but was permanently back in Cameroon, although she had plans of doing a job for the UN which would involve to travel again. She was also an owner of a cyber café in Bamenda. Another man in his late thirties had lived in the US from an early age, but was back in Cameroon because he was able to take a responsible and well-paid job with Camtel through the influence of his father. Another man of the same age had been studying in Italy and had been back in Cameroon for a few years, working in a hospital. He had plans to go to Europe again for further education. A man in his forties had lived for many years in Holland, was divorced from his Dutch wife, and he was currently building up a business related to agricultural products, together with Dutch business partners and Cameroonian friends back in Holland. Frederik, who was in his late thirties, had lived in Holland and had returned in order to build up a spare car parts business. Loveline, a young woman in her thirties had studied and worked in the UK for seven years, and had returned permanently in 2009 in order to take over her family's business. Among those who were very mobile, going back and fro between Cameroon and their current country of residence, was for example an elderly man who had built a bar in the city centre, was supporting young musicians in Bamenda, and was called "the German" because he had lived in Germany for nearly

---

furthermore, how situations and views on life projects – here dreams of return and being involved in diverse activities in one's society of origin – which are tied to economic prosperity and juridical status – are also correlated with age. It thus renders insights into the narratives of success and position of youth in society, which are also followed up by migrants themselves, see later in this chapter.

[399] In the indications of the quotes I integrate here also dates of the interviews, because with some interviewees I followed up their life stories over long periods of time and met them several times – in particular Paul, René, Godlove, Louis, Alice, Alexandre and Dieudonné – the last two mentioned were Francophones.

forty years and had German citizenship. A young man from the UK was spending a couple of months in Bamenda – and was continuously extending his stay – because he was building up a cyber café, but encountered difficulties finding reliable staff to hand it over to and return to the UK. Another young man, Fred, had come back in order to develop a poultry business in Cameroon, after which he intended to go back to the UK. A lady, who was based in the UK, spent several months per year in Cameroon due to her NGO. Furthermore, I talked to two women and one man, who had been working in China as English teachers. Others were business-men, who were regularly travelling for a few weeks per year – for computer and mobile phone spare parts – mainly to places such as Dubai, China, but also the US and Europe. Two of the interviewees I had known and interviewed in Switzerland I met again in Bamenda. René, who was in his late twenties, had to return because his application for asylum had been denied. I met him in Bamenda in 2008 and 2009. In the meantime he won the DV lottery, obtained the visa for the US and went there in 2010. Paul, in his 40's, was involved in business and NGO work and travelled to Cameroon several times a year, so I had the opportunity to meet him several times in Bamenda[400]. Additionally, in Cameroon, I led informal conversations with people who were about to leave, such as a young man who went to Denmark in order to play football, another who went to Ireland to settle a family affair, 25 year old Christian went to study in the UK, and Edwin, another young man, was able to join his wife in Norway. One young woman went to Germany for studies, another migrated to Canada to do a course in nursing, and another woman in her 40's had won the US green card and travelled to the US with her teenage son. There were many more stories and encounters with people about to leave[401].

---

[400] Travelling out is related to making one's living, but also coming back is most often – with the exception of the lady who had come back for retirement, or René, who returned involuntarily – related to probing into a business venture, be it that people intend to stay permanently or just for a period of time in order to set up a business.

[401] When following up their stories, some were then successful, and many others not.

An anxiety to migrate, and felt hardship of being a migrant

As I have already stated[402], regarding their migration ventures, people seem to inform themselves only partially through media, but they are guided in the main by the impressions they obtain through their personal social relationships to Cameroonians in diaspora. They also count on the help and support of others who have already migrated, sometimes partly being ignorant of juridical systems, politics and laws governing immigration[403].

In the cases of the interviewed migrants, four of the migrants came to Switzerland through marriage, and five came for a job or to study, another three came upon an invitation letters of relatives and friends in Switzerland, of which one and another five – who came so to say in a clandestine manner – had then to apply for asylum[404]. The migrant's choice of the immigration destination, is in this sense either dependent on social ties, or it depends on their general urge to migrate, applying for different visas in Cameroon, or their dependency on migration brokers. Regarding the interviewees' planning of migration, in the case of Switzerland, some people said that Switzerland was not the country they had aimed to reach, but rather Germany, France, the UK or the US. It seems that migratory paths are not always easy to plan. As one migrant expressed: "People do not think about the long term planning, getting a visa is difficult enough, they sometimes only think about that and inform themselves accordingly. They do not think, for example, that in Switzerland you are not allowed to work when arriving and that your chance that your application for asylum is approved, is very low" (Alexandre. Interviews Diaspora. 17.04.07). Four of the interviewees stated that it was a coincidence that they had come to Switzerland, two of them through a recommendation of migration brokers who helped them to get there[405] – they were then applying for asylum upon their

---

[402] In particular in chapter 2 and 4.

[403] See chapter 1 about Cameroonians in diaspora and figures regarding Cameroonians in Switzerland.

[404] Of those who had applied for asylum, I have not always followed up their stories. In the case of Alice, she married a Swiss national: when I met her, her status in Switzerland had already been secured. René went back to Cameroon and went to the US two years later, Godlove was still waiting for the decision on his case by the end of the year 2011.

[405] Not everybody has relatives abroad, a few Cameroonians came to Switzerland without having any contact, as they said, through a migration broker to whom they had

arrival. Most said that they had no idea about Switzerland when they arrived, apart from that what they had learned in school about Switzerland in a very superficial manner. One migrant said: "When I saw pictures from Swiss in school in geography lessons, I thought, this looks so beautiful, and it is the most peaceful country in this world, the standard of living is good, and people are nice… but I never thought of coming here at first" (Godlove. Interviews Diaspora. 19.05.09). In this sense it is not the knowledge about the destination, which is decisive. Some migrants said that they were just following an anxiety to move in order to change their lives. A young man who sought asylum said: "I had to migrate at all costs. And now I will stay here at all costs. I need to make it, now that I have finally reached here" (Blaise. Interviews Diaspora. 06.03.08).

Depending on these migration histories, migrants have different starting positions. However, the issue of a difficult integration into social networks in the country of migration destination is a common experience of migrants, even though varying in intensity. Even though all of the interviewed migrants had very different personal backgrounds, education, gender, age, state of residence, and time spent in Switzerland, or abroad in general – they all emphasized the experience of "struggling". The notion of "struggling" relates to imaginaries of what I have called the "flipside of the West", as well as to positive counter imaginaries of their home society. Interviewees either expressed their struggle in a retro-perspective, that they were able to get through these difficult times and had established themselves after some time, or they were relating to it in order to describe their current difficulties, which was of course particularly true for 5 of the interviewees, who had applied for asylum. This period of struggling is seen as a liminal state. Their first, pressing and burdensome concern is the insecurity of their permit to stay. Most of the migrants said that they never expected life in diaspora to be as hard as they found it to be. "I was very ignorant. I realized that life in Europe is hard. But I have to go through it. If I can't, I can go back to Cameroon and relax" (Godlove. Interviews Diaspora. 19.05.09). Several migrants said, that their experiences in Switzerland were difficult,

---

to pay a considerable fee. In the case of Alice, she said that her broker promised to bring her to the UK, but left her at the airport in Zurich, which is a hub for international flights.

but would also add to their life in the sense that they would have matured through them: "I am feeling old, since I am here in Switzerland. Old in the sense that I went through so many difficult experiences. Other difficulties in life will not make me lose balance anymore" (René. Interviews Diaspora. 06.11.07). One of the main issues of struggle in Switzerland – apart from the difficulty of making one's livelihood – are feelings of loneliness, and often a prevalent feeling of rejection. "Swiss people are so cold, it is sometimes not easy to support" (Aboubakar. Interviews Diaspora). Or René remarked: "You can come closer to the Swiss, but it is always myself who has to do the first step, and also (I need to) be ready to accept rejection, but often they like it when I speak to them and we chat. These are good experiences, but it takes effort, to always make the first step" (René. Interviews Diaspora. 28.02.08).

Contacts to fellow Cameroonians and migrants in diaspora

In view of such struggles, being in close contact with loved ones at home is of great importance. In new life situations, however, new relationships also gain importance. When looking at ego-centred social networks of migrants, these often reach out worldwide, also to fellow migrants, relatives and friends who have also migrated, or connections from the individuals' previous diaspora life in other countries[406]. In particular young migrants stated that they were in contact with friends and former class mates, who also lived abroad, working or studying, for shorter or longer periods, well established or in insecure situations. These relationships can become important to migrants in diaspora: often of the same age, hierarchically equal, having the experience of the difficulties in diaspora, exchange in such relationships is valued as uncomplicated. "It is with them that I exchange myself and talk about frustrations and problems, rather than with my parents. They are able to understand because they have similar experiences" (Daniel. Interviews Diaspora). Or: "They can understand my situation, because they went through the same difficulties. With them I can exchange things, people in Cameroon do not understand" (René. Interviews Diaspora. 06.11.07).

---

[406] These contacts are maintained by using email, messenger chat, Skype and other internet based communication technologies, with people who have equally access and skills using them. Some of the elderly migrants, who I have interviewed, were only internet literate in a limited sense, and depended more on phone calls.

Additionally, regarding face-to-face contacts, the connections to other Cameroonians are highlighted as a way to keep "a sense of home", to keep contacts to others, who have been known prior to migration back in Cameroon, or meeting new Cameroonians who somewhat share "the same fate". Most migrants are involved in local associations of Cameroonians, to a greater or lesser extent. The groups are often informal and pursue different aims, ranging from purely social and leisure, up to involving aims related to development. Some Swiss NGO's related to Cameroon I have come across through my interviewees, are "CASA-Net" – Cameroonian Skills Abroad Network - an international network of Cameroonian diaspora organizations, "Ambassuisse", a Bassa cultural group, and "Association Oyili", a local women's group. Furthermore, there exist groups aiming at development projects in Cameroon, as "Hilf Kamerun", "Hilfsprojekte für Kamerun", "Hope Foundation" and "Bridgehead Foundation", founded by Cameroonians and Swiss partners. Other organizations have socializing as their main aim, including leisure activities, social events and sports – most of them informal. "Njangis" – how credit associations are called in the Western Provinces of Cameroon – are often integrated into such groups. The value of "preserving Cameroonian culture" in diaspora – also for the offspring born in Switzerland, is also mentioned. As one interviewee stated, who had been in Switzerland for more than 15 years: "Many Cameroonians rather think of Cameroonian culture as negative, backward. It is only in diaspora when they gain interest in their culture and moral values. You cannot adapt to another culture in a foreign country, when you do not know where you are coming from" (Alexandre. Interviews Diaspora. 20.06.07). Sometimes, social ties and arising tensions reflect differences along the lines of ethnic groups, or in the case of bilingual Cameroon, language. I have heard of some groups, which were splitting up, producing new set ups and dividing again, due to such reasons. This is also lamented, as an interviewee expressed: "Cameroonians cannot even stick together across language and ethnic boundaries, when they are in diaspora" (Louis. Interviews Diaspora. 30.03.09).

Even though some Cameroonians are active in maintaining relations with fellow Cameroonians abroad, for example within such organizations, there are also those who distinctly refused to involve themselves in such groups, or did so at a certain time and were disappointed precisely because of the tensions that occurred: "One comes here and one can see that life is not so easy. In such a situation, everybody becomes a competitor" (Marguerite. Interviews Diaspora). Some Cameroonian migrants expressed certain mistrust towards fellow Cameroonians: "They will judge you, where do I come from, what ethnic group, what province, and so on. I do not want to be judged like this, I am Cameroonian and I am here in Europe. So that's it" (Godlove. Interviews Diaspora. 23.03.10). In this sense some interviewees emphasised their preference for mixed social groups. Among such social groups there are sports clubs, and also churches, which provided an opportunity for being part of such a community: "My church is like my family here" (Godlove. Interviews Diaspora. 23.03.10). The contacts among fellow Cameroonians in Switzerland are to a great extent mediated by phone, rarely by internet. People live in different cities, and meeting is subject to time and financial constraints - transport costs - and other obligations.

An option of return - the urge to "make it" and fruition of one's achievements abroad

The great aim of migration is to achieve something. As the term bushfaller relates to: when one goes to the bush he is expected to return with game, one does not come back empty handed. To return without having achieved something would be embarrassing, since the imagined perfect conditions of succeeding abroad are given – not to succeed is attributed to a personal failure of the migrant. A non-migrant in Cameroon expressed these expectations in very lively terms: "(As a migrant) you should not sit and expect. Life is not ABC. Some stay for twenty years in the US and are empty. They do not even have a structure (house or business) in their own country. If you come back empty everybody turns their back on you. Why are you there, for what? Then nobody would even welcome you, that's our tradition. You must do something" (Valeria. Interviews 2010/11).

"Dreams of return" may arise among migrants in all life situations. However, a return depends especially on the estimation of whether or not they have acquired an economic base, and if they are able to maintain a positive presence in their societies of origin[407]. Often migrants think of returning as soon as they have "succeeded", in the sense that they have been able to save money for a starting capital, and having invested in business or a house, during their stay abroad. Subsequently, the newly arrived migrants do not think of return, despite the difficulties they face. Even in an insecure state and in the view of low chances to obtain a residence permit in Switzerland in the case of asylum seekers, hopes are held high, as one interviewee expressed: "I cannot go back. I cannot go back now that I am here at last. I am sure I will find another solution, if they reject me here in Swiss" (Blaise. Interviews Diaspora. 06.03.08). A migrant in Germany, a student, stated that he wanted to stay until he had gathered some wealth, and that his presence abroad would figure highly in the estimation of his family: "I realized that I give my family hope. Even though I could not give much in return until now, as long as I am here, they have hope. If I do not send money back today, I will do it tomorrow, if I do not send tomorrow, I will send the day after tomorrow. But when I come back, they cannot have hope anymore and they will criticize that I did not use the chance, that I had given up. I do not want to take their hope for a better life away" (Daniel. Interviews Diaspora). Meanwhile, these migrants tend to reduce their social networks, due to high expectations of relatives and friends in Cameroon, and their feeling of not (yet) being in the position to respond to expectations, or to invest. Thus, due to the difficulty of making it in diaspora, dreams of return remain, but plans are postponed to a later point in time, such as retirement[408]. Alexandre, an elderly Cameroonian said: "I always wanted to return. But life has kept me busy. But I want to return to my village in old age - in order to do this, I have to maintain my presence there: I have my plot of land and keep relations with the extended family. I need their respect and support, when I come back to farm my land" (Alexandre. Interviews Diaspora. 26.06.07).

Alienation, which could have arisen in the years living in a different country, is a factor, which can influence decisions regarding return. Having the knowledge of both worlds raises contradictions and induces

---

[407] Also their plans to possibly marry and starting their own family in Cameroon, play a role here.

[408] Here it of course also depends on migrant's marital and familial state in the country of residence.

personal transformation, as some migrants said. "They miss a part of my life, they cannot imagine my life over here. This leads to that I feel sometimes grown away and detached from them (the family)" (Maximillian. Interviews Diaspora). One migrant critically said, after four years in Germany: "When I was in Cameroon lastly, I realized that I have become somehow German. I could feel it when discussing with people, about the world, life chances, or politics. However, when I spent time with my family, with the people I know for so long, I felt home again. I can still go back I think, that is my plan, to go back after having finished my studies and start a business" (Jean. Interviews Diaspora). Most migrants expressed that they could imagine living in Cameroon again, on the condition that they had achieved something on their return. However, most of the returned migrants in Cameroon stated that it was not easy to readjust, after having lived abroad for years. 33 year old Loveline, who had come back to take over her family's business, complained that it was easier to focus on work in Europe, that people would talk behind her back, and about the role of women in local society. However, she as well stated that she felt comfortable to be close to her loved ones: "home na (is) home" (Loveline. Field notes 2010/11. 25.11.10).

Returning home can also mean a cutting of resources from the perspective of those in Cameroon, a cutting of transnational flows, and also of status that arises from being integrated into social networks of such flows and ties. Regina, a former migrant in Bamenda, teacher and owner of a cyber café, stated: "My family would have liked me to stay longer abroad. But after finishing my studies, and having worked for a certain time, I decided to come back and try to build up something here. I had then the impression that some of my family members would have preferred that I stayed abroad, they feared that the sending of money would stop, when I was struggling here to set up something. But it was my decision and my life" (Regina. Field notes 2009/Migration-Stories).

Overall, the desire to return, as an often firmly uttered statement, seems to be a blend of a feeling of social obligation towards those who stay back, of emotional attachment to loved ones, and also of the prospect to "live a good life" in Cameroon with the background as a migrant – regarding financial means, as well as status and recognition.

Imagination and solidarity - migrants between pressure and demonstrating success

Migration is indivisibly bound to the notion of "success": therefore "bushfallers" seem to be ideal targets for projections. The easiness of being connected coupled with the images of abundant wealth in the West, lead to a doubly enforced tendency of demanding from migrants. The disparities between ideal and reality, between imaginations of non-migrants and the life realities of the migrants abroad, have often to be borne by the migrants[409]. Explanations and expressions of subjective realities from the part of the migrants – as well as attempts of non-migrants to understand them – can only partly overcome these differences in perception and expectations. However, the migrants also often align themselves with ideals of support giving and solidarity. In this situation migrants contribute their part in maintaining imaginaries and narratives. In this sense, migrants balance between taking social imaginations of solidarity and success as templates for their life projects, also orient themselves to their personal social mobility along partly alternative pathways, and as well, they negotiate and evaluate their social practices according to these conditions.

F60: Staying connected through calling each other
F61, 62: Local call boxes and price signboard, Yaoundé and Bamenda

---

[409] See figures F60, F61 and F62. Whereas migrants abroad might limit their contacting to the most close social ties, in Cameroon the availability of phones is an expression of an expectation of connectedness.

Strong and weak social ties - the desire to support and solidarity as a burden

"Giving something back" is viewed by most migrants as the highest imperative. These felt duties to support and showing solidarity – or refusing it – arise from different dimensions as a basis of justification. One is the actual financial indebtedness, as for example a few aspiring migrants in Bamenda said, that some relatives had borrowed a certain amount of money, which they aimed to pay back when they finally fulfilled their aim of migration, and start earning money abroad. Then it is also a moral sense of indebtedness to the family, which is not only be supportive regarding the actual migration venture, but also invested in education of the aspiring migrant for example. The other dimension relates to general claims towards migrants, based on mere fantasies of migrant's apparent state of being in an economically better state. Regarding the first dimension, migrants take their duties seriously – it relates most often to strong ties of emotional and/or social closeness. Regarding the second dimension, it relates to what I have called claiming relationship and is rather concerning weaker social ties. Migrants are also reminded of their duties by communication technologies, or in the local context, as Edwin, a migrant to be in Bamenda, expressed: "They are reminding you that they did you good in the past, and that it is now your turn to return the favour" (Edwin. Field notes 2009/Migration-Stories). In his case, relatives and friends were turning up in his family's compound, recalling that they had sponsored him to go to school, when he was a child, and now he had grown up and was about to leave Cameroon.

Regarding the most close and intimate social ties, migrant's responsibility for their physical, social and emotional wellbeing is highly valued. For these people, many migrants invest huge amounts over time, which is seen as both, a burden but also a moral duty, which has a satisfactory component when fulfilled. The interviewees stated that they were happy being able to support their family back home. It was also uttered in Bamenda that taking over of responsibility for one's family – as part of the notion of "good life" – was one of the main motivations to migrate. Regarding notions of solidarity, some migrants emphasized a kind of implicitness, or naturalness to support each other within families[410]. For migrants in diaspora, the family is a node of belonging: "Family is it what is tying me to Cameroon, family is home" (Louis. Interviews Diaspora. 30.03.09). In most cases close family ties provide a feeling of closeness and emotional support in the difficulties in diaspora. "After having communicated with my mother and my brother, almost every Sunday, I feel good and I really feel relieved to hear that they are doing well, and I feel prepared to face everyday life for the next week" (Maurice. Interviews Diaspora).

However, also concerning the closest social ties, expectations need to be constantly negotiated, between meeting expectations and maintaining good relationships, and not to overburden oneself: some newly arrived migrants stated that they would rarely send remittances and that their families would understand that they first had to build up something for themselves, before they could send money to them. Many migrants expressed, that it was less the close family but rather friends and weaker ties who were deliberately claiming relationship with them, who put them under pressure. "My close family knows me and they know my situation, they do not put me under pressure. When I have something to send, I will send it to my family. Friends are more intrusive, bothering me with their claims. It just leads to, that I minimize my social ties to the closest ones" (Alice. Interviews Diaspora). Nevertheless, some feel burdened by responsibility, feeling that they can always run into danger to be looked at as non-responsible when not corresponding to demands for support. Another migrant said: "When something happens in my village, people say, good, we have somebody from our village, who is abroad, we can ask him for support. Often I cannot say no. Like now, a member of my family died, and they collected money for the burial, about 100 Swiss francs altogether. But the burial expenditures will be about 1000 francs. So they expect a substantial contribution from my side. I sent them 500 francs yesterday. This is much for me, I earn small money apart from being engaged in my studies" (Maurice. Interviews Diaspora).

---

[410] I am aware, that some might have over-emphasized the importance of their role in their families to the researcher.

Reproducing the myths of greener pastures – migrants between not being understood and playing with imageries

The reproduction of the myth of greener pastures occurs through mediated interaction – and also non-interaction – between migrants abroad and non-migrants in Cameroon, but also on a local level. It is effected through verbal – including what remains unsaid – and nonverbal means, as the appearance of the bushfallers when they return to Cameroon, pictures sent, and narratives told. Assessments from the part of migrants range from feelings of not being understood, disbelieved, even ridiculed, feelings of shame, up to the desire for openness and conveying "the truth".

Among Cameroonian migrants who were less settled in Switzerland, who were in a prevalent state of insecurity, and who did not dispose over strong social ties in the country of residence, it was often emphasized that they mostly suffered from not being understood – their life stories being less convincing than the strong imageries in the minds of people with no migration experience. Hence, silences also come in as an issue. René, a young asylum seeker stated, that he had nobody to share his concerns and problems with, since he would not want to disconcert his family back home by telling them his worries and daily difficulties, and it would be difficult for them to understand or to imagine a system functioning so differently from the one in Cameroon. "My father asks me again and again, if I had found a job. I have tried to explain that I am not allowed to work while waiting for the decision of my application for asylum, but he cannot understand these kinds of regulations. Nevertheless, I was able to earn some small money and sent back small amounts, that makes them being content, and they do not ask me too many questions" (René. Interviews Diaspora. 01.07.07). Often migrants do not even talk about their experiences, by anticipating non-understanding, and they want to avoid further misunderstandings. Or they disclose these issues only towards certain people, with whom they evaluate it as vital to share, resulting from pragmatic intentions in order to justify their behaviour – for example not being able to send money – and to maintain an ambiance of familiarity and closeness in the relationship.

Yet another point here is the perceived disinterest, and the feeling that their problems are not taken seriously: "I was even disappointed by close friends, of whom I thought they could understand my situation, because I have explained to them. But the next thing is that they ask me for money for this or that" (Maximillian. Interviews Diaspora). Migrants complained that they were not listened to, and that others would not believe them when they tried to tell them about their difficulties. "Some people in Cameroon have weird ideas about the life in Europe. Some even think that one does not even have to work here, since one is at the source of money. But these are people who are not well educated. If I tell them that even here in Swiss not everybody is fine, they tell me I am a liar. I really do not like to be called a liar when telling the truth. So I rather keep quiet" (Godlove. Interviews Diaspora. 23.03.10). In this sense some also uttered the fear of being ridiculed by non-migrants, in the sense of attributing it to their personal failure when facing difficulties. Or their explanations would be turned against them, such as when they are accused that they do not grant the triumph of migrating to others, when they deny helping others in visa issues. Thus, some migrants adopt strategies to keep a neutral position towards questions, desires and demands for helping others who also desire to migrate. "When they ask about Swiss, I say, you have to come and see yourself. I do not want to tell them that it is paradise, because it's not true, but I also do not want to tell them, no, it is not good, you should not go there, because they will think that I want to prevent them from succeeding" (René. Interviews Diaspora. 06.11.07).

Another issue is holding back information, as an expression of a desire to keep some dignity in the face of the expectations of people back home and the circumstances of constraints in the country of residence. As in the case of Godlove, who expressed that he wanted to convey a positive image as an answer to the embarrassing questions about what he was doing in Switzerland and his being unable to explain the conditions of his life here, especially as an asylum seeker, but also, that he did not intend to invent stories: "I had a job I could do from a Beschäftigungsprogramm (occupation programme), in a butchery. I was very happy to be occupied, even if I hardly earned anything. I asked a friend to snaf (photograph) me when I was working. The picture was very nice, me standing there in the white butcher dress and the big knife in my hands. But when I wanted to send the picture, I was informed that we could not work there anymore, the migration authorities (Amt für

Migration) had decided this. So I never sent the picture. Until now, I have never sent any picture at all" (Godlove. Interviews Diaspora. 19.05.09). However, migrants also highly value the imaginations and expectations of their relatives and friends at home, and in order to correspond to these expectations, they are not willing to expose their real life circumstances out of shame, for example if they are doing work that is below their social status. Another aspect of the perception of not being understood was expressed by a young asylum seeker, who rejected fuelling imaginations of people in Cameroon. He said that even though he had pictures of himself he could send to his family, he would hardly do it. "When they see the pictures I am smiling on, they will say: you see, he is fine, he is very happy, because he lives in paradise. I do not want them to imagine this, because I have many difficulties and often I do not feel well" (Blaise. Interviews Diaspora. 25.05.08).

However, some of the interviewees gave examples of how they tried to "illuminate", or exert control over non-migrant's perceptions and imaginations, about their state and situation in Europe. As 50 years old Alexandre stated: "When I am in Cameroon and ask young people who eagerly struggle to migrate, what they want to do in Europe, they say: putting up a small business – no you cannot just do that, there are regulations. I will clean the streets – no, these jobs are institutionalized, I will make my life playing African music – sorry, but you can never make a living. I really feel bad, it is not that I do not want to let them dream of a better life, but we should address the problems in Cameroon." (Alexandre. Interviews Diaspora. 17.04.07). Or, as in the case of an interviewee who was using the convincing force of pictures, in order to "enlighten people about the real circumstances of life abroad"[411]: "They (the migrants) send pictures where they sit in the armchairs of their bosses' offices after closing time, they stand in front of a sports car parked in a street, or make a walk in a nice quarter and snaf (photograph) themselves in front of a nice house. I wanted to do it in a different way: I sent pictures of me cleaning the trams. Some people thought this was a bad joke, because they had seen me as a big man in Cameroon, but I wanted to convey the truth to them about the life of African migrants in Europe" (Paul. Interviews Diaspora. 22.04.09).

---

[411] However, the interpretations of such pictures can also produce ambivalent impressions.

F63, 64: Images of success: decoration in a youth's room, DVD issued by a Pentecostal church

Migrants also related to the co-responsibility of fellow migrants, who would not stick to predicaments of transparent communication and information, and would deliberately convey false images and impressions, as a main contribution to the reproduction of people's imagination of "greener pastures" at the expense of migrant's reputation. One migrant said: "The reason why migrants pretend these wrong images is excessive pride. They are viewed as already successful when only leaving the country, so they think they need to stick to this image" (Paul. Interviews Diaspora. 22.04.09). The interviewed migrants would of course highlight that they were different, delineating themselves from those taken as negative examples. These narratives about co-responsibility of fellow migrants are connecting directly to the narratives in the Bamenda context about the irresponsibility and immorality of "bushfallers". As a non-migrant in Bamenda said: "It is not fair towards people in Cameroon, when they convey wrong images of the life abroad. It is not fair because people will continue to dream dreams that have not much to do with reality. And it serves just to maintain their own status" (Titus. Interviews 2009/F) (compare Tazanu 2012:137ff). And another migrant criticized the bushfaller's comportment in Cameroon: "The fact that they go for visits in Cameroon, displaying wealth by driving a big car, wearing an expensive watch and spending a lot of money is adding up to people's wrong ideas about life abroad" (Jean. Interviews Diaspora). By these interplays and the lack of transparency, the "mystery of white man's kontri" is reproduced, enhancing both, difficulties for migrants to deal with expectations, but also the opportunity to play with imageries and enhance their status in their society of origin[412].

Negotiating expectations – through "managing" communication media

Due to the availability of the internet and mobile phones, it has become possible to maintain close relations in the country of origin, due also to the fact that these media have become available and affordable in Cameroon. Migrants in Switzerland use mobile phones and internet in different ways than non-migrants in Cameroon. Some of the interviewees had internet at home, or they used it at work. Most of them however used their phones, be it landline phones or mobile phones. Peculiar is that they mostly called from private spaces, which is related to the differing set up of communication technologies[413]. That the phone is emphasized, has to do with the nature of the social ties maintained at home, rather small social networks of strong social ties, which includes, for example, parents, and other – also elderly – relatives, who are less internet literate. Furthermore, about one third of the interviewees were in their forties and fifties, and some of them less internet literate themselves. Migrants are also expected to call, and they emphasize emotional closeness, which is effected in calls, whereas youth in Cameroon maintain rather vast social networks of weak social ties of friends and class-mates: Only few of my interviewees in Bamenda had close relatives abroad, and non-migrants are also more inclined to use less expensive means of communication to get in contact with migrants, and rely on migrants calling them.

I suppose that the accessibility of New Media has enhanced the claims posed on migrants, based on the condition of being more easily accessible and available. The growing prevalence of attention, when being easily contacted and being connected to people in Cameroon can also become a burden and a psychological stress. Social ties need to be

---

[412] See figures F63 and F64. Youth's dreams of a better life, expressed by idols of global youth culture or status symbols – here also with an allusion to a Born Again Church.

[413] Which meant for my research – contrary to the Cameroonian context – that I observed hardly any migrant's use of mobile phone and internet, but I had to rely on interviews. Compare also Tazanu's description of differences of migrant's and non-migrants communication environments (2012:108ff).

dealt with in new ways by migrants abroad, which is mostly done by regulating connecting practices through New Media.

Regulating communicative exchange and expenditures for communication

Expenditures for communication are in general dependent on the economic or emotional valuation of a social tie. Most often, when "investing" into communication and calling abroad, people expect, or at least hope for, a certain positive outcome from the investment, at least in the long run. In general, it is expected that migrants abroad are the ones who bear the communication costs, since they are seen as economically viable, and, as it is estimated, it is "only small money for them"[414]. Costs are also expected to be taken over by migrants, when contact initiatives come from people in Cameroon. It can be done deliberately by beeping the person abroad, or, in order to express more valuation and politeness, to make a short call, conveying the wish for a call back. Some moral notions of conduct related to age, position, and gender, are thereby taken over, however in other cases not, since the shift in position through migration is evaluated more strongly (compare Tazanu 2012:186ff). The more frequent the interaction, the better established the social tie, as well as the higher a migrant ranks regarding the evaluation of his or her economic viability, the more it is expected that migrants abroad bear the main costs for communication, and the more migrants are ready to do so concerning the respective close social tie.

However, like non-migrants, most migrants related to financial constraints in communication especially regarding making calls. In particular for example for asylum seekers, staying in regular contact with those at home is rather difficult. Depending on migrant's social networks linking them to their home societies, the activities they are pursuing back in Cameroon – such as business – and on the availability of financial means in general, some spend considerable amounts for communication purposes. In order to keep these costs as low as possible, they try to favour less expensive means of communication,

---

[414] The wide spread idea also prevails that costs for calling from abroad to Cameroon is less than the other way round, an impression, which is related to relativity of expenditures rather than to effective expenditures.

such as internet based means such as Voip or Skype for calling. Further possibilities are using call cards, SIM cards from providers with cheap offers for international calls, such as Lebara or Lyca[415], or using pre-dialling numbers[416]. To balance the expenses for communication, but also to deal with demands, which are expressed towards them through New Media of communication, migrants develop different strategies: "You are forced to find a way to deal with communication with people from Cameroon. Otherwise you would be constantly confronted with calls, beeps, emails and demands" (Paul. Interviews Diaspora. 22.04.09).

Expenditures for communication are appraised by migrants on the basis of an evaluation of social ties, conduct in communicative interaction, the importance of issues addressed, and the importance given to what I call a space of self-determination, by regulating the use of communication media, temporal, or possibly spatial frameworks for mediated social interaction. It is about when and how to be available or not available for others – here specifically non-migrants in Cameroon.

Sorting out valuable social ties by regulating practices for connecting

When regulating practices of connecting and most importantly, being contacted, it is not about rejecting communication in general, but about an evaluation of social ties, which are important for migrants. These might be measured by an evaluation of social and emotional closeness, and conduct maintained in mediated communication[417]. Being connected through New Media of communication is on the one hand evaluated as facilitating connecting, and thus prone to becoming a burden for migrants with whom everybody wants to claim relationship, but on the other hand, mediated connecting can also be dealt with and negotiated through a range of strategies. Reducing the number of contacts is of course valued negatively by people in Cameroon, in particular the ones who do themselves not belong to the circle of intimate and highly evaluate social ties of the migrant. "I know that some people in Cameroon think that I am neglecting them, but I cannot

---

[415] See figure F65. The same is visible in Cameroon, see figure F66.

[416] They have to dial a number and then adding the country code and personal number of the person to be called. Some migrants also use call cards (compare Tazanu 2012:129ff).

[417] Tazanu (2012:176) speaks of creating a family within the family. Compare to the previous chapter.

be constantly available for them, I have my work, my family, and many things to do, and I cannot spend hundreds of (Swiss) francs every month just to say hi to everybody" (Alice. Interviews Diaspora).

In this sense, some migrants in Switzerland, but also those with migration experience I talked to in Bamenda, expressed that they restricted being contacted prior to migration: since they change their contact number when going abroad, migrants have the chance to consciously valuate whom they would give their contact number to[418]. Some stated that they inculcated to those who had it, that they were not supposed to hand it out, which is generally observed by non-migrants, since to dispose over an exclusive contact with somebody abroad is also seen as a resource, which is not carelessly handed out and shared. Migrants often use net calls when calling people in Cameroon, to pay less, but also at times for not being forced to disclose their personal phone numbers. It is often stated by youth in Bamenda, that they dispose over email addresses of classmates and friends who migrated, and not over a phone number[419]. An email address is thus a convenient means to potentially stay in contact without the feeling of being urged into excessive expenditures, related to calls, or disturbances by constantly being beeped, however it is restricted to those who are internet literate. Media of communication are used here in order to classify different contacts, in evaluations of closeness or distanciation.

There are other strategies in order to regulate networks in the sense of enhancing desired, and controlling unwanted contacting and communication. A strategy, in order to control by whom and when being contacted, is keeping separate phones. Almost all migrants had extra phones – and extra numbers – for contacts from Cameroon and for domestic contacts and other people. Migrants involved in business had separate phones for these contacts respectively. As a tendency they use different SIM cards for international and domestic calls, some have

---

[418] This is enhanced by habits of only telling the closest family members and friends that one travels abroad, also due to fear of jealousy and its potential malevolent implications, as I will relate to later in this chapter.

[419] Some people in Cameroon stated that they did not dispose over phone numbers of even close friends who were abroad. It makes them feeling more dependent on the responsiveness of migrants to call them. It also includes complaints to migrants to hand out their phone number. A refusal is then seen as a lack of trust and consideration.

double SIM cards in their phones for the purpose of separating networks, a tendency which I had also observed in Cameroon. Another strategy to enhance easy communication with close social ties – and at the same time keeping the costs for calls under control – is to provide people within a close circle of intimate social ties with phone cards – and/or mobile phones – in order that the callers set priorities related to calls themselves. Again another strategy is to set up close ties with key persons, or persons in key positions, for example within families, who act as key informant to other members of the family.

Setting up rules of conduct – beeping, emails, and priorities in communication

Regarding demands, migrants not only restrict their contacts and try to manage them by the use of different media, but also, they insert rules of conduct they would stick to as a principle, in the sense of setting priorities regarding issues addressed in communicative exchange and ways of contacting them. This consists, very commonly, in the restriction of beeping. As a migrant expressed: "You need to set clear rules. If you start to call back everybody immediately after receiving a beep, people have these expectations, and claim on it" (Godlove. Interviews Diaspora. 19.05.09)[420]. Some migrants have clear guidelines set for their relatives and friends in Cameroon regarding contacting, which are individually communicated. One migrant said that he would not react to beeping as a principle, as he felt bothered and put under pressure when beeped. "Even my mother is not allowed to beep me. She should call me" (Paul. Interviews Diaspora. 22.04.09). For some migrants, beeping is seen as legitimate, in the view of inadequate finances for those trying to make contact from Cameroon, but however, it needs to be controlled, differentiating between close social ties and more distant ones. Another migrant said that only the members of his closest family, such as the mother or the brothers and sisters were allowed to beep. Concerning less close social ties, "I restricted this way of contacting me because I realized, when I was always calling back in the beginning that I spent too much money on unnecessary calls, which I would better have used to send to people directly to solve their

---

[420] When together with Tazanu in Basel or Freiburg. I learned about his strategies of controlling contact and demands.

problems" (Louis. Interviews Diaspora. 30.03.09). Migrants also stated that their reactions to beeps had changed with time, in the sense that they did not feel anxious any more when receiving beeps. "In the beginning, I was anxiously calling back when I saw many miscalls of the same numbers in my mobile. And every time I called back, it was nothing important, just greeting and often asking for money. So now I do not bother anymore. To beeps I do not react at all, and it has almost stopped now. When people have something important they know that they have to call me. I have also a good excuse, I tell them sometimes that my husband (who is Swiss) would feel disturbed by the beeps" (Alice. Interviews Diaspora). By not responding to beeps, migrants hope to be able to force non-migrants to "invest more" when they want to get in contact, in order to turn down less serious requests (Wilding 2006, Wiles 2008), and to avoid unnecessary call backs.

Some interviewees expressed that they had "educated" people in Cameroon how to address them through communication media, and what rules of conduct they needed to observe thereby. "They know that I do not react to beeping. If something is important, they should call me for a short call, tell me what it is about, and I will call them back. If they call me for things, which I evaluate as less important, I do not call back. This is very strict, but I found I have to regulate the expectations for support and expectations for what purposes I am calling them" (Louis. Interviews Diaspora. 30.03.09). Three interviewees said, that they had "educated" most of the people who were internet literate, to send them an email and describe an issue, their problem, why they needed support, or render information. Then they would check the message and write or call back depending on the importance of the issue. This would give them time to clarify the issue concerned in order to decide whether and in what ways to respond to the issues brought to them. By such strategies of regulating communicative exchange, migrants try to sort out – from their perspective – important from non-important issues addressed in communicative exchange, in view that they are the ones taking over the cost for it. "I do not call people just to greet, there is always a reason for contacting. I cannot spend my money by doing useless calls" (Paul. Interviews Diaspora. 22.04.09). However, migrants have their circle of close social ties, regarding whom regular calling in the sense of unfocussed exchanging and keeping each other up to date is seen as a priority. Regarding this intimate circle of social ties, migrants expressed that they called regularly and were willing to assume the cost for this communication. Migrants' setting priorities communicative exchange then disfavours people who are considered as less close.

In this sense, migrants try to enforce their priorities regarding conduct in mediated communication, what issues are to be considered "important", or notions of social and emotional closeness, and the means of communication favoured, in order to establish moral conditions for mediated communicative interaction (compare Tazanu 2012). From their perspective, it can be considered as strategies to maintain their dignity and exert a certain control. From a non-migrants perspective in the Cameroonian context, some expressed a certain understanding, but more often people comment in a mocking sense, that migrants impose their will on them, and that they have become self-centred and detached from people in Cameroon.

Restricting being available – setting up times and places for mediated sociality

Regarding not being ready, or not being able, to be available and reachable all the time, many migrants raise the issue of availability, which is taken for granted and expected by communication partners, as problematic. In this sense, switching the phone off is not always a good solution and can be perceived as an offence[421]. Regarding expectations of availability of migrants, for example internet related media, such as email and chat, are easier "to control", as some migrants explained, since they were not always online and would not feel as disturbed as "when you have your phone with you and it is always ringing" (Alice. Interviews Diaspora). Not reacting to beeping and determining the ways of being contacted, keeps migrant's agency intact regarding how and for what reason being contacted by people, thereby delineating a sphere of self-determination in space and time for themselves.

Rules of conduct are in this sense inserted regarding times of being available, by restricting time frames for phone calls, as for example not

---

[421] Indeed, time frames for calling are less restricted in Cameroon, also depending sometimes on people profiting from certain offers, for example from inexpensive calls between 22 p.m., and 6 a.m., limitations in the sense of "no go times", are not so distinctive and thus not in line with temporal conditions of daily life courses in "the West".

answering the phone during the day but only in the evenings, or only checking emails at a certain time of the day. Or in order to avoid exhausting beeping they would take out the SIM card for Cameroon and only insert it to be reachable, for example, in evenings or at the weekend. Here, the main reasons are that migrants have other obligations such as work, and their times of sociality reserved for the social ties in their immediate context. Of course, most migrants emphasized, that they had a certain regularity of getting in contact with close people in Cameroon, some had reserved time frames, such as Sundays, weekends, or certain evenings for certain social ties. In this sense, migrants are very conscious about having their "times of sociality" with people in Cameroon, whom they consider as close. However, not being reachable creates tensions, due also to the fact that not being available for communication is uncommon in Cameroon[422]. As Alice expressed: "When I switch off my phone in evenings, since I am tired from work and I want to spend a peaceful time with my kids and husband, friends from Cameroon sometimes reproach that I have adopted white man's habits. But I do not care about that, they do not know my life here, so they cannot judge" (Alice. Interviews Diaspora).

The restricting of times for mediated sociality is also related to place. For example, migrants said they could not make private calls at work[423], and only in the evenings or weekends, rather confined to their homes. Furthermore, when they call they often call for a slightly longer duration compared to calls usually effected by non-migrants. It means that they want to be cosily adjusted at home when making a call, not as people in Cameroon calling from different – also very often public – places, from call boxes and cyber cafés. Additionally, for most migrants, they use internet from their homes and not from public cyber cafés. It means that as a tendency, migrants are in their communication habits more confined to specific places than non-migrants. Their sense of being

---

[422] Also, when not being reachable, people in Cameroon usually suspect and interpret it as a non-willingness to be contacted from the part of the migrant, and not as a possible failure of technology, such as, for example, lines not passing.

[423] This was of course also dependent on people's occupation and profession in Switzerland.

mobile and reachable at any instance and place[424], changes in the Western context, where activities – also mediated communication – are more temporally and spatially regulated. Due to the fact that non-migrants would not understand such everyday life conditions in white man's kontri, as some migrants stated, misunderstandings and tensions can arise regarding not always being reachable.

However, the impression that non-migrants in Cameroon just wait for migrants' calls, always happily ready to accept their attempts to communicate, could also be deceptive. Even though in general, it is expressed, that "one should never miss a call", non-migrants also have their times when they said they are busy and not ready for a call. An interviewee complained that his sister from the US would call him in the night without being conscious of the time difference (Olaf. Interviews 2009/R), and another uttered that she could get tired of talking to her brother when he called from South Africa and wanted to "talk for hours" (Andrea. Interviews 2009/O). I heard such statements several times, that migrants tend to talk elaborately, contrasting the way of effecting domestic calls in Cameroon, which are kept rather short. This tendency is by some people interpreted in a negative sense, that migrants have the assumption, that people in Cameroon are idle, have nothing to do, and are only anxiously waiting to be called. Migrants taking non-migrant's responsiveness for granted, is seen as a devaluation of non-migrants' integrity.

---

[424] As also visible for example in figure F67, call boxes, telecommunication publicity, and so on, visible and perceivable in the public sphere, where much of mediated communication also takes place.

F65: Telecommunication publicity in Switzerland, Basel
F66, 67: Connections are visible in the public sphere, Bamenda

As I have shown so far in this subchapter, migrants are constantly trying to balance their social ties in Cameroon, between the desire to support and maintain emotional closeness on the one hand, and on the other hand to deal with exaggerated expectations on the part of (too many) people. As I have described, on the basis of being connected through New Media of communication, this balancing is effected through regulating, connecting and disconnecting practices in these media, in a range of different strategies. In this sense, such strategies might lead to tensions, conflicts, and disappointment, since both sides are inclined to complain about a non-observing of positive conduct by their communication partners. From the perspective of non-migrants, it is the migrant's being non-responsive, non-reachable and non-supportive, and expressing their superiority, and from the perspective of migrants it is related to non-migrants' exaggerated expectations, claims on duties, and not observing rules of politeness. Observing of positive conduct is furthermore taken on to a face-to-face level, when migrants visit Cameroon.

Managing expectations on a local level: inconspicuously visiting Cameroon

Whereas New Media restrictions are used by migrants in order not to be too exposed to expectations and demands of a wide circle of people, when being present in Cameroon, avoidance is more difficult. For all the interviewees – who could afford to travel and were permitted to do so according to their juridical status – the main difficulties were said to arise when they went to Cameroon for a visit. It depends of course very much, for what purpose migrants return, if it is for a longer term, just for holidays, solely to visit their family or for a specific purpose, and how regularly they return. On the occasion of a visit, people with whom migrants are not in contact during their absence, can now hook up with them - situations, which could involve a great deal of pressure, demands for support, and reproaches that they have not kept contact and had forgotten about their people[425]. In order not to expose themselves to too many expectations and embarrassing situations, some migrants expressed that they had strategies of avoiding certain people, who also did not have their phone numbers they used when entering the country. Phone numbers are in this sense also a means of disconnection. In Bamenda, they would for example not stroll down the main roads, where recognition and meeting people is most likely. Or they would avoid staying longer in the same place, some said that they would stay most of the time in their village, whereas others preferred to instead stay in town, for example living in a hotel. In order not to worsen the situation, migrants said that they hardly announced their coming to Cameroon, New Media were used selectively to announce their coming to a few people only.

Migrants on visit or back in Cameroon are likely to be overwhelmed with expectations and claims. Moreover, also before their travel, if people know about a planned visit, it was mentioned that migrants were asked for gifts, such claims are conveyed through New Media. People ask for computers, cameras, mobile phones, and all kind of electronics, alluding to the prestige and desire for consumer goods coming from abroad and the idea of their easy availability[426]. The demands are expressed in an apparent disregard of the costs of these items, and how they can be transported, based on the claim of the tradition of giving presents on the occasion of a visit. In this sense, also in Cameroon, many migrants feel continuously under pressure regarding expectations for financial support, and distributing gifts, expressed in the often

---

[425] When I came to Buea together with Tazanu, in 2009 and 2010, I could witness how he tried not to be recognized.

[426] I heard that migrants at times deal with demands for bringing gifts by buying them in Douala instead – also because some products are cheaper, not to exceed luggage weight, and to avoid problems at the customs.

addressed question: "You bring me wetti from white man's kontri?"[427] (What have you brought for me?). Raymond, an elderly migrant, who lived in the US with his wife and children for five years, came to Cameroon for holidays for the first time since his departure. He stated that he had got used to demands while abroad, but when he came back and was recognized by so many people, he was overwhelmed with expectations that he would solve financial problems and everybody was seeking advice from him, "It was a difficult encounter" (Raymond. Field notes 2009/Migration-Stories). Or Alice said, that the first time she returned to Cameroon, two years after her departure, with her Swiss husband as a newly married couple "I realized that from now on it would not be easy. I went to the market, and prices had risen for me. I realized that everybody was thinking: ah, she brought a white man, now everything will be fine for us" (Alice. Interviews Diaspora). In particular migrants who have returned in order to stay, are forced to find ways how to deal with such expectations and projections posed on them. René, who had temporarily returned to Cameroon after his application for asylum in Switzerland was denied, stated that he tried to behave as inconspicuously as possible, but he discovered out that somebody who had been abroad always attracted attention and expectations. Even decent behaviour could lead to misunderstandings and entail certain evaluations: "When visiting Cameroon, people you visit do not know that you have saved money for these six weeks for two years. (...) Now that I am here... when I invite them for a drink, they could think that I am rich, ... but also they expect me to invite them. All what I can do is trying to keep a balance in everything what I am doing" (René. Field notes 2009/Migration-Stories).

Many migrants also stated that they were aware that people mocked them. Migrants are often not open towards a wider circle of people, regarding the purpose and under what conditions they have returned. They thus quickly become a target of ridiculing narratives[428], concerning

---

[427] The question: "You bring me wetti from white man's kontri?" is notorious and also serves as mocking statement, referring to people's strong desires for consumer and imported goods, causing a burden on migrants, and disappointment for themselves, since it is never possible to consider everybody.

[428] Narratives about why they have returned, whether they "have been repatriated" – a common notion - because they have possibly committed a crime abroad, or about what they intend to do in Cameroon. Additionally they are judged according to what

migrant's possible failures, or their undefined status they have as returnees, who have not yet taken a position. Raymond, who had returned for holidays, stated: "They are all talking behind my back. This guy, what has he been doing, what is he doing here now, what has he brought for us" (Raymond. Field notes 2009/Migration-Stories). It can thus also involve migrants' re-evaluation of their social ties. Whereas some migrants tried to get along with their previous friends as much as possible, others stated that they had to choose their "friends" anew, according to their attitude towards them as a bushfaller, upon their return. Frederik, a young returned migrant, who after a few years in Holland wanted to set up his business in Bamenda, said that the relation to some acquaintances had changed: "There are some guys I will not meet now, since they are constantly asking me favours which I do not want to fulfil. I do not want to be considered as a "wallet on legs". But I have plenty friends who just take me as the one I am" (Frederik. Field notes 2010/11. 03.01.2011). Here social distance arises when people treat migrants differently from previously, which is at times a painful experience. Another migrant, who had returned, described this social distance as a position imposed on migrants in order to render them into the responsibility to care for others: „My friends are people like me, we go out and share the bills. Friends are people I do not pay for because I am a bushfaller. I hate this concept of bushfaller. My good friends are not people who like that I dominate them. They have the same mind-set like me" (Isidore. Interviews 2010/11/EC).

However, whereas some migrants apparently try to behave as inconspicuous as possible, others seem to enjoy a certain attention given to them. Especially those who just come for a brief visit around the Christmas period[429], seem to belong to this category. This impression does not only derive from wide spread narratives in Bamenda, but also from my own encounters in the streets, at events, in nightclubs, bars and other "leisure" social spaces[430].

---

they have already "done" in Cameroon during their absence. Compare later in this chapter.

[429] Compare to chapter 3.

[430] Cars are a very common means to state economic viability and status. I have met bushfallers with conspicuous sport cars, hummers, jeeps, and other cars, which attract a lot of attention in Bamenda, when cruising through the main streets. Also in night-clubs, bushfallers buy rounds of whiskey for their acquaintances, and so on.

Keeping a foothold in one's place of origin – negotiating "virtual presence" at home

The facilitated connections through New Media of communication open many doors and offer opportunities for migrants, in order to keep themselves up to date with what is happening back home, performing activities, running businesses back in Cameroon and maintaining a certain presence back home. Not least, it enhances mobility, offers opportunities to "reach out and back", and it enhances flows of goods and finances, to be easily transferred from here to there. For migrants, claiming their belonging back home involves a long-term maintenance of their virtual presence among their social group. Maintaining good relations with their kin and family is, apart from the emotional bonds and moral duties, also an issue of being authorized and making a valid claim to their rights. The continuous negotiating of expectations from relatives and friends by migrants is also a negotiating of their own authority and status in their society of origin. First, migrants need to establish a material base for it, investing in a house in the village, planting crops, or in business: it is about rendering proof that they are viable enough to play a role as a distributor of wealth in their family and community. Secondly, it is about creating a feeling of their presence among the local community and also being able to exert control despite their physical absence. And thirdly, it is related to maintaining a good reputation and moral legitimization of their activities on a local level.

Investments in one's home society and struggles for power

Migration has always played a role regarding transformations of social and political influence, domestic migration towards towns and cities within Cameroon as much as transnational migration. Regarding the Bamiliké, Den Ouden (1987:19) states, the influence of the urbanites in their villages of origin increased in the eve of colonial rule and early independence, due to their new ways of achieving success in towns, but also when the chiefdoms were losing influence and independence towards centralized state authorities. Traditional power in the chiefdom decreased in influence, as the "nouveau riches" in towns were involving in their villages of origin more strongly, in terms of money, investment and claiming of land, and through such influence, also entering traditional hierarchies. Geschiere states (1997:101) that the real authorities in the villages were now the "absentees", the civil servants dwelling in the city. Gugler's (compare 2002) study of the relations between people in Ibo villages in Nigeria and their young descendants in cities in the 1960s and a restudy in the 80's, has shown that despite more offspring being born in the cities, involvement in the villages had even increased (compare to Trager 1998)[431]. These roles are also strongly taken over by migrants, especially since the possibilities to "make it" within Cameroon, have decreased.

Regarding the interviewed migrants, who were involved in diverse transnational activities, it depended on their situation in Switzerland[432], and also their legal status and mobility, as to whether they were financially viable enough to become involved into activities of supporting others, in a business, or other ventures. Generally, older migrants, from their forties and migrants who were in a cleared situation of residence and working allowance, were likely to be involved more strongly in activities back in Cameroon. In this sense, migrant's involvement back home seems to depend also on their age and status, personal history and relationships, social pressure, "dreams of return", interest and gender. It seems that expectations regarding sending back remittances are applied to both gender, due to a general high evaluation of their social position of being abroad. Nevertheless, male migrants always emphasized their responsibility to serve their families in the long run, in the sense of setting up "something durable" in their home societies, investing in a business, or building a house. As one interviewee – a non-migrant - stated: "(As a migrant) you have to make known to the family that this is your own progress, (something) that people can see and appreciate. You need something in material form. Some kind of security so that relatives go there and survive (…). What is money, sending money, (the relatives will) just eat it, and forget about it" (Valeria. Interviews 2010/11). Thus, in particular valid for male

---

[431] I was not able to consider hometown or village organizations, or what role remittances play on a local level.

[432] However, it depends also on the burden in their working life: I spoke to a few migrants who were financially well established, and were regularly supporting their families back in Cameroon, but were not engaging in any business and had hardly time to visit Cameroon regularly.

migrants, the underlying idea is not only to fulfil duties, but also to profit from confined rights in their home society, in terms of taking over "a responsible position", which is also related to respect and status in their community of origin.

There are however also the migrants - male and female – who try to avoid too much involvement: These migrants do not invest so much in order to acquire a certain position in their home society, either due to the lack of means – often correlated with age – and/or due to a stronger focus on their families and investments in their country of residence. They are involved on the one hand in a fragile balancing of meeting the most pressing obligations, by sending monetary remittances to a smaller circle of people, but they disclaim exerting control and influence, which leaves them in the position that they have actually no say in their community of origin.

Keeping one's position through a material base and activities, and the fulfilment of a good life

Many migrants seem to have rediscovered their roots, their base of belonging, their villages, and give these places of origin a high evaluation. Activities in their home societies give migrants the opportunity to "keep the position" and stay virtually present in their society of origin while being abroad. These activities are to be seen as a continuous work in order to maintain validity of being a member of the community despite one's absence (compare Geschiere & Gugler 1998: 310,311). Migrant's maintaining of relationships and keeping foothold at home is evaluated in different ways. It relates to strong bonds of solidarity and obligations towards their families, as well as to investments and validation of status. It also relates to the desire and the need to install something durable, a business, which could generate income, in order to satisfy the community's constant demands, which are often not easily followed up by migrants, as well as investing in their future. Notions of "giving something back to society and community" are in this sense rather practical and not just philanthropic or nostalgic reflections (compare Frei & Tazanu 2011).

As I have mentioned before, migrants' mere support of individuals cannot solidify a basis for durable influence and power in their home communities, but rather, in order to state their presence on a local level, this can be done through material investments and taking part in local activities of income generation, accumulation and redistribution (compare Tazanu 2012:230ff). Many migrants use land as a means to state their presence, and to claim their rights to land in their villages of origin. Three migrants said that they owned property – one or several houses in their villages of origin on their family's land, houses – which were used by either the family of the migrant or rented out. Another three migrants talked about their plans to build a house on their land, in order to claim use of it. Some migrants were farming their land and organized their family's labour on their behalf, in order not to lose these rights, which needed to be validated by the land's use[433]. 50 year old Alexandre said that he had built a house, claimed a piece of land and went to Cameroon in order to see how, and for what he could best cultivate the land. People in the village did not take him seriously and even ridiculed him. However, with the help of some friends and elders who advised him, his plantations ran well today and turned a profit, which benefitted his kin in the village. Some young men said they had a business venture in growing and selling cash crops in their absence. Some migrants stated that they wanted to make a contribution to the development of their communities back home, be it through regularly supporting relatives and friends, but also by sponsoring chosen ventures, rather than individual persons, in order to enhance sustainability. Furthermore, the satisfaction they had in such ventures, especially when they also played a role themselves, was mentioned. A migrant said that he contributed when they built a new school in his village of origin, another had founded an NGO related to giving care to orphans, and others play a role in an NGO from afar[434]. Other migrants – most often in addition to land use – were involved in business activities in Cameroon and had their business enterprises running back in

---

[433] According to Kaberry (1952:41,45), in rural areas, where much land is still clan land, formal property belongs to the fon, or the head of the village as his representative. He is distributing land and assigns the right for usage and profit from its usufruct. Only when land is cultivated, can use-rights be claimed upon by people.

[434] Of course they are at times also "forced" into such activities, since founding of NGO's is wide spread, and having a "viable member from abroad" is always desirable. Since facilitating measures to found NGO's have been taken by the government, founding an NGO seems to be looked at as an opportunity to obtain access to external funding.

Cameroon[435], investing a lot of time in their ventures and their work of coordination from afar, including – as valid for a few of them – regular travel to Cameroon. Others rendered assistance for business, as Aboubakar, who stated that he was collecting old computers for his brother who wanted to build up a cyber café, or another migrant regularly collected loads of second hand clothes in order to ship them to Cameroon. We should not forget that migrants depend deeply on the help rendered by non-migrants, relatives and friends in Cameroon, to effect business ventures. As for example in the case of Francis, one of my interviewees, who travelled to Douala to receive a car loaded with cartons of wine bottles, sent by his brother in Belgium. He had the task of selling the wine in Douala, then he drove the car to Bamenda to sell it to a specific person, and sent the earnings to his brother[436].

Being involved in income generating activities in Cameroon can contribute to the migrant's income in their country of residence, supporting their own livelihood, and in order to benefit and satisfy their community. Many migrants, but also local business people in Cameroon expressed that if they had the necessary financial capital, influence and contacts, it was easy to set up a business and generate money in Cameroon. Regarding business ventures, migrants also believe in profiting from the status attributed to them, as Paul uttered, who was frequently travelling between Switzerland and Cameroon: "It makes it easier to build up a good network. People collaborate easily, they look at you differently, (…) they look up to you, because you have been abroad" (Paul. Interviews Diaspora. 23.03.10). Two (male) migrants in their 40's expressed their desire to enter politics in Cameroon, as their contribution for "the development of the country", but also as an investment in status work and entrance to business opportunities: "Now I am coming to an age when it is meaningful to enter politics, to use my influence and status. I also hope to profit for my business, it is a good way for gaining attention" (Paul. Interviews Diaspora. 23.03.10). For

---

[435] See figure F68, an example of a migrant's business in Bamenda. Especially the internet field is an interesting option.

[436] Only few individuals mentioned that they profited from investments of migrant relatives, probably because it was not directly benefiting them, but rather the wider family. Often profits from migrant's businesses are pooled into family duties, such as the support of the parents, which then however, also relieves the siblings of certain duties for example.

male migrants, being a member of local men's societies is widely the case, sometimes induced by the community itself, since having a viable member is seen as positive for the association[437].

F68: Example of a business initiated by migrants
F69: Money transfer institute, Bamenda

In this sense, some migrants invest effort in maintaining their virtual presence on site: it can mean that a migrant could be inclined to slip in a more responsible and powerful position through the support and opportunities he or she is providing to members of the community in Cameroon. In order to be able to set up a base for such achievement, to create and maintain a foothold in their community of origin, migration is seen as crucial – sometimes inevitable, but successfully returning to Cameroon is seen by many migrants as the final stage and fruition of their struggle. As I have already mentioned in this chapter, many migrants see the ultimate goal in returning to Cameroon, in order to be fully able to profit from their efforts performed in both sites, work in their country of residence, and their investments at home, to profit from their material investments, efforts in business and social networking, as Paul said: "I have seen in Switzerland, that you work hard but you never become rich. When I have saved enough money, when returning to Cameroon, I can lead the life of a rich and highly regarded man" (Paul. Interviews Diaspora. 22.04.09).

---

[437] Origin and belonging are important also in relation to politics and regarding influence and an economic base, and therefore, gaining access to traditional hierarchies and prestige. A title in the traditional hierarchy, or other local institutions, can be used by migrants as a springboard to achieve access to state power, or claiming titles and rights in their societies of origin - Political power, associations and positions are still largely dominated by men in Cameroon.

A sense of control and maintaining migrant's credibility

However, maintaining their presence while being physically absent also poses many difficulties for migrants. When they intend to play a role in their society of origin, they depend on the help of locals, and on strong social ties. Migrants' authority in attempts to exert influence on site depends on the frequency of contacts, intensity of engagements, scope of support and social integration in their society of origin. Investing and redistributing wealth on a local level might result in achieving status and power in return for investments, but only when able to exert a certain control over such processes. Migrants engaged in investments, business and ventures in their communities of origin, have to cope with the problem of being physically absent, trying to compensate this drawback through continuous exchange by New Media, and also regular travel to Cameroon.

Even if it is considered a moral duty to support one's home community, which is gladly fulfilled if possible, most migrants stated that it was important to them, that the money they transferred was used for the purpose brought to them and considered valid (compare Tazanu 2012:198). They stated that they only wanted to provide for "important things"[438]. Financial support from migrants is often demanded for issues such as sickness, renovating houses, education, or social events[439]. From a migrant's side, it was emphasized that they solved the immediate urgent problems of their families. However, migrants also complained about the notion of non-migrants, that support must always be effected swiftly, since their problems are also most often urgent (compare Tazanu 2012:223,238). Some migrants emphasized that they did not ask for any proof for investments, in order to acknowledge trust towards people in question. Or they claimed that they would be informed about what was going on in the family, in the sense that they would know, for example if the child for whom they sent school fees, would go to school for the term. Nevertheless others mentioned that they asked the beneficiaries to render accounts – also depending on the kind of investment and the amount of money concerned. Some mentioned that

---

[438] I have related more elaborately to financial, material and other forms of support in chapter 4.

[439] It could then be sent instantly through, for example, Western Union, see figure F69.

they expected to receive pictures of the "investment", for example when a house was renovated or a business started, as a kind of proof and update, or they reassured themselves on the occasion of their next visit.

The narratives of migrants being cheated in practices of investments are numerous. Such narratives are often related to in a mocking sense by non-migrants: for example such that bushfallers have no idea about business practices in Cameroon because they have been "too long abroad". There are indeed many examples of business ventures by migrants or using migrants' capital, which have failed due to a lack of experience, collaboration and lack of ability to anticipate and react to difficulties. However, among close relations also, issues of misusing financial means that have been sent by the migrants, create much disenchantment[440]. As one migrant stated: "You send them money for something important, such as sickness in the family or money for school fees, and then you hear through somebody else, that a few days after the transaction they had a new TV set in their living room" (Alice. Interviews Diaspora). In such cases migrants feel that some people try to exploit them by all means. In mocking narratives in the Cameroonian context, migrants are often described as "stupid" and "foolish". Thereby it is uttered, that migrants, who would just spend their money on any venture or relatives without making sure whether the amounts are used accordingly, would contribute negatively to the reputation of migrants in general, in the sense that non-migrants could conclude, that migrants seem to dispose over abundant wealth since they "do not care" what is happening with their investments. Thus, migrants evaluate their own comportment as decisive regarding their reputation in their communities of origin in this regard. Not letting things slip, but trying to keep control is therefore a precautionary measure with wider implications. As an interviewee expressed: "I do not want to be seen as a fool. It is important not to lose credibility in the eyes of the people in your community" (Louis. Interviews Diaspora. 30.03.09).

---

[440] It is also important which means of communication is used, such as conveying a code for Western Union for a financial transaction for example. It is seen as less adequate to convey the code orally, because writing it down on paper involves an additional risk for the intended recepient. It is necessary for migrants to first call the recipient of the transaction in order to verify that the phone is secured with him, in order to secure that they would exclusively receive the message in question.

Migrants, in cases when they have been seriously cheated by those in whom they had been investing, have to deal with the difficulty of reacting accordingly. A solution to effect consequences is to possibly stop flows of support, since it is seen as not easy or impossible to withdraw from close social ties, even if the people concerned have misused migrant's trust. Transnational flows of support and investments can therefore not only have positive effects, but are also a source for arising mistrust, tensions and disconcertion even among close family ties and communities.

Exerting control on the base of being informed, and evaluating the trustworthiness of social ties

Being well informed is the main condition of not being duped and not losing credibility, since good information also provides a sound basis from which to validate situations and take decisions. Staying informed is mostly effected – apart from reading online local newspapers – by gathering different views of different people, as people in the migrant's community of origin have differing interests and therefore tend to represent biased perspectives. Many migrants said that they were in contact with different people from their community, family but also beyond, in order to have different views and sources of information. In order to exert influence, some migrants rely on an intermediary or a representative on site, who acts on their behalf. The characteristics of a trustworthy intermediary are described as someone "who is not very much involved with family issues", someone who is located within the local community, but does not have a central position, and not too many personal interests are involved, to obtain a more "neutral" appraisal. This could be a more distant relative or a long-term friend for example, who seems to be trustworthy. As Louis expressed, a migrant who had been in Switzerland for many years: "My intermediary is my cousin from the maternal side. She is outside of the quarrels among the powerful, she is a kind of neutral, and she is a female. People do not even know that I am in frequent interaction with her, and if they knew, not everybody would like it" (Louis. Interviews Diaspora. 30.03.09).

Such precautionary measures regarding an appraisal of the trustworthiness of social ties, are also seen as necessary where business is concerned. It is a wide spread idea that it is difficult to do business with, or give a job to a relative, because control over the person is limited. Relatives cannot easily be dismissed from the job, since they are family members, and they would sometimes not whole-heartedly engage in their task. Thus, by choosing less personally "freighted" relationships, migrants imply a strategy in order to avoid direct obligations and expectations, which could hamper business or other endeavours. Often in Cameroon, as it is stated, members of the family would not understand migrant's decisions in this regard – from their part, it would be somebody from the family, in whom they trust and eventually invest. Thus, regarding contacts and investments or supporting certain individuals, the basis of decision making for migrants sometimes relates to criteria other than kinship and hierarchies. This can be an issue, which raises further discontentment between migrants and non-migrants.

Evaluating situations, and balancing interests and activities

Migrants emphasize the power of their statements, support giving and involvement in their communities of origin, due to the high attention given to their actions. Some migrants – above all older and male – stated that they exerted a position on the evaluation of their respective age, added to the social position and status value as a provider. For these reasons, migrants said that they were often asked for counselling and to be in a position to give advice, they needed to be informed what was going on. "I want to be in a position to say something useful. People expect that from me. In order to do so, I need to be up to date" (Alexandre. Interviews Diaspora. 26.06.07). Some migrants said they assisted regarding business, migration ventures, and family issues.

Apart from such possibly positive influences migrants can have on site, migrants' activities and their investments, particularly in business ventures and people, based on their appraisals of situations, can also lead to strife and disconcertment in the community. These tensions can be related to the migrants themselves, their actions and involvements, and can be expressed by members of the family, but also they can concern feelings of other members of the migrant's community, who see themselves as being left out, and not profiting from the migrant's investments[441]. When money – or investments of any kind – is involved, it is likely to raise attention, leading to a struggle of power between people and groups. Another issue, which some migrants stated they needed to be careful in their actions, was their involvement in family affairs, which are difficult to evaluate from afar. To involve migrants in local disputes is also a means of forcing them to take position, which can in turn then be used against them from the perspective of other members of the community. An interviewee described that he would not want to be involved too much into local quarrels: sometimes people beseeched him to take sides, which he tried to avoid. "I realized that I have a lot of power in the village, and I can influence the ambiance to a high extent, regarding my own activities there, so I always try to consider the outcome of my activities and try not to fleece certain people." (Maurice. Interviews Diaspora). However, as it was stated, it was also a matter of being self-assured regarding one's own actions, despite the objections of certain people. As Alice expressed: "It is impossible to always please them, there are always some who are not content with what you do and what you do not do, according to them…" (Alice. Interviews Diaspora).

In this sense migrants relate to the responsibility they have concerning peace and harmony in their families and communities. It is thus about keeping a sense of moral legitimization of their actions in the setting, by carefully evaluating their activities. Many migrants relate to discreteness regarding activities in their communities, in order to not provoke malevolence among people and towards themselves. It is related to as a negative example that those who celebrate themselves and openly declare whom they support and whom not, contribute to tensions and disputes.

---

[441] Some migrants also expressed their uneasiness about an extended circle – of mostly relatives – who would claim from their close family members, pressurizing them, or mocking them, questioning the visible investments of their migrant relative, and so on, and also because they lacked the opportunity to reach out to the migrants directly.

Negotiating social and moral positions and status among migrants and non-migrants

In the view of profiting from each other, the local community from investments and the migrant from opportunities to enhance his or her status and power in the community, both parties have an interest in maintaining relationships. In this sense, the rural base of communities is exposed to such struggles for influence (compare Geschiere & Gugler 1998:114,310). I have related to the need of migrants to exert control over support giving and investments on site, and the power they could set up relating to their status of being a provider in their societies of origin. However, non-migrants have an interest to integrate migrants into social systems of duties and obligations. From their perspective, migrants can also easily slip out of control. I have related to complaints about migrant's disconnecting, non-responsiveness, non-availability, and non-supportiveness[442]. In view of the sometimes ambiguous relations between migrants and non-migrants, I will relate to trust and mistrust, levelling practices and jealousy, ranging around discourses on moral legitimacies of power[443].

Trust and mistrust – enhancing and obstructing social collaboration

As I have related in the previous chapter, in interpersonal mediated communication, trust is seen as a favourable, advantageous and even indispensable capital. Notions of trust – and mistrust – come to the fore here, when related to social cohesion and common basis for exchange and collaboration, on a societal level. As Miztal expresses, the motivation to trust is either a result of a strong positive personal bond for the object of trust, or a "result of our belief that we have "good rational reasons" to trust" (Miztal 1996:21). In social interaction the fulfilment of others expectations is based on a shared knowledge of what seems to be appropriate in a certain situation and in certain social relationships, by taking into account the consequences of exertion or omittance of the same (Miztal 1996:24, Luhmann 1973:9). Placing trust – in a person and/or a situation – is a reflexive process. In daily social

---

[442] Compare above all to chapter 4 and 5.

[443] Compare to discourses on moral legitimacies of success of bushfallers and scammers, chapter 3.

336

interactions related to trust, people depend on past experiences related to other people and situations, which influence what Miztal calls "mood of collaboration" in the sense of the willingness to take risks to trust. A society's ability to secure trust among social actors and to enhance a good climate for collaboration can be valued as favourable for common action and development, as collective social capital (Miztal 1996:142,143). According to Luhmann, mistrust can serve as a functional equivalent[444] to trust (Miztal 1996:74, Luhmann 1979:75, Hjarvard 2002:78), in this sense contributing to inhibition and impeding social interaction and collaboration. Neither trust nor mistrust are universal, individuals tend to adopt one or the other strategy in certain contexts. As Miztal (1996:188) relates of Poland during the communist rule, the greater the degree of the informal sector, economic insecurity, bribery, and corruption, the greater the negative impact on social solidarity. A low level of economic and social security is here connected with low levels of social and personal trust (Miztal 1996:197,198). In this sense, a certain level of social trust could enhance personal trust, since the interpersonal relations are less fraught with expectations. In the case of Bamenda, or in general Cameroon, I would argue that mistrust is very prevalent in society[445], towards formal institutions, but also in interpersonal relationships, even though it could be considered a society relying strongly on personal trust rather than on opportunities for the building of institutional and social trust. In an environment where interpersonal social ties are directly related to social capital as opportunities, it is possible that trust becomes a scarce good. Thereby, related to a high competition for scarce resources (compare Fuh 2010, Geschiere 1997), a feeling of mistrust is not only prevalent vis-à-vis the elderly, high ranking and powerful, but also vis-à-vis migrants, and generally fellow youth who seem to be more privileged.

---

[444] The contrary of trust would be an "existential angst" or anxiety of existence (Giddens 1990:100, Miztal 1996:92).

[445] Undoubtedly, trust or mistrust is also a highly individual issue. Some people told me that their "motto" in life would be trying to trust people because they also thought that "everybody deserves a chance to be trusted", and that relating to prejudices would not be correct, as the level of gossip was high in the local sphere.

The notion of "eating" as expression for mutual accusations and moral legitimacies

Social disconcertion between migrants and non-migrants refers to struggles of power related to social and moral legitimacies. Thereby, accusations can come from both sides. From the perception of those who remain in Cameroon, migrants or bushfallers would be inclined to employ a sense of superiority: "Distance is a first factor, the other aspect is the superiority aspect. Unconsciously the person abroad feels superior. There is a hierarchy in it…" (Kenneth. Interviews 2010/11/EC). As an interviewee said about her cousin who had come back to Cameroon: "The relationship has changed, before it was like, you go there, sit down, laugh, create fun, now I just greet her, and that's it, we do not have to discuss something. When they leave Cameroon, they have their own type of people to meet and interact with, when they come here they only move in their class of people" (Lizette. Interviews 2010/11). It leads to the accusation towards migrants of deliberately demonstrating their superiority. In this sense, negative local narratives about the bushfallers seem to arise from their perceived distanciating. From the perspectives of migrants, non-migrants are trying to deprive them from status by their attempts to exploit them economically. As expressed in the saying: "Kontri fashion go kill you", migrants are exploited on the basis of claims, which are "naturally" imposed on them, because they are or have been abroad.

These differentiating perspectives can be related to mutual accusations to be "eaters". The notion of "eating", "to chop " in Pidgin English, is prevalent in colloquial expression in a wider African context (compare Bayart 1993, Mbembe 2001, Jua 2003, Argenti 2007, Geschiere 1997, Hansen 2003). In the sense of "eating money", the notion is related to practices of corruption and deceit, wealth creation and accumulation in general[446], alluding thus to the ambiguity of the notion of success. "Eating" happens always to the disadvantage of someone else: if someone has much of something, be it children, wealth, success in business or a high rank, somebody else's possessions would not be satisfactory (Rowlands & Warnier 1988: 123, Eyoh 1998:341). "Eating" could in a context where power and accumulation takes on

---

[446] Including politicians or traditional rulers, illicit business people, and also migrants.

disconcerting forms, easily be related to occult forces and witchcraft. In view of an opaque means of accumulating, the commitments to moral obligations of redistributing wealth also, would be not secured anymore: migrants can be seen as corrupted by a decadent morality of "the West" and self-centred tendencies. However, "eating" can also stand for notions of solidarity, literally, people in powerful positions offering food to their subordinates on the occasion of festivals and celebrations. As Hansen (2003:219) states "(…) eating together could be understood as a metaphor for consuming and sharing commodities other than food; such as money and power". (…) Understood in this way, the eating metaphor reflects closeness or distance between people" (Hansen 2003:219), and it has to do with claiming relationship, and keeping social control over the ones who access realms of opportunities. Whether migrants seem to slip out of control, or they are successfully integrated in networks of redistribution, the notion of "eating" can be used as reproach and accusation from both perspectives. Either the migrants are eating, leading a pleasant life while forgetting their family back home, attracting jealousy from the less privileged, or migrants feel that their relatives are eating at their expense as profiteurs from remittances (Page 2007:236). In this sense the blame for the emphasis on materialistic terms, is mutually related to by migrants and non-migrants.

"Levelling" practices, jealousy and subordinating migrants

Migrants are in this sense judged according to their investments and activities back home, and not always in a positive sense. Compliance with social norms often comes into conflict with notions of individual success, as we have already seen in the previous chapters, where I have related to notions of a "good life", and it's different dimensions. Migrants are always prone to being negatively judged by people in the local context, depending if and to what extent they are likely to be integrated in the social network of profits through them. In the background of what I call "mocking success"[447] in this setting, are the allusions to potential success and opportunities. Migrants can be accused of attracting attention by conspicuous consumption, also as a sign that they might circumvent notions of redistribution and solidarity. Migrants'

---

[447] Following the notion as it is used by Tazanu (2012a:114).

demonstrating of success in material investments and local influence can invoke feelings of aversion and jealousy among other members of society. At the same time, however, it is expected from the part of non-migrants, that migrants adhere to certain images of success. One issue is the general suspicion of failure of migrants, when they do not show the signs of success: As in the example of Fred, a young migrant, who had returned for a couple of months from the UK to build up a poultry business. Those who did not know him closely said that he had "returned empty": he had apparently no job, no fancy car, nor had he brought a wife from abroad.

All these negative narratives would lead to a subsequent devaluation of the migrant's status (compare Tazanu 2012). In this sense, mocking success serves to equalize social positions. Moral claims serve to level out differences in wealth, as strategies in order to "cut the other person down", in an extreme case through killing or harming the person who is seen as more successful (compare Geschiere 1997, Bayart 1993). "Some people try to destroy the success of others, in order that the successful come down to the level of the non-successful, except if people can profit from the successful person directly" (Edwin. Field notes 2009/ Migration-Stories). Levelling strategies can take on different forms, and are often related to occult forces and witchcraft. As Geschiere points out, kinship ties are under siege, as they have to bridge increasing inequalities and social and spatial distance (Geschiere 1997:213, Geschiere & Fisiy 1994:326). Witchcraft is seen in the first instance as arising "from the intimacy of the family and the home" (Geschiere 1997:10,11). The intimacy of family relations bears a hidden danger, in the sense that the social ills and moral illegitimacies start from the very intimate realm of the family (Geschiere & Fisiy 1994:326). One cannot escape neither from family members, nor from dangers of sorcery within these realms of intimacy. This could also be transferred to the relations of the migrants and non-migrants. Physical closeness, which is attributed to the danger of being afflicted by witchcraft through co-present kin, can be substituted by medial contact. In this sense, practices of Nyongo are also called "the long arm of the family" (Nyamnjoh 2005:246): practices of control towards the ones entering the affluent realms of West, in order to keep them at disposal of their family and kin. Related to new communication technologies, I would argue, that particularly because these media allow enhancing and maintaining of social relationships mediated over distance, they contribute their part to social ties strongly influenced by tensions, accusations, and new degrees of jealousy.

Bushfallers are constantly in danger of being afflicted by spiritual attacks, such as witchcraft or other destructive actions through jealousy, before they leave, during their presence for holidays for example, and even while abroad[448]. Migration ventures are therefore sensitive issues handled with great secrecy. I have come across many stories of people leaving and only telling their closest family members that they intend to leave. Ivo expressed his disappointment: "A friend left for the US, a very close friend. I had not seen her for days, then I met her in chat. She was telling me that she had gone to the US, she arrived two days ago, I was the first to know… I was very disappointed, why she had not told me. It was aching me, because I thought that we were close friends who trust each other" (Ivo. Interviews 2010/11). Even apparently close ones could not be considered above suspicion. Edwin, who was about to join his wife in Norway, said that he immediately felt alarmed when some people with whom he had not much to do with, invited him for dinner to their house before his journey. He was then torn between the intention of being polite and accepting their invitation, and the fear of being harmed or poisoned. Several other people, not only migrants, said that they would preferably not leave their drink alone[449] when socializing – even among friends. The "poisoning" of family members due to competition for resources is a wide spread issue, and sometimes goes beyond families, to non-kin, such as jealous neighbours, and so on. "There are stories of people being killed, poisoned and rendered ill, in this sense, also people I know" (Edwin. Field notes 2009/Migration-Stories). In the case of a girl who was studying in Europe and returned

---

[448] This jealousy is not only turned towards migrants themselves, but also towards the entire group of people who profit from bushfaller's attention (compare Tazanu 2012:238ff). Not only where migration is concerned, but related to scarce resources in general, the notion of jealousy is prevalent, it could be generally related to somebody's success, in business or education, or somebody's assets of land, house, and so on. As an interviewee expressed: "It seems that jealousy has originated in Cameroon" (Titus. Interviews 2009/F).

[449] For example by either finishing their drink first, or taking the drink to the toilet when they need to go there. I have to admit that I found these statements "shocking", since it expresses a constant fear, even among friends.

for a holiday and died a sudden death, she is said to have been killed through witchcraft by her mother's co-wife, who was jealous that not her own children were studying abroad. Or a bushfaller's girlfriend hired a hit man to kill him, because he wanted to split up. Another friend was involved in finding out about the sudden death of his sister in Ireland, which he was convinced had been unnatural: According to his perspective the sister had been poisoned by a jealous co-migrant. Some migrants said that parents came up to them with issues such as finding a job or a wife for their sons and daughters. "There must be a recipe for success, they think. They want to know the recipe. If you cannot give one, they are disappointed and think that you hide it from them and that you want to keep your success for yourself only" (Clovis. Interviews 2009/R). Because they cannot fulfil the expectations of so many people, and have to turn expectations down, migrants fear people's jealousy turning into harmful acts against them. Some bushfallers prefer to seek spiritual support, arming themselves against malevolent spirits and the actions of others. Another issue is the fear of being physically threatened, robbed and beaten up (Tazanu 2012a:121). Raymond, who had come for holidays, said: "I do not believe in witchcraft, but I fear that some people could try to harm me or destroy my property" (Raymond. Field notes 2009/Migration-Stories). For this reason, many bushfallers, during their presence in Cameroon, try to take certain security measures, and it is another reason for migrants not announcing their return, staying in a hotel[450], and avoiding certain people.

Migrants and status - continuity of subordination or new powerful positions?

As I have described in this book, in local discourses, migration is seen as *the* way of social mobility, which includes the opportunity to achieve social recognition. However, when referring to young migrants, regarding their struggle in diaspora, it seems rather to be a continuity of the subordination of youth, than an emancipation and turning away from traditional notions of social solidarity (compare Jua 2003, Argenti 2007, Nyamnjoh 2002, Fokwang 2008).

---

[450] Status and a desire for privacy and liberty are also motivations for not staying in the family's compound.

This is so in two aspects, firstly, young migrants usually live under doubly precarious conditions abroad, being subordinated, and to an extent exploited, in the society of sojourn – going through the hardship of adapting and settling down, being left alone and suffering from non-acceptance of members of the society of sojourn, and doing low status and low paid work. Furthermore, they are burdened by demands and claims by their families and people of their society of origin, based on notions of servitude of youth towards their parents and extended family. The second aspect alludes to a continuity of notions of power linked to seniority and status, and also age – and even gender[451]: Migrants who "have made it", through work abroad and establishing themselves, having reached a certain level of economic viability – and also a certain age in their life, try to not only serve their families, but also to exert power on their investments on a local level. They rise in status as important providers, and they re-establish themselves in their society of origin by entering social systems of rank and social hierarchies within circuits of power, adopting a role of gate-keepers and big men vis-à-vis the masses of young aspiring migrants. The narratives about young migrants, who are apparently circumventing their duties, and the narratives about youth, who – by demonstration of prestige – give the impression that they do not stick to moral norms of social conduct and do not occupy the subordinated place foreseen for them in society, are continued also by youth themselves. This mocking of success is often related to notions of jealousy and a common suspicion towards those who have gained attention through their falling bush – be they present or absent on site. Thus, success achieved in white man's kontri is deeply ambiguous.

From another perspective, however, youth in fact deploy new social spaces in a sense of a social emancipation. We have seen that this can be employed in a local context, among youth groups in youth related spaces, often based on "reaching out" to powerful realms of potentiality, such as New Media, scamming, youth culture, and so on. In a transnational space it is a fact, that through migration ventures young people also can gain attention by assuming roles of gate-keepers, or at least viable contacts worth pursuing from the perspective of non-

---

[451] Establishing oneself in one's community of origin, can be considered as more important for male migrants.

migrants. As I have showed in the course of this chapter, migrants then try to adopt strategies in order to create spaces of self-determination by negotiating social ties and duties.

Conclusion: Between accord and disappointment – liveness through New Media as a means to effect claims and rights in social relations between migrants and non-migrants

As I have tried to show in the course of this chapter, being in contact through New Media enhances the opportunity for close and meaningful social relations and support, but also raises the chances that mistrust, tensions and disenchantment can arise. Migrants try to balance the notion of a "good life" in their actions, balancing between meeting social responsibilities, acquiring status and taking part in consumption, between accumulation of wealth and individual success or solidarity and redistribution. Thereby, norms and moralities of solidarity and sociality are transferred to the mediated realm, where they are super elevated in a transnational context: they come to the fore in mutually high expectations, because New Media seem to offer a high potentiality of maintaining contacts despite physical dislocation. However, sociality in New Media of communication must limit social and emotional cues[452], and bridging the gap between differentiating life-worlds and strong imaginaries often fails. This inevitably leads to disappointment and disagreement. In order to handle responsibilities, but also claims and demands going beyond their scope and acceptance, migrants adapt their practices of social networking, and their use of New Media of communication. However, New Media use is not only at the migrants' expense, in the sense that expectations of being reachable and available overwhelms them, but they can also be adapted to serve their interests, by controlling flows of communication and support according to unfolding conditions.

Migrants' attempts to maintain a "virtual presence" in the local context involve practices of media use, and also being present through representatives, through regular visits and material effects on a local level, could be subsumed in the notion of creating and maintaining liveness. It mainly relates to liveness as work and effort. It involves

---

[452] Compare to the previous chapter.

strategically canny ways of contacting and communicating through diverse media, as well as consulting regular news, in order to stay up to date, and also a conscious and careful evaluation and balancing of actions. This is closely connected to other dimensions of liveness, as the sensory experience of closeness in the sense of fostering good quality social relationships. From a migrant's perspective, liveness as a potential includes the dealing with high expectations and claims in reflective ways, in order to maintain positive social relations, but not to overburden oneself. It also relates to migrants' aspirations of being able to profit from their investments and involvement in their societies of origin by securing a position of status and power. Even though New Media of communication are viewed as enhancing a greater level of involvement, migrants also see their limitations, and that of mediated liveness: "You lose touch relatively quickly. If you really want to continue to play a role back home, apart from communication you need to go there regularly" (Louis. Interviews Diaspora. 30.03.09).

Thus, high expectations are prevalent on "both sides", among non-migrants and migrants alike, towards each other and towards the endeavour of migration. New Media of communication serve here to maintain and manage these expectations, which are transferred to a mediated realm of social interaction. At the same time, New Media contribute strongly to creating the high expectations towards each other and towards success and sharing of success, since they allow close contact, emotional closeness and mutual control. In this sense, liveness enhanced through New Media use, contribute to negotiations in social relationships, about moral conditions of social interaction and communication, which are symptomatic of social transformations.

7

Self-Reflection and Fieldwork Methods

This chapter is dedicated to fieldwork methods and self-reflection. Even though the researcher is always a part of the field, a fact which contributes to the "outcome" of research, regarding the topic addressed in this book, this is even more strongly the case: my research and my own being are interrelated in many ways. As an outsider, coming from "the West", I often experienced imageries and narratives about my own origin – "the world" I come from, and the kind of "lifestyle" I must lead there – very directly, mediated by my own physical presence. Thus, I was in the situation of being myself being "target" of link-up practices and expectations. The topic of research is also interrelated with my personal experiences of New Media use. This familiarity and the close relatedness of my own presence to my topic of research might have been decisive in the perspectives I have adopted in my research. Due to these interrelations, I have decided to integrate this chapter only in the end of the book, in order to allow the reader to connect the descriptions of the previous chapters with the issues I intend to address here. In this chapter, I will reflect on my position as a person and researcher in Bamenda. With this background, I want to relate to applied fieldwork methods, interviewing and participatory observation, the collaboration with my "research partner" Primus Tazanu, and what influences our perspectives had on research and findings.

Using New Media in my research – as a working tool and research topic

When in the field, New Media were my topic of investigation and my tools to organize my daily work. When in Switzerland, New Media served to maintain contact with people in Cameroon.

Linking up, coordination work, and providing contact coordinates in the field

As a tool for coordination for meetings and link-up, the mobile phone was my most important working tool, whereas the internet gained more importance when I was absent from the field.

When I was in Bamenda the first time in 2003, mobile phones were only about to being widely used or available - neither myself as a "latecomer" in adopting mobile phone technology, nor people I was interacting with in Cameroon had a mobile phone. I remember that it was more difficult to coordinate appointments, and meetings were less spontaneous. I would however not conclude that meetings were less likely to work out, but the handling of spatial and temporal coordinates was less flexible. When returning to Bamenda in 2008, mobile phones were already considered as common. Undoubtedly, the bulk of the coordinated meetings and the number of handled contacts had considerably increased with the mobile phone. Re-meeting people and following up their activities, as well as cultivating social ties had also become much easier. Likewise, the urge to maintain contacts regularly, by link-up calls for example, had risen. My expenditure for phone credit was higher than for accommodation, food, and domestic transport expenses combined[453]. While in the field, the mobile phone was a tool to coordinate instant sociality on a local level of presence availability, and the internet rather served instead to maintain transnational ties[454].

F70, 71: Using New Media in the field: local coordinating by mobile phone and transnational communication by internet

---

[453] See figure F70. Less often I coordinated meetings by internet. It depended however most often on coincidentally meeting a person online in chat, since email as non-simultaneous media was less feasible to coordinate meetings.

[454] See figure F71.

The internet and mobiles phone set the coordinates of being accessible, thus, I was always asked for these coordinates myself. Being very mobile, my phone number was my address. I adapted to habits of instant socializing, such as adding contacts by sending or receiving beeps in the presence of the person with whom I intended to exchange phone numbers. I also got used to – most often receiving – beeps, and accordingly interpreting them. For some people it was likewise important to have my email address, since they knew that my Cameroonian number was only a temporary one, or they wanted to have my foreign number, which I had to restrict and explain when not giving it out. In this sense, handing out my contact, phone number, email and chat ID was something, which was part of my daily experience, and which gave me the opportunity to "observe" people's networking practices directed towards myself[455].

Linking up through New Media during my absence

Regarding link-up with people in Cameroon during my absence from the field both media, internet and mobile phone, were relevant. With internet literate individuals, I was mostly in contact through email and chat. Communication with less internet literate people depended on the mobile phone. My Cameroonian friends influenced my everyday use of internet and mobile phone for transnational sociality. For example, I received many invitations to various friendship sites, which I usually followed, and I started – to a certain extent – to be active on these sites. Since many of my friends in Cameroon had Facebook accounts, I was also inclined to use Facebook more regularly. I stored and shared pictures on Facebook, which was motivated by my friends in Cameroon, who wanted to have access to my pictures[456]. Regarding chat – which I had hardly used before – somebody in Bamenda created an account on

---

[455] I gave out my contact coordinates freely – which was at times deplored by some friends of mine – where my temporary Cameroonian phone number was concerned.

[456] Furthermore, some of my face-to-face friends from Bamenda interlinked me with their friends on Facebook, also with a few migrants, including some of those in Switzerland. Some people from Bamenda also contacted me on Facebook especially when they saw that I already had a wide range of contacts in Cameroon. In one case I chatted a few times with a young man, who I then met in person when going to the field for my next stay. In this sense, online contacts could even turn into face-to-face contacts, and face-to-face friends and mere online contacts could intermingle.

yahoo messenger for me. My dealing with time for socializing on the net, apart from the usual checking and responding to emails – transformed in the sense that I started to take time to chat. I also listened to recommendations of Cameroonians in Switzerland regarding cheap calling offers for Cameroon, and tried out several of them, from different call cards, prepaid calls, Skype or Voip through the internet, or buying special SIM cards for this purpose, helping me to economize and control my communication expenditures.

I took mediated communication serious in the sense that I tried to answer emails and write to people with whom I had not communicated for a while. There were some people with whom I was not communicating often, even though I had spent much time with them in the field – this was mostly the case with those who were not particularly internet literate. Likewise, there were people with whom I had not too much contact while in the field, with whom communicative exchanges increased through New Media, especially the internet. With the exception of some people who were calling me regularly, in most cases I was expected to call. I often received beeps and short calls. I made regular calls especially to people who were not particularly internet literate. However, even though I tried my best I was susceptible to allegations from people who felt neglected. I had the impression - similar to how it might be in the case of migrants – that I was the one to blame when communication was not "smooth", and that I was held responsible to maintain regular communicative exchange. It was interesting to see how people who I had interviewed about their use of New Media were actually using these media to pursue our relationship over distance.

A favourable research topic – New Media's presence in the local setting

Researching mobile phones and the internet in Bamenda turned out to be very interesting. Almost every situation could serve to obtain information, or to learn about practices and narratives about these media. Especially the mobile phone as a personal and mobile device could arise as a topic and in action at every place and every situation in everyday life. Be it at an event, a marriage, a church visit, be it in an encounter and conversation with somebody, be it in a taxi or somewhere

350

in the street, a bar or restaurant: the phone could ring at any moment, or was deliberately taken out and presented openly, or "played around" with, and talked about. Regarding the internet, researching internet use was also fruitful in particular in Bamenda, with all the cyber cafés spread in town. Overall, issues related to my research topic were easily observable and to be encountered virtually everywhere, which made it in turn sometimes difficult to systematically and analytically examine these phenomena. A wide range of people could be my potential interview partners. The topic of research was by most people perceived as unproblematic, and they were usually ready to talk about it. Interestingly, it happened to me on a number of occasions, that somebody just came up to me, started a conversation, asking me – without knowing me and the purpose of my stay in Bamenda – if I could accompany him or her to the cyber café in order that I could indicate some sites and render some information how to go abroad. The research topic was in this sense directly interrelated with my interaction with people in the field.

Having multiple identities in the field

My physical presence – being white – but also my skin colour in combination with other factors, such as gender, age, and conduct, has influenced my fieldwork. It had an impact on my emphasis on the topic of New Media use coupled with mobility and imagination, and my appearance influenced people's ways of responding to me. Furthermore, mutual interpretations and projections influenced relationships over time when mediated through New Media during my absence from the field. The different aspects of my personality, which were attributed to me, were a complex aggregate of collective imaginations, but also depended on people's individual interpretations.

My presence in the field site: Being known and anonymous – but always noticed

Since Bamenda is a rather small town, but also an urban place, it offers a good mixture of being known, but also a certain degree of anonymity. Sticking out is a common experience for every ethnographer who comes from outside, whether you are recognized as an individual or not.

When passing in streets and areas, where I was often present, it was convenient to meet people on a daily basis at their fixed places, where they had their businesses[457]. At such places, I could spend time greeting people, observing interactions, and listening to narratives and conversations about the topic of my research. Such sites and related people were integrated into my "mental map" for Bamenda, as nodes of interaction and sometimes places where I also went to when, for example, an appointment had not worked out, or at times just to be there and have a rest, when I felt exhausted from rushing around. Such sites were also a number of cyber cafés. In some there was a high turnover of customers, where people were working. In this sense some cyber cafés were places where I could spend time – more or less – anonymously, and using this opportunity to check my mailbox and observing what was happening around me. In other cyber cafés I was known by the employees and by most customers. Thus, some of these cyber cafés were places for me to spend time, to relax, to chat, and to regularly meet certain people. In this sense, the urban setting of Bamenda offered both, moving more or less anonymously - however always noticed – and regularly meeting certain people at certain places.

Being measured with different yardsticks – being a "white man"

Directly related to my presence, my appearance and conduct, I could find out about people's images of "whites" in their statements and comportment towards me. "Whites" are perceived as complicated, difficult, and demanding, as far as their needs for comfort are concerned. "They only live in nice and expensive hotels and go to expensive restaurants", or: "They do not want to try African food, because they do not like it and they fear that they will get stomach problems", indicating an imagined "better environment" they are used to. In that sense, it is often attributed to "whites" that they are unable to adapt to African culture, related to their conduct, their "being distanciated", or that they "only move in groups", which is related to their being anxious, insecure, and only willing to interact among

---

[457] Which sometimes provoked me into taking a taxi or a motorbike – called Orgada – for short distances when I needed to pass quickly when in a hurry to reach another place.

themselves[458]. I saw no reason to take these sort of statements – about "lifestyle in the West" or about "whites" – too personally, but I rather took them as general statements. Since I was eating African food, used public transport, and was interacting with locals, some people referred to my comportment as differing from these images of "whites": "You are different from these other white man", or, that I was "adaptive"[459]. However, my comportment was always judged according to images of "whites". Being white - or coming from "the West" - served as the most important marker to categorize me as an outsider[460]. Asking for the nationality was used as a way to link up, by asking "Where are you from" – or: "For husai you kommot?" as one of the first question right after asking for my name. When I defined myself as "a Swiss", some general reactions were made such as: "Swiss is the most peaceful country in the world", or Switzerland was seen in connection with wealth and banks, where African rulers hide their money.

Very obviously, as a "white man", I was measured with different yardsticks – which also meant being seen as an outsider. It was at times peculiar, how some people – who did not know me personally – apparently viewed me awkwardly just due to my colour: also because it collided with my own subjective estimations of feeling "not strange at all" in Bamenda. I had the impression that being an outsider was a matter of perspective, degree, and of a specific situation. Being an outsider does probably not mean that one is not accepted. The people I interacted with often integrated me verbally, be it to express appreciation, to please me, or to stress relatedness: "You are one of us", "You are a real African", or: "You are a true Mankon"[461]. I also profited from being perceived as an outsider. I could refer to my status as an outsider in order to explain my position, to excuse mistakes, and to – at times – express my "right to be different". I could address issues and

---

[458] Compare to the imaginaries and imageries of and about "white man", in chapter 2.

[459] Such statements and compliments could possibly also be used strategically to create a "positive relation" with me.

[460] Some seemed to already include their judgement about my nationality. Sometimes, I was automatically seen as an American by some people, which had to do with the fact that many "whites" – and specifically young women – in Bamenda are related to "Peace Corps", an American organization in the development aid sector.

[461] It refers jokingly to my living in the part of the city being a territory of the Mankon ethnic group.

obtain interesting explanations, which were related to my status as an outsider, and therefore, ignorant. Being an outsider could also have influenced the fact that most people acted very supportively towards me, since I was not part of local networks of competitive relations.

Being a "white man woman"

My identity as a woman in the context of my field research has been an important issue. For example it is perceivable in the proportion of interviews, which I conducted with men, that it was easier for me to contact men, which is related to activities of young men of – often in direct and determined ways – getting into contact with women in general, and a white woman in particular. However, I was also contacted and addressed by young women, in direct ways also, but in general concerning women I had to rely more often on my own initiative to contact them. The fact that I was moving around alone, and that I was apparently in the field without a family, made it difficult for people to guess my marital status. I was often asked, after my name, where I stayed in Bamenda, my nationality and my purpose of being here, and if I was married. I was rather inclined to introduce myself as being married in first encounters with men, whereas I preferred to stick to the truth concerning those I knew better. The fact that I was in a long-term relationship, which from my perspective proved its seriousness, was not always taken at face value. However, I tried to be transparent regarding my situation: that I was talking to men in terms of my research, and that I was not on the lookout for a boyfriend or husband, often served as explanations, which were accepted. I experienced intrusive situations hardly, which went beyond what I considered acceptable. In only two cases I experienced conduct, which I could not accept, which lead me to cut contact.

My identity as a woman also differed in relation to social interaction with different men. Their reaction towards me – according to my impression – was sometimes balancing between seeing me as a "white man" or as "white man woman" initially, depending on the individual, and also – relatively often – shifting from the latter to the former with time. Depending on the age difference, in many cases, it was clear from the beginning, that I was the senior in the relationship: I was seen more as a white man, a sister or friend, at first, which included generosity from my side. However, in other cases I was treated as a white man woman at a first. In a few cases it involved using the opportunity of an interview appointment to make "a date" out of it, by for example inviting me to a "good restaurant" or drinks, according to the notion that men ought to pay for women in such situations. From my perspective, I profited most of being a woman where my security was concerned. I always had somebody who accompanied me home, brought me right to the entrance door in order to make sure that I was safe, or see me off into a taxi. It gave me the impression, and it was also at times stated, that some men wanted to prove that Cameroonians were generous and hospitable, and that they were gentlemen, in this sense referring to me as a woman and a white at the same time. In a way I also took advantage of being measured with different yardsticks, in the sense that I of course tried to stick to certain modes of comportment attributed to females, but sometimes I deliberately went beyond it. Going out in the evenings, apart from the relaxing aspect, was also interesting in order to spend time in sites of youth sociality. Thus, I heard a few what I would call rumours about me, which had an ambiguous touch, for example one friend – religious and strict –told me anxiously, that I had been seen in an irresponsible place – a bar – together with a group of scammers[462], drinking beer and smoking! However, by other people it was understood as "the nature" of my work: "It is your job to meet all kinds of people"[463].

Being a researcher

I was often asked about the purpose of my stay in Bamenda, and I explained what I was doing as simply as possible. I had the impression that some people did not understand that I was doing a PhD. They repeatedly related to my occupation as a student in "school", which might be explained by some people's lack of understanding of formal education, especially where higher education is concerned. Sometimes I had the impression that it also related to my identity as a woman, and how I interacted with people, which did not so much allude to my

---

[462] The designation of "scammer" was used for all seemingly immoral young men, who were seen as performing conspicuous consumptive practices, drinking, smoking, "making noise" – attracting attention – in a sense of attributing delinquency to them, which had nothing to do with their working actually in the scamming business.

[463] See figure F72, conversing with people in every situation.

position as a doctoral student[464]. I also purposefully tried not to emphasize a position of high status to minimize social distance[465].

In general I had the impression that many people were not so much interested in what I was exactly doing. It was known by most people that I was "doing research", though. Many did not even ask what kind of research I was involved with: it seemed to be a well-known notion that "whites" in Bamenda could probably be researchers[466]. Some people however expressed their confusion about the kind of work I was doing, by asking for example: "What are you really doing? I only see that you are socializing the whole day", and similar questions. Or they asked, whether I was "free" or "on duty". I then tried to explain that I could hardly separate work and non-work, and that this was the particular character of the work I was doing in Bamenda. Some people supposed, that I was doing "the real work" then over there – back in Switzerland – when I would be occupied with bookwork. Even though many people were unclear as to what kind of work I was doing, they always related to me as "being busy"[467]. "Being busy" served as an evaluation that I was working hard, and that I was serious in my work. It referred to my various activities going on: mainly related to being with many different people every day and to my "being mobile"[468]. It was indeed true that I was constantly rushing around on public transport, sometimes from one corner of the city to the other. In this sense, being mobile and busy influenced people's notion of my work.

A few times people took interest in my work[469], and deliberately wanted to know what I was doing in detail, and how I was going about

---

[464] In the Cameroonian context, being a PhD-student – a doctor to be - was seen as a position of high status, which – as I experienced a few times – also showed in how such people were addressing others, demonstrating their superiority.

[465] For example, I only handed out my business card a few times, usually to elderly, higher ranking people such as business people, owner of cyber cafés, and so on. I hardly handed out a card to one of my young interviewees.

[466] In a similar sense, the notion of "Social Anthropologists" was known by a number of people, referring to the fact that the Northwest Province has always been an area of intense cultural and social study.

[467] Which is also another imagery concerning the character of "whites".

[468] See figure F73. Being mobile sometimes also included travelling.

[469] As an example, and also out of reasons of transparency, I sent an article I was writing about scamming out to three people, two of them scammers themselves. However, I did not obtain much feedback on it, they just gave their "consent". Two people, who were, as they stated, themselves interested in "social issues in our society",

fieldwork: among others, my fieldwork assistant. In a few cases people wished to know, how I would use the information they were giving, this was specifically the case concerning people who were involved in scamming activities. A few times, I was seen as a messenger: "When you write a book, you should tell your people how we live here", or: "You could give them an insight so that they can understand the reasons and motivations of our youth going into criminal activities" (Peter. Field notes 2009FS. Information/Scammers).

Being – more or less – young in the setting

Even though most of the young people I interacted with realized that I was older than them, I was not often asked about my age also, and often my real age was underestimated. Some of my interviewees were of my age as well, or slightly younger, however many were considerably younger, in their early or mid-twenties. For some age seemed deliberately not to play an important role, as it was stated, maybe so as not to lose the opportunity to address me as a potential girlfriend. In other cases it was clear from the beginning that I was senior, then I was instead addressed as a "sister".

I guess that the difficulties in estimating my age was strongly related to difficulties of categorizing my status, position, and so on. In the Cameroonian context, women of my age – mid thirty – usually dressed differently, "responsibly, as a lady", than I did: I was most often dressed casually[470]. That I was also a student, and that I did not appear with my family, might have contributed to estimations of my age. From my perspective, I had the impression that the social distance in the sense of age was not too great between myself and my acquaintances. The fact that I was unmarried and childless, seemed to render me in their category, something which could however also be seen as ambivalent. When some people knew my real age, the question about why I was not married and had no children – in particular the second issue - arose[471].

---

expressed their interest writing something with me - maybe it was seen as an opportunity to profit from my "connections".

[470] In the Cameroonian context, teenagers were dressed in simple trousers or skirt, and T-shirt, however, global youth fashion may blur such differentiations to an extent.

[471] Many expressed hereby their knowledge that "it was different in the West" regarding age for marriage and having children, but hinted at "the urgency of having children for me".

In this sense, some people expressed their feeling that my life-style was not in accord with my age in their opinion, even though it seemed to be accepted in the sense that I was measured with different yardsticks.

Being a "friend" and a "sister"

Most often people related to me as a friend or a sister. Denominating me in these ways related to people's claiming relationship, which served to integrate me into social networks of mutual solidarity and relatedness, in degrees of closeness, from friend to sister. Some people expressed that I was "responsible" for them – a notion, which was common to express relatedness, but which made me feel a little uneasy. Moussa, a young man and close friend stated: "You are like my elder sister, I will always listen to your advice… You are responsible for me as a junior" (Moussa. Field notes 2009. 24.10.09 (compare interviews 2010/11). Or when I had proved my solidarity towards certain social ties, expressions such as "you are a real friend in deed", or "you are a real sister", were used in the sense of acknowledging my given support and loyalty. Treating me as a friend or sister by treating me generously, inviting me, and introducing me to their social network, was also used by my acquaintances in order to ensure my loyalty in the future.

In view of people's claiming relationship in the sense of seeing me as a "resource", it is maybe not always easy to speak of "real friendship". However, I think that it was a lesson in order to realize that the notion of friendship could be used in a wider sense, and not to stick too much to "Western" notions of a "pure" relationship free from claims and duties[472]. I think that being friends can deliberately have an aspect of support, however, to an extent it should be reciprocal. Valeria, a young woman, who I consider a good friend, said repeatedly that she was not content that she herself hardly called me, because she had no means for credit: "I am not satisfied with this situation, it should be a two way traffic, not only in one way"(compare interviews 2009/O and 2010/11). Concerning a few people, in return for my loyalty and occasional support, I had the impression that I could count upon them, and I also came back to them when I needed their help. In general, by maintaining

---

[472] Also I think that people's motivations are seldom unambiguous, it is probably always both and not either or.

connection when I was not present in the field, some people invested work into the relationship with me. Being connected to each other was also supported due to the fact that I have repeatedly come back to the field, which seemed to contribute to people's certainty regarding our relationship, since they knew that I was coming again. And of course – I have to admit that too - I myself had also an additional motivation to keep my relationships "smooth" – also beyond the closest friends - in the view that I would come back to the field.

F72, 73: Fieldwork situations: conversing and travelling

Being seen as "a resource"

Being seen as a resource was an important part – and also the most problematic one – related to my research and my presence – and also absence - from the field site. It could range from profiting from, most importantly, financial and other forms of support, up to profiting from my presence in the sense of raising status. It happened a few times, that unknown people just came up to me saying: "I want to be your friend". Such instances made me realize that I could be seen as a resource, a feeling, which ranged from being indifferent to being disappointed about it. Overall, I think being seen as a resource is a normal experience when doing fieldwork, and I think it is important never to forget that I was myself also coming to the field "to get something".

Even though I accepted being seen as a resource by people to a certain extent, my tolerance and also my responsiveness to their claims depended on how I felt related to people, their comportment towards me, the nature of their claims and situations in which claims came up, and also of course my ability, to respond to their claims. My support giving was related to my estimations of the person, our relationship, and the person's economic situation, estimations, which were based on my experiences and knowledge of the person I had gained in face-to-face situations. I was conscious of whose demands I had declined, and whose I had already previously agreed to. At times, I would have supported people if I could, but I tried to be reasonable because my small PhD salary was hardly sustaining myself[473]. However, the motivations and estimations from my part were often driven by subjective and emotional evaluations. Additionally, regarding support giving and stories of people's financial struggles, there is always the risk involved "of being taken for a fool".

My willingness to support was also related to what I have called positive conduct in communication. I had helped people out in financial terms several times when being present in the field. However, I realized that people were more often asking me for support when I was absent from the field site, which I interpreted in the sense that it was easier for them to ask me via mediated communication, and that in the context of face-to-face presence, it was rather their turn to prove their relatedness, their generosity, and their loyalty towards me, in order to come back to these investments later on, in the sense that the relation to me could come to fruition at last. I appreciated it, when people were investing in communication in order to stay in contact with me: a few people were regularly calling me, and without ever or rarely asking me for support[474]. I was usually aware that I was expected to bear the costs for communication, which I was ready to do, but in this sense it was also up to me, if or after what time I would decide to call a person. When somebody would start beeping me continuously and sending me an email that he had a very important thing to discuss with me, and then when I called, he only conveyed his demand, I felt abused and disappointed, and I became very reluctant to invest in such calls.

I experienced rendering support as disappointment in a few cases. In one case, a young man was trying to convince me to send him money

---

[473] This reason as an explanation for turning down a request seemed to be accepted by most people. Also, their motivation seemed to derive from the expectation – as some uttered – that in future, my life circumstances – and ability to support – would considerably change, since once my PhD was concluded, I would "pick a well-paid job".

[474] Which was maybe also a part of their strategy in the hope of a positive outcome.

for a specific issue. I explained to him that my financial situation was difficult, however, in the end I sent him a minor amount of money. When he then was asking me for more money, after only three or four days, I was upset. In a few cases, people never thanked me for sending money, a comportment, which I did not view positively. I found that the reaction to my support, no matter how small, was an important part of conveying appreciation for my support giving. Another issue was people's willingness to accept my reasons when not being in the position to support them[475]. When they showed understanding, and continued to interact with me by writing and calling, this was then a reason for considering them next time. One person was repeatedly asking me to call him, which I did a few times, and each time, the person complained that his relationship with me should turn out favourable for him – during these years we had known each other, I had repeatedly sent him money – that I was responsible for his wellbeing, and that other people were asking him, why he was in financial difficulties, even though he had a white friend. Or when I was in the field, a young man – with whom I was not very close - called me and in a weak voice told me that I had to meet him at the pharmacy. When I rushed there, he had bought drugs, which I had to pay for, without leaving any other option open to me.

Apart from financial support, people were asking me for other favours. It was common to ask for material gifts, which I should bring with me when returning. When coming back to Cameroon, I always brought heavy loads of gifts along. Especially around the time people knew I would come back, there was an increase of demands, also very general, "bring me some nice thing", or when I was in Bamenda, they asked: "What have you brought for me?". Many people asked for valuable items, such as camera's, laptops, shavers, MP3 players, mobile phones, and other electronics. I often reacted to such demands by asking them how much they could spend, making it explicit that these were no gifts in the scope of my ability. I brought laptops several times

---

[475] Some people said that they appreciated my seriousness, since, as they stated, Cameroonians would rather have the habit of not turning down the request, but to make promises, which they would not keep.

to people, also experiencing the difficulty of getting back the advanced money[476].

Another dimension of support giving was providing information, which was most often related to migration ventures. It encompassed help provided on a local level, when for example going with somebody to a cyber café to browse, or on a translocal level, information rendered by email or calls. Providing information about migration ventures was also an opportunity for me to give something back and to show my appreciation. I had been asked certain questions so often that I had prepared a compilation of suggestions, such as links to important sites, in order to convey such information just by copy and paste. Such links included informative sites, of some information related to procedures for obtaining a visa, applying to a university, and what people needed to consider when they intended to migrate. Of course, my advantage here was that I had gained certain knowledge through my research, and that in turn such demands were also interesting for me.

Since I was generally seen as knowledgeable in terms of internet technology, I was at times also asked to provide help in such terms. For example I created email addresses or Facebook accounts for people, or showed them basics regarding emailing and the use of messenger chat. Twice I was asked, once by a female and in the other case by a male friend, to create them an account on a dating site, and several others - mostly men - asked me to recommend dating sites, which I considered as trustworthy. Similarly, a considerable number of – male – friends wanted me to find them a wife or a girlfriend in Switzerland[477]. I was asked to create a site for an elderly man, who was involved in NGO work, a problem I solved by creating a Facebook account so he could present his NGO project to a wider range of people. Valeria, a close friend, who was not particularly internet literate and was sent to work in a village with no internet access, had asked me if I could regularly check

---

[476] I could never retrieve back the full amounts I had spent. I learned from this later on and did not promise the laptops to anybody particular, but sold those I brought to the field to the person who could provide the money.

[477] Because they said they were disappointed that they never had positive results from dating sites. I tried to convey that it was not easy and that I was lacking female friends – of their age – who were looking out for a man online.

her mailbox, which I did over the duration of a couple of months[478].
Again another dimension of support was my integration into NGO's. I
had accepted proposals to become a member of two NGO's, where my
function was related to a position in "international resources".

Another part of "seeing me as a resource", consisted of profiting
from the status, which apparently derived from being with a white
person in the Cameroonian context[479]. Some people deliberately wanted
us to go to places where we would meet other people: they insisted
taking me home, or visiting them in school, at the university, and so on,
in order to show me around, to their parents, neighbours, friends and
classmates. Some people also openly expressed the issue of status related
to my presence: "It made me so proud when you came to our school.
Everybody has seen you there with us" (Moussa. Field notes 2009.
24.10.09). Or: "They treat you differently when being with a white. I
receive a lot of attention now, because many people have seen me going
with you to the cyber very often" (Valeria. Interviews 2009/O). It also
included that some people wanting to have a picture of the two of us,
then I had to go with them to the photographer to make a picture in the
studio, which they would put into their albums. I was at times left feeling
bewildered when, for example, parents of young men were calling me –
as if they would already consider me as their future daughter-in-law. Or
twice I had to call people's bosses, just "to say hi" to them, or three
times I had to "solve problems" on behalf of friends, by talking to
higher ranking persons. However, I tried to see such contributions and
activities as a means for equal exchange, in the sense that both of us, I as
well as my friends, could profit from each other's company.

Transnational research collaboration

In this part of the chapter I will write about my experiences and
reflections regarding the collaboration with my research partner Primus
Tazanu. He is attached to the University of Freiburg in Germany and is
a member of our research group related to the programme "Passages of

---

[478] I then informed her by text message when she had received an important
message and also answered the same for her then when we agreed over the answer. I
did that when in Cameroon and also when I had returned to Switzerland.
[479] The interrelation of being white and high status can create ambiguous feelings.

Culture", sponsored by the German Volkswagen Foundation. He is Cameroonian by nationality, with life experience in Cameroon as well as in Europe. In his PhD he addressed similar questions in regard to internet and mobile phone use, in the context of migration and transnational social relations.

Reflections about nuances of being "insider" or "outsider"

I could call myself an Anthropologist influenced by the "classical" stance of Anthropology regarding fieldwork in a foreign society, from the perspective of an outsider, spending time in the field in order to gain an insight in the interdependencies between cultural background and the topic of research (Hannerz 2003, Berry & Clair 2011, Robben & Sluka 2006). Contrarily, my research partner Tazanu[480] was doing his research in his own society (Hannerz 2006). Since the era of the crisis of representation, anthropological methods have transformed, and statements of anthropologists in the "classical" sense as "cultural and social outsiders" to the society where they work, are complemented and compared to statements of anthropologists working in their own society. However, it is not that simple, as Tazanu has spent years of his life in other parts of the world than Cameroon, meaning that he was also perceived as an outsider in Cameroon to an extent. Our topic of interest, tracing transnational connections, also brings with a blurring of field sites, and makes it necessary to adopt a two-sided perspective. Regarding our field sites, both of us did part of our research in Europe – Switzerland and Germany respectively - among diaspora Cameroonians and in Cameroon – Bamenda and Buea respectively. Regarding research in Europe among diaspora Cameroonians, Tazanu's background as being a migrant himself, was an important issue. Even though new to his town of residence, Freiburg in Breisgau, when he started doing his field research there, he had an understanding and years of experience of what it means to be an "African" in a European society. Regarding myself, I do not have an equivalent migration experience myself, nor the experience of being perceived as an outsider in this society.

---

[480] Prof. Dr. Judith Schlehe from the Institute of Social and Cultural Anthropology in Freiburg i.B., Germany, calls it a "tandem partner", the collaboration of two researchers with a different cultural background, as "tandem model".

Nevertheless, in the field site in Cameroon I also gained an impression of what it means to "be different"[481].

When visiting each other in our field sites in Cameroon - Bamenda in the Northwest Province and Buea in the Southwest Province in Tazanu's case – in 2009, the "being an out - or an insider" again took on a different meaning: it was the first time for me visiting Buea, a town, which Tazanu is very familiar with. When he in turn visited me in Bamenda, it was a different experience, since he was not familiar with this town. I had spent time here, and felt at "home" in Bamenda, I showed Tazanu the town and introduced him to people, which made me feel like a kind of "insider" in this context and circumstances. Visiting each other, apart from the opportunity to discuss and exchange, was interesting in order to experience the other in one's site of fieldwork including the other's familiarity with the site, relations to people and "going along in the setting".

As a researcher one is also classified by people. Regarding differences of origin and attributed background, by markers such as skin colour – in the sense of being black in Europe or being white in Africa – one is immediately and often automatically classified as an outsider. In that sense, regarding the fieldwork context in Cameroon, as a white – European and Swiss – I had a clearly visible outsider status, whereas Tazanu's identity was not obvious at first sight[482]. Only when knowing him, or when interacting with him, did people become aware of the part of his identity, which made him to a certain extent also an "outsider". Tazanu had a special status in Cameroonian society, being a bushfaller, somebody who had been migrating, studying and working abroad. However, since this status was not necessarily apparent, he could play with roles by playing with appearances and comportment that were attributed to a certain status. I did myself not have such opportunities, for people in Cameroon I was at first a white – yet my role was not clearly defined and left up to people's imaginations. I also used my

---

[481] Of course an experience with different implications: being in the setting temporarily, and usually addressed positively, whereas as an African in Europe experiences could be differing and often as well negative.

[482] For example, he said that it subsequently made a difference, how he was dressed, whether he was dressed casually or „responsibly", or if he used his car or not for example, or if he was speaking Pidgin or "good English", how people reacted to him, and if they viewed him as a "bushfaller", or as a local, in the first instance.

identity in flexible ways in attempts to reduce social distance or profit from my status as being an outsider. Related to imaginaries of white man's kontri as a place of abundant opportunities, and considered the fact that Tazanu and I were linked with these realms, both of us were considered as viable connections.

Gender aspects – different practices of contacting and being contacted

Gender aspects, regarding how researchers are perceived, as well as the researcher's gendered comportment, are important issues in field research, and also interesting to compare regarding our research collaboration, here our being male and female, coupled with skin colour. In this sense I was attributed transient roles, as a woman, a white or white man woman. The reason I interviewed more men in the field I attribute to the fact that it was easier for me to make contact with men, but also because, for example, in public cyber cafés there was a dominantly male clientele present. In general, I had the impression that men were more likely to see themselves as eligible to comment about societal issues, and also they were apparently more inclined to represent themselves as individuals with specific attitudes and opinions related to certain topics. At times, it happened that women told me when I asked for an interview, that they did not see themselves as "competent" to comment on such issues. Tazanu also stated that he experienced difficulties at times in approaching women, which was in his case however due to other reasons: he said that he sometimes found that women shied away from him depending on their family or marital status, when they thought that he was showing interest in them. However, at times women showed interest in him when it transpired that he was a bushfaller. For both of us, gender was connected with people's hopes of profiting from a relationship with the researcher. Another issue, which concerned gender, was the competition aspect, which was relatively prevalent and openly expressed among – in particular – young men. I was told by another – white – colleague, who had worked in Bamenda, that he was sometimes addressed by young men in a way that indicated social ranking and position, in verbal expressions such as: "A de pass you" (I am better than you, I am surpassing you) referring to certain issues. I have never come across such direct statements of superiority towards myself, since I was treated differently by men, and women expressed feelings of competition in less demonstrative ways – rather, for example, such as frowning at me.

It seems that gender roles are a crucial factor in fieldwork, a fact, which might be more or less consciously experienced – or profited from – by the researcher, gendered conduct seems to intersect and influence our experiences and interpersonal encounters in the field. Even though in certain situations I probably did not fit into local expectations of gendered behaviour, I was excused if I stepped out of line, because I was a "white", and measured with different yardsticks. In general, it is difficult for me to separate the experience of being a female and white.

Integration of researchers into social networks of solidarity

An important aspect is duties and expectations of people towards researchers: this relates to an integration of the researcher into local social networks of relatedness and respective duties.

I myself am not part of social networks defined by family and common history, or other grids of belonging. I could not be morally held responsible and judged on the basis of fulfilling my duties, as my relationship with people was based on friendship "out of free will". Furthermore, I was belonging somewhere else: even though not really feasible, I must belong to a social network and family "over there"[483]. At the same time, however, strong imaginaries regarding my origin contributed to my being seen as a potential resource for everybody, who could make a connection with me. However, I had the impression that the longer I was in the field and the more often I returned, the more I became involved in relationships of – quite one-sided – support and duties. Also inclusive statements, of "you are one of us", or the continuous insuring of mutual support expressed by the very common saying "we are together", or simple questions or intentions like "I want to be your friend", were ways of creating connections, including me in social networks[484].

Regarding being part of social networks with a more clearly defined role, Tazanu of course had different experiences than I did, when it

---

[483] My family was back there, it was very obvious that I had one, but it was not considered too important at the moment and situational context, even though I always had pictures with me, since people asked for them.

[484] Compare to the earlier subchapter in this chapter.

came to issues of support, duties and expectations. In his case he was part of social groups that were defined and reassured by referring to duties within so called "natural" relationships, such as relatives and family, based on a continuity of social interaction. During his absence, he had maintained his presence in the sense that he was connected to people through communication and different activities such as sending remittances, responding to expectations, and negotiating his position back home. Imaginations also concerning his life in Europe, financial situation, work, and familial situation, served as justification of his ability and therefore duty to contribute to the wellbeing of a wider group related to him in one way or the other. In this sense, people did not consider the fact that he might also be integrated into social networks with related expectations in Europe. From what he told me and what I experienced while spending time with him in the field, he had to deal with a direct and sometimes even intimidating course of action: when people asked for support or a favour, they insisted on certain advantages by considering that they had a legitimate right to profit from opportunities through him.

Responsiveness of potential interviewees towards the researchers

In general the topic of research on the use of New Media and migration was not seen as delicate. People were quite open towards me, in the sense that they seemed not to have the impression that opening themselves up could become inconvenient for them, or that they thought that I could be a competitor in the field, and thereby withholding information. People were in general very responsive, even such "vulnerable groups" as the scammers[485]. Since I was not integrated into local networks of competition, I might have had easier access to certain information, than Tazanu. Tazanu certainly had a better insight as member of the local society, but as a Cameroonian he had also to deal with issues of jealousy, being envied that "he had made it" and sometimes people would not collaborate with him. Nevertheless,

---

[485] Their carefreeness was also related to their relative safety of operating – on rather small scale – in an environment of corrupt policemen and employees at money transfer institutes, with whom they had "smooth relations". It depended here also very much on being integrated into social networks. In this sense, it had a strong impact when I knew one or two people of a group, in order to be accepted by the others as well.

Tazanu evaluated research among Cameroonian migrants in diaspora as more difficult than research in Cameroon, where the aspect of being seen as a potential door opener was also highly valued. In turn, migrants' reactions towards him were at times ambiguous: on the one hand he was seen as being one of them (Tazanu 2012:39ff), but on the other hand he sometimes experienced suspicious reactions from fellow Cameroonians, because they might have thought that since he was new in Germany, he wanted to hook up with them in order to profit. Myself, I also had the experience that people in Cameroon were generally more open than Cameroonian migrants in diaspora[486]. Of course, how potential interviewees react to researchers in the field, and subsequently the researcher's access to information, depends also on our characters to a large extent, how we deal with situations, and the individual contacts.

Different perspectives – different findings

So far I have related to differences of my own and my research partner's opportunities of contacting people – how they reacted to us in relation to our research work we were doing – how we could access information, and how different images and perceptions were related to us and our potential of being "useful" contacts – as a bushfaller or a white man (woman). We experienced certain similarities in the sense that we were – to different extents and in different ways – seen as outsiders and insiders alike, and also as potential resources. Another dimension – but interrelated with these opportunities and peculiarities our personalities were offering – is the content of information, we both were encountering, and their interpretations, on which we were building our findings. Here our personal experiences seem to have caused us to emphasize on specific issues or to take specific perspectives related to our going about research[487].

---

[486] Migrants who had applied for asylum were more inclined to be reluctant to render information.

[487] Compare to Tazanu's description and reflection about our collaboration and cross perspectives (2012:48-50).

Different starting points in research – "reaching out", versus a migrants' perspective

Triggered through my own presence and response to me, I have accentuated people's practices of reaching out and linking up as a kind of "outgoing" perspective. It is reflected in my research and writing, that my perspective and experiences as a white outsider in the site of fieldwork has influenced my emphasis on the motivation and attempt to migrate[488] in relation to New Media use and social networking: People emphasized their difficult life circumstances towards me. Their complaints, as well as some people's desperate decisiveness to "leave this country", have left a deep impression on me. Imaginaries and imageries of "the West" and a "good life" have thus become an important part of my book. In this sense, I have generally laid a stronger focus on the Cameroonian context, and New Media user's tendency to align their practices of media use with "reaching out". Regarding local attitudes towards migrants – apart from negative connotations - I often came across people who uttered understanding regarding migrant's life circumstances. I think that this was a part of people's desire to demonstrate that they were knowledgeable regarding life conditions of migrants abroad. Similarly, I think it is not a coincidence that Tazanu put a stronger emphasis on aspects of ridicule, devaluating, and negative perceptions of returning migrants, impressions, which might have arisen from his personal experiences. He thus laid a stronger emphasis on migrant's perspectives.

Different field sites, and differing emphasis on media

Obviously the field site also had an impact on our research. The differences between Buea and Bamenda reflect our different emphasis and approach to the topic. Buea is a university town. Apart from its symbolic stance, as the former German capital, Buea is rather small regarding its number of inhabitants[489], and the central urban core mainly consists of one major double lane street through its centre, and a few small adjacent streets. Tazanu has accurately called Buea an "urban

---

[488] However, I do not think that I had come across an emphasis on migration and reaching out to opportunities, merely due to my presence. Various narratives and discourses circulated, which dealt with migration and bushfallers.

[489] According to the 2005 census, Buea has about 90,000 inhabitants, compared with Bamenda's 270,000 inhabitants. Compare http://en.wikipedia.org/wiki/Buea.

village" (2012:44). Compared to Buea, Bamenda could be considered as more urban, with different quarters, different hot spots of social life, and the range and number of businesses is estimated to be higher than in Buea. Furthermore, New Media in regard to services and businesses are more numerous. In Bamenda I met comparatively high numbers of internet users, due to the prevalence and density of the cyber cafés in town, whereas in Buea there are comparatively few[490]. This might explain the greater emphasis on the internet in my research – also coupled with the greater emphasis on the "reaching out" quality of New Media use, for which the internet is an important tool. My starting points for research were to a large extent the public cyber cafés. The prevalence of young and internet literate clients in cyber cafés in the urban area has also lead to the tendency to put a stronger focus on youth and the urban sphere, than Tazanu did in his own research.

Not talking about what is self-understood, or enlightening an ignorant researcher

It is obvious that interviewees always deliberately shape the messages they want to convey to the interviewer: in this sense they influence the researcher's interpretation and writing and therefore contribute to the shaping of cultural representations. It has to do with cultural and social distance or closeness between people, here between the interviewer and the interviewee. People would address or avoid, emphasize and neglect certain issues towards a certain interviewer. Towards Tazanu, people might not have talked about certain issues, because they presupposed Tazanu's competence due to his being a Cameroonian. Towards myself, people were more inclined to explain and thus verbalise issues, since they saw me as non-biased, clueless, and ignorant – and to be possibly influenced. It could be an advantage that as an outsider, one is inclined to reflect on issues, which seem to be a given for an insider, however, I suppose that people were interested in

---

[490] I have related to the high prevalence of the scamming phenomena in Bamenda. However, it was also found in Buea – often pursued by students from Bamenda, being involved in such activities back in Bamenda during semester holidays. Furthermore, even though I had tried not to relate to the issue of scamming too much in my book, when being in the field, this topic – interrelated with other practices of internet use - had been taking a part of my time. It was also a topic of narratives in the field, among youth. Furthermore – I have to admit – the topic fascinated me.

rendering certain impressions towards me as an outsider. It was then more difficult for me to evaluate such "selective performing" and put it into context, since I am lacking a cultural and social background as a basis on which to judge. I experienced such strategically reflected performing as peculiar with regard to two aspects: performing a positive impression and demonstrating equality.

Performing a "positive impression" towards the researcher

Positive representations of traits of character were made on individual or collective levels. This could be in narratives or normative statements referring to the comportment of other people, in positive or negative ways. It was then most often verbalized explicitly by my companions, that they estimated their own comportment towards me as morally correct.

Very often, individual comportment and the demonstration of positive traits of character were explicitly related to what was then denominated as "culture" or collective positive traits of character and comportment. Collective statements were made on different levels, such as relating oneself to a specific social group – such as, for example, certain youth from a certain quarter - or "people from Bamenda", "Anglophones", "Cameroonians", or sometimes "Africans"[491]. Another dimension of performing a positive impression consisted of delineating oneself from other - groups of - people who were perceived as different, by portraying them in a negative way. For example, I often came across comparisons of Anglophones and Francophones. Here, also the topic of "Anglophoneness" or the delineating of "Anglophone culture" from "Francophones", and "Francophone culture" came into play. People often demonstrated and narrating the good qualities of Anglophones, by asking me, what experiences I had had in the Francophone part of the country: "What was your impression, how were you received in the Francophone area (in Yaoundé, in Douala, etc.)?", sometimes in a suggestive manner: "Don't you think that Anglophones are more open, friendly, welcoming… than the Francophones?". In this sense, Anglophones often related to themselves as morally superior, but

---

[491] Less often people related to ethnic groups, but it happened at times when I was expressing my being familiar with certain attributions of different ethnic groups in the setting – for example concerning Bamileké, Mankon people, and so on.

marginalized and disadvantaged compared to Francophones. Using negative stereotypes seemed to serve to address perceived negative tendencies, as critique, but also related to moral values of "true" Cameroonian, Anglophone, or Bamenda "culture".

It also included the "negative examples" of those who did not – according to the speakers – correspond to these predicaments, by relating to "we groups": We - the group of youth I was interacting with – are nice people – but beware of ... - mentioning another group of people, such as "dangerous" youths, noise makers, scammers, and so on. Often such negative narratives were also made concerning specific individuals. It was sometimes striking to what extent some people would relate to others in negative ways, whereas my own experience of the same person was a completely different one. These narratives often included explicit warnings directed to myself, to be careful, not to be too trustful, and how I had to avoid these dangers, such as going home early, only trusting people who I knew well, and so on. In this sense, by pointing out to dangers, people conveyed their own image of being trustworthy individuals, whose advice I had to follow in my own interest, and under who's protection and responsibility I would not have to fear inconveniences. It was several times stated by people, that they wanted to prevent me from having any negative experience in Cameroon[492]. Interestingly, people also lamented that people in the West had mostly negative impressions of Africa, as they would only see negative news about African countries and people: war, poverty, catastrophes, criminality, and illegal immigrants, and that they wanted to prove that it was not just like this. Some people stated explicitly that they wanted to "prove that we are nice people, we are friendly, hospitable, open, welcoming and generous", in a way rendering the proof to me as a representative and spokesperson of my own world. In this sense, I could at times not help the impression, that people as individuals or collectives, tended to perform a positive impression, by portraying others negatively[493], apparently in attempts to strengthen our relationship.

---

[492] Here it meant that I could be robbed, intimidated, harassed, and so on, experiences, which I – luckily – never made.

[493] It was related to by saying: "Cameroonians like "kongossa" – rumours, gossiping.

Demonstrating "equality" towards the researcher

In general, a mixture of positive images related to people's own positive estimations of "culture" was demonstrated towards me. It contained images of "traditional culture" as well as allusions to "being modern" alike, by emphasizing certain issues and leaving out others[494]. People's demonstrations balanced allusions with modernity and accuracy on the one hand, and on the other hand deliberately shaping "African culture" and values. Such images served to do both delineate African, Cameroonian, or Anglophone cultural and social values positively versus the negative side of "modernity": such as alienation from families and kin, lacking responsibility and morality, and so on. In this sense, an "alternative African way" was presented. On the other hand, it served to blame African/Cameroonian negative traits, such as corruption, fraud, jealousy, as well as an overstatement of ethnic origin, traditional hierarchies, and witchcraft beliefs, as reasons for "backwardness". Both uses of imaginaries were demonstrated towards me, often mixed up in arguments and tied together as different sides of the coin. When showing interest in various topics, it seemed that many people had no difficulties in addressing "an African way of life", relating to traditions, life in the village, to topics such as witchcraft, and so on, in the sense that these issues were part of their lives, too. Thus, on the one hand, such discourses seemed to serve as an attempt to deliberately shape traits of African culture by emphasizing difference[495]. Thereby people were stating that they were versed with both of these "cultural worlds". On the other hand, others delineated themselves from such issues, which were by them seen as backward or "primitive", and not put on an equal level with traits of culture, which were attributed to "modernity".

People often seemed to carefully evaluate which attitudes they should adopt towards me. I assume that for example practices of consumption commonly served as demonstrations of economic viability

---

[494] Of course, conversations in the urban setting, regarding my topic of research, often related to "modern issues."

[495] Thereby, such aspects of "culture" were then sometimes portrayed by expressing that I was not seen as competent to judge of them. A good example here relates to witchcraft and occult powers, which were issues often addressed in urban narratives, especially related to scammers and the internet in general. "For you it is no issue. But for people here it is a reality. But you do not have to fear, it does not inflict you" (Festus. Interviews 2009/F).

and success towards one another. However, some youth's pointing out their being "equal" seemed to also have an important stance in their comportment towards me – as a white man and representative of a realm of "modernity". Youth's demonstrations of their "being up to date" could also include for example addressing me in a deliberately adopted American accent, or showing that they were versed in "white man's technologies", being knowledgeable regarding "life in white man's kontri"[496], and they would emphasize connections to good friends or responsive relatives abroad. Demonstrations of such attitudes were related to their imaginations of what would please me, what would be convenient for me, and what I probably expected. Such allusions could also serve to bridge people's insecurities of how to address me, in order to be on the "safe side": some youth did not consider themselves to be "knowledgeable what whites want, need and what they expect", and they were curious to experience how it was to interact with "whites". This was a few times verbally expressed towards me. In this sense, I had the impression that a demonstration of "being equal" was emphasized by some, because they felt inclined to do so towards me.

Fieldwork methods

In this section of the chapter I want to relate to the methods of fieldwork. I have already related to phases of work and different field stays in the first chapter of this book. In this subchapter I intend to describe interviewing and participatory observation, as well as related chances and difficulties.

Several field stays, addressing questions, analysing data, and the writing process

I was lucky to have already gained an insight into the topic of my research in the field site during my internship in Bamenda in 2003 (Frei 2003, 2005), and I could rely on already existing contacts, which made it easier to reconnect, when coming back after five years, in 2008. In an exploratory phase of fieldwork in Bamenda at the end of 2008, I could investigate the situation of New Media and migration in Bamenda.

---

[496] In the sense that some emphasized that they were aware that "life in Europe or America is not always easy".

When coming back from the field I adapted my overall research questions, breaking them down into more detailed questions to address three dimensions: Firstly, the interrelation of the use of New Media and the imaginations of mobility and migration of youth, and how New Media use was adopted to purposes of pursuing such ventures. Secondly, how the use of New Media and imagining was reflected in the local sphere, in materialization and practices of media use alike. Thirdly, how interpersonal social relations transformed when being mediated by New Media of communication, regarding notions of sociality and solidarity.

During the second and longest field stay from June to November 2009, I concentrated on conducting interviews with New Media users as well as participatory observation in diverse New Media social spaces. When back from the field, I needed to find a suitable theoretical concept, which could guide me through my work. Eventually I adopted the notion of liveness[497]. My aim was to write the main part of my book as a draft, before going back to the field from October 2010 to January 2011, in order to round up and to go into further detail into central topics[498].

Regarding the analysis of literature revised, as well as the data, which I collected during my field stays, I adopted my own "system" of ordering and classifying documents and notes. Concerning literature, I had been copying important sections of books or articles, and stored my notes on my computer, ordered in subsections under specific subtopics. This system worked well, even though it sometimes meant going through a lot of paper, when I had to readapt my writing. Regarding the data collected, I worked without additional programmes for citing. I had stored the notes of my interviews in folders ordered in different subsections. I had highlighted sections in my notes, concerning different

---

[497] This decision was also influenced by discussions with Tazanu.

[498] I only went back to the field-site a fourth time when I had almost concluded my writing, for 5 weeks in September/October 2011. This stay was dedicated to clearing some detailed questions, and to reassure myself regarding quotations and images which I intended to publish later on.

guiding questions, in order that I could compare interviews, even though not all of them were following the same general order[499].

Regarding a diary, which I wrote whilst being in the field – above all in 2009 and 2010/11 – I wrote down everything elaborately which had happened each day, split it up into description and additional interpretations from my side. I then copied sections belonging to certain subtopics of my research, and included them in separate documents, where I collected information, citations, interpretations and descriptions regarding specific subsections, such as for example mobile phone repairs, narrations of bushfallers, notions of solidarity, and many more[500]. When analysing this material, I had to repeatedly read through what I had written, which was probably unsystematic, but proved to suitable for recalling and "processing" what I had collected, when turning to the writing.

I tried at different stages of my work to integrate my interviewees into the process. Once when I was writing a paper for a conference, I sent the draft to three people, and at least one returned some interesting input. I also sent drafts of chapters of my book to a few people. I of course also gave parts of the chapters, which coincided with the results of the interviews which were partly lead by my field work assistant, to him accordingly, in order to see what he could contribute to it. In his case, our discussions on issues of the research were fruitful and he freely gave his input. Nevertheless, a general problem was that some people, even though I had "selected" them carefully, said that my writing was "too much" for them, expressing a reluctance to give more insightful comments, only very general ones. Overall several people expressed their interest in my work, when it was ready, and that they expected to have access to it in one way or another. I plan to create an internet site, in order to publish some excerpts of the chapters, with a blog included.

---

[499] Compare to practices of interviewing, later in this subchapter. Also, since I did not have many interviews recorded which needed to be transcribed, I followed this system of ordering my notes taken in the field.

[500] I did that partly in the field, and partly when back from the field, when I had time to analyse my data.

Collecting background information in an exploratory phase of my fieldwork

When starting my research, I collected background information about the operation and organization of New Media sites. In 2008 I worked on a mapping of New Media sites, such as cyber cafés, computer and mobile phone repair workshops, and respective selling sites, including the suburbs[501]. Furthermore, I compared Bamenda with its more rural environment, suburbs and villages, and I spent three weeks in Yaoundé, investigating the cyber cafés and talking to people. I spend shorter periods of time in Douala, Limbe, and Buea, and visited other cities and villages.

I lead informal conversations with key persons in the field of New Media, such as teachers, government employees in the field of education, owners of computer training centres, principals of schools, cyber café owners, owners and employees of computer and mobile phone repair workshops, dealers of electronic equipment, computers and mobile phones, employees of telecommunication companies, as well as vendors of credit at call boxes. I also visited schools in the area, in order to talk to ICT teachers, and to see the school's media centres. I talked to managers and employees of Money Gram and Western Union, and telecommunication providers Orange, MTN and Camtel. However, my experience when talking to such "officials" was that they were giving me "official versions", which was in most cases not too insightful. I learned more by conversing informally, with employees for example. Furthermore, I was continuously collecting narratives, by writing them down accordingly, and by reading through the headlines of local newspapers, in order to collect anything related to my topic of research. In such a case I bought the newspaper and collected the respective article. In this sense, I was informed about current issues and discourses.

I intended to make a small survey, with the help of employees in cyber cafés, about the composition of the clientele, regarding estimated age, sex, and the amount of airtime they bought, and how long they stayed in order to browse. However, I gave up this idea along the way. I realized that it would be too time-consuming and probably not generating really enlightening insights. Furthermore as per my

---

[501] I also tried to backup this mapping later on in 2009 in order to integrate the transformations.

estimations, in cyber cafés with a high turnover of clientele it would have been difficult to pursue my idea, and in such cyber cafés with a rather stable clientele I found it to be not insightful enough. I also had much information derived from spending time in cyber cafés systematically, and the information I obtained from interviews with internet users.

A visual back up of my research

Taking pictures in the setting of my fieldwork was not in all aspects absolutely crucial for my topic, but of course it could support my work considerably. In general, I was rather reluctant to take pictures in a public sphere in the presence of unknown people. It was a different thing when I was together with friends. Often I was asked if I could take pictures, since many people knew that I carried my camera along with me. In general, with portraits of known people there was no problem, but some people would for example not like to be pictured, as it was partly the case with scammers. A few times I gave my camera to friends, as a favour in order that they could use the camera for private purposes. Three times I gave my camera to somebody with the purpose that the person should take certain images for me. I did so because I felt a time constraint, because I could integrate another person's view on the environment, and I felt that when a local was taking pictures, it could be less conspicuous than when I did.

However, unexpectedly, I gained a deeper insight regarding the visual coverage of my issue of research. Tazanu and I were integrated into a filming project of the sponsor of our research, the Volkswagen Foundation, portraying our research group and in particular covering our own fieldwork about New Media and migration: we had the task of filming in Cameroon during my field-stay in 2010/11[502]. Furthermore, we did filming work among Cameroonian migrants in Switzerland and Germany, as well as during the VW Foundation Workshop we held in Nigeria in September 2010. We also covered the making up of the different sequences by taking pictures. We tried to film inconspicuously, since we had no filming permit, as it turned out to be too complicated and too much effort for the small filming work we intended to do in Cameroon. It was an interesting and tedious experience, and also

---

[502] See figure F74 and F75, filming in Yaoundé at the university, and when we were travelling for this purpose.

regarding the question of how to address our research topic in a simple and "consumable" form adapted for a broad public to view our work on the internet[503].

F74: Doing filming work in Yaoundé
F75: Travelling around for this purpose together with Primus Tazanu, here on our way to Fontem, in the South West Region.

Conditions of interviewing in the setting

I have already referred to the posed questions in interviews in different field stays, notably 2009 and 2010/11, in the first chapter of this book. In chapter 4, I have described samples of interviewees regarding the 52 interviews, which I conducted specifically for the topic of New Media use in 2009. I also conducted 26 interviews - half of them were conducted by my field assistant – in 2010/11, adding another 4 in September 2011, with emphasis on practices of social networking and notions of solidarity. I conducted a further 17 interviews regarding migrant's practices of transnational communication in Switzerland, and additional informal conversations with people with migration experience in Bamenda. I have related to this sample in chapter 6.

The "convenient" issue about my research topic was that almost everybody was a potential interview partner. Finding interviewee partners was not a problem. Most often I got to know and interviewed people coincidentally, or I was introduced to somebody by a friend. A few times I addressed somebody, introducing myself and what I was doing purposefully, asking if I could interview him or her. Often I was contacted by people myself. In a few cases two friends of mine – who

[503] See the site www.sciencemovies.de

were both working in cyber cafés – were organizing meetings for interviews for me.

Nevertheless, the interviews with people whom I did not know before were, from my point of view, less fruitful. People's background and life story was missing to a certain extent, and also their trust in me was limited, when I just met them once. In that sense I preferred to conduct interviews when I had already known my interviewees beforehand and had interacted with them several times. In about three quarters of the 52 interviews on practices of New Media use, I had conversed with the interviewees at least once before[504]. Meeting people more than once also gave me the opportunity to follow up questions, which had not been entirely answered, or details, which I could not follow up. I also considered it to be useful to browse the internet together with my interviewees. Practices of New Media use, I had the impression, seemed to be integrated into habitual patterns to a certain extent, some of them were hardly verbalized, which made it at times difficult to obtain satisfactory answers. Some issues I could only follow up when witnessing people's internet or mobile phone uses. Such as contents of conversation, New Media habits, style of writing or talking, as well as people's address lists or their photo albums online or in their phones.

Some of the interviews I conducted in a cyber café, but only a few of them, since not every cyber café provided the opportunity to talk to somebody in a calm environment. In one cyber café where I lead a few interviews, I could use the back office, twice I conducted an interview in a cyber café spontaneously, when there was a breakdown of current, and other people left. In most cases I met people at drinking spots somewhere in central town, or in other cases in their homes. Leading interviews in a public space also involved interruptions at times, from incidents happening around us, or interviewees met people they knew, and so on.

---

[504] Regarding this sample of interviewees, which I had conducted myself, I was already familiar with all of them.

Interviewing – recording or notes, and different levels of formality

I preferred to lead interviews in not too formal ways, my questions rather wrapped up in a more informal conversation[505]. I think that such interviews, which are quite informal, provided more insightful answers. I had my questionnaire in mind, but how the interviews were structured also followed the development of the conversation. Furthermore, when meeting people several times, the asking and answering of questions was distributed over time. While interviewing I took notes in my note pad. When just spending time with people and interacting with them informally, I, in some cases, preferred not to take out my note pad and make notes, but I used every opportunity to at least write down a few key words. I always tried to write notes in detail right after a conversation had ended. Then I sometimes went to a café, or sat down somewhere at a friend's business place. People I knew around central town were used to me performing this activity of taking down my notes. They were making statements such as: "ah, you are working", or "just do your work first, after we can talk". Sometimes I also took notes in cyber cafés. I got very used to working in this way and when I tried to follow up the conversations going along my collected key words, or more elaborate notes, I had the impression that I could easily recall the course of the conversation.

When first coming to the field I had been planning to record interviews, a plan, which I did not entirely follow up. I felt inclined to put back the issue of recording interviews, concerning the large part of interviews conducted in the field. The difficulty I had with recording was probably rather a subjective reluctance. I guess that with most of the interviewees it would not have been too much of a problem, apart from a few individuals and more "vulnerable groups". These groups were namely the scammers, who could fear that I was a "spy" - of the CIA, as it was mentioned a few times - and with migrants in Switzerland I also felt reluctant, moreover with those who had an unsecure juridical status. As some interviewees uttered, in conversation only, "all what we talk is gone for good afterwards" (Festus Interviews 2009/F), whereas when recorded, it was "fixed" and one had to ask what was done with the

---

[505] The bulk of informal conversations I did not count as interviews, they are partly to be found in my field notes, and additionally I have stored them in sections of different information and topics.

material. I felt that the combination of being white and recording could be sensitive, as in Cameroon recording has a touch of something official, and thus some people might have a certain mistrust about it. Another issue was that some people expressed their doubts as to whether they could say something, which was useful to me, because they did not see themselves as being knowledgeable regarding certain issues. I then tried to explain that it was about their personal opinions and experiences[506]. When recording, some people had the tendency to lay greater emphasis on "telling me the truth", talking about "how things are", and less about their own ideas and feelings. I feared that the presence of a recorder would make the interaction too formal and influence people's comportment and answers too strongly. I tried to circumvent this by mostly recording interviews with people I was already highly familiar with.

However, regarding not recording interviews, I am aware, that certain details might be forgotten, overlooked or I could not remember a statement in the exact way people had uttered it, even though I very often noted specific statements in a detailed way to how people had addressed it, in their words and notions they were using. During the fieldwork stay in 2010/11 I then decided to record a number of interviews[507] with people I already knew, and with whom I had already been leading interviews, or had informal conversations about my topic of research before. I explained that I intended to record in order not to forget details, to use the records as personal notes, which were then much more elaborate. Since the people I recorded trusted me, most of them had no objection, and the interviews were in this sense to a large extent informal and relaxed[508]. The interviews were also quite lengthy, and lasted one hour as a minimum, and sometimes up to two hours.

---

[506] However, some uttered their desire to "enlighten me" about certain topics.

[507] Apart from conducting interviews with an emphasis on solidarity I recorded and added a few interviews to the bulk of interviews with the emphasis on New Media use, which I conducted in the fieldwork stay in 2009.

[508] But even among well-known people, of whom I considered that they trusted me, there were a few who said that they would not like to be recorded, but they wanted to preferably "chat in a relaxed way".

The collaboration with a fieldwork assistant[509] was fruitful in order to include different perspectives of the interviewer, and especially useful were our subsequent discussions, listening to his completed interviews, commenting and evaluating them from our own perspectives. It turned out that our styles of conducting interviews were slightly different. My fieldwork assistant's interviews were in the first interviews very formal and when listening to them I had the impression that concerning many issues, I would have liked to ask further questions. By discussing and talking about these issues, in the later interviews his way of asking questions was more flexible and adaptive according to utterances of the interviewees and the course of the interview. My fieldwork assistant knew his interview partners, but not all of them were close to him, and in this sense his interviews were also much more formal than mine. Of course, less formality and more flexibility in the way of how questions are asked and adapted to an extent to the specific interview partner, is sometimes at the expense of strictly observing all parts of the questionnaire, as I realized when comparing my own and my fieldwork assistant's recorded interviews. His questionnaire, which he adapted for his own use[510], was very elaborate, many questions were complementary, which meant that they were answered at times by initiating just one question. In my own interviews, I often received answers to questions contained in long narratives, which were given to me by interviewees after having initiated it by asking a broader starting question. I agreed with my assistant that he did not have to fully transcribe the interviews, but rather note down the main points. He laid emphasis on writing down his own interpretations of the answers of the interviewees, accordingly[511]. These interpretations were based on what he found he needed to explain to me. I listened to the interviews and added my own comments and parts of transcription in this sense, or asked questions.

---

[509] I knew the person before, but he was not very close to me. He had been conducting interviews for researchers from Social Sciences before, and was also interested in "societal issues" in general. I of course paid him for his work.

[510] I had given him the central questions which needed to be adopted, but I left him some freedom to adapt them according to the setup, order, and addressing details in specific sections of the interview. Also, we had discussed the questionnaire before he used it, as well as after the first two conducted interviews.

[511] He added such explanations in cursive writing to the statements of the interviewees. They concerned most often statements about "Cameroonian culture", based on his own assumptions of what knowledge I might lack.

According to my emphasis on informality, informal conversations gave me the main insights and also helped to find specific questions, which needed to be asked in more formal interviews. Informal conversations could be lead everywhere, and most were unplanned. Furthermore the topics I was interested in were widely discussed in the public sphere, in different places and induced in different contexts, among people I knew, or strangers, and I could listen or actively participate. Often, I was included in conversations by people in one way or another, be it that they also addressed me, or that they were directly asking for my opinion. It was a good way to – often by coincidence – obtain interesting information about New Media, the internet, the "scammer issue", about bushfallers and migration, and so on. Related to such informal conversations I usually did not take down notes on the spot, but rather afterwards. One drawback regarding informal conversations was – especially in situations of everyday life – that the issue of discourse was quickly changing, people were speaking disorderly, and the context was sometimes not favourable, distracting and noisy. I never purposefully instigated group discussions, but on several occasions I witnessed conversations about interesting issues related to my work, arising without my influence.

My interviews were face-to-face narrative interviews. In two cases of migrants in diaspora, I asked questions which I had failed to address through email, because I could not meet them again. In a few cases, I also asked further questions by email to interviewees in Cameroon, when I had returned to Switzerland. However, since I kept returning to the field, I had the opportunity to re-address questions. At times I addressed specific topics related to my research, when conversing with people in chat, in order to see how my communication partners would respond. Thus, my use of the internet regarding interviewing was rather sporadic and not systematic.

Participatory observation - opportunities and difficulties

I did participatory observation in a range of what I call "New Media sites". I chose a range of ten different cyber cafés, according to their size – numbers of computers – location, and demographic and turnover of customers. I went to cyber cafés with people in order to browse, to converse or conduct interviews, but I also spent time there to observe.

Not all cyber cafés provided a space where one could sit for a longer time, without appearing strange to customers[512]. Thus, I was often browsing myself, buying two hours airtime and meanwhile observing what was happening around me, interactions between people, who were coming and going. In some cyber cafés I knew the employees or owners. While browsing, I always had my note pad on the table, where I could note down observations at any time. In such ways, I tried to spend time systematically in the chosen range of cyber cafés, at different times of the day[513]. I tried to spend more or less a whole day within the regular opening hours in each of these cyber cafés, at intervals[514].

During my fieldwork in 2010/11, I was systematically doing participatory observation in "social youth spaces". These appeared to be good places in order to observe youth's social interactions, how New Media were integrated into face-to-face interaction, and how social interactions were influenced by them. I had chosen a sample of social spaces, such as two different drinking spots – a "responsible" place, where elderly people also met, and an "irresponsible" place where the average age was much lower and people coming here were often seen as scammers and hustlers, and different night clubs. Furthermore, I spent time at the swimming pool at hotel Ayaba, where in the dry season many young people met for socializing and bathing, and took part in meetings at the occasion of one "veteran club" youth organization, and some spaces in the public sphere in the streets. Apart from that, I tried to gain more insights into how New Media were used in more private spaces, in people's homes. In such youth social spaces, the notion of participatory observation was most valid, since it also included active participation, socializing, and conversing. Difficulties regarding participatory observation could be the tendency of being unsystematic, since much

---

[512] Regarding participatory observation in the field the aspects of the visibility of being an outsider, regarding my appearance comes into play. Even when unknown in the corresponding site, I was always observed and never unnoticed.

[513] I intended to witness night browsing, but I did not do so, because I doubted whether it would be useful to spend a night in a cyber café with people who would be occupied working, also, I was not sure if everybody would like my presence, and since the cyber cafés were closed during night browsing, I would not have been able to leave the place until morning. Here, also the issue of keeping up with sleep in order to use my energies for the day came in.

[514] Also I paid attention to weekends, when the operation was slightly different than compared to weekdays.

depended on coincidence, what was happening in a place precisely when I was there. I thus sometimes doubted the use of systematic observations – however, it meant spending a certain time span with a certain regularity in a place in order to get an overview of what was happening there.

Whereas habits of communication and performances related to mediated communication could be covered well by observation, a part of New Media related activities are not observable, but take place in a virtual space. For example contents of conversations are more difficult to assess. Since contents of conversation are intimate and private, me as a researcher I could only obtain a glimpse of it, when people were deliberately showing, or letting me take part in communication processes. In this sense, it was fruitful to browse with people. Similarly with phone conversations: even though one can easily listen to calls in the public sphere, when not knowing the person communicating and being able to integrate additional explanations with the conversations, understanding could only be partial. It also raises the ethnical question as to whether or not it is correct to "eavesdrop" on phone calls. It is necessary to address some issues by interviewing people.

"Hanging out" as a research method

As I have stated before in the subchapter concerning participatory observation, I spent more or less systematic time in youth social spaces. I will call such methods "hanging out", referring to Kusenbach's (2008) "go-along", a less systematic method, relating to groups of people in a certain milieu. "Hanging out" corresponds to participatory observation, in confined social spaces or places, where a researcher spends time with people, flexibly reacting to context and incidents in the environment. It is relating to actively being there, interacting, conversing and observing at the same time. The aspect of the participatory is highlighted in the notion of "hanging out"[515]. When actively taking part, the attention is directed towards certain activities, to certain people and conversations, and the range of what is possibly observed is reduced. However, consciously directing one's attention to certain issues, for example observing how people use their mobile phones, can contribute to the understanding of the situation in enlightening ways. Informally asking

---

[515] See figure F76 and F77.

questions is also a part of widening the view and deepening the understanding of what is going on, in addition to observations. A certain background knowledge is however necessary to be able to make relevant conclusions regarding observations. It relates to a certain familiarity with the situation and people – also including regularly spending time in situations, and a kind of socialization into the setting, in order to integrate the observations meaningfully into existing inventory knowledge.

F76, 77: "Hanging out" as a research method, or participatory observation, with the emphasis on "participatory"

What I call then "hanging out with scammers" was a specific experience. This was related to their non-expected openness, and the sense of casualness with which they accepted me "hanging out" with them. There were two groups of scammers in town with whom I was regularly interacting. Even though I did not interact with all members of these groups – apart from small talk - my presence was apparently not contested. Everybody greeted me, making jokes, casually relating to my presence. One group where I occasionally spent time was not confined to a specific cyber café, but these young men were co-habiting, and others in the neighbourhood were also interacting with the group. They had a computer with internet access in their house, where a number of the group members were working in shifts. When entering the house, there were always one or two people sitting in front of the screen, and some others in the bed next to the computer sleeping and dozing, or watching TV and videos. At the street corner there was a bar, where they hung around, and in the neighbouring house there was a room with a snooker table, where they used to play snooker. Young women from the neighbourhood and some of the boys' girlfriends were visiting here and then. Sometimes I just met group members by coincidence or I went to their house to greet when passing by. Three times I spent extensive time with them, twice for a whole day. I was not used to spending the day with them hanging out, drinking, eating, smoking, discussing, and playing snooker: in the evening I really felt exhausted from this type of "fieldwork". With the other group of scammers I was regularly "hanging out" in their cyber café[516]. Concerning these groups, apart from doing a few interviews with individuals, I was avoiding formality, which would have created mistrust, and I also would not have been able to use a tape recorder. In this sense, hanging out with the scammers was a special kind of being with people[517].

Fieldwork in Switzerland – Cameroonian migrants

In Bamenda I used the opportunity to talk to people who had migration – or travelling - experience. With some I planned our meeting and we talked elaborately, with others our meeting and conversation was mere coincidence. I was interacting with returned migrants in different social spaces.

Regarding the interviews I did in Switzerland, the sample of interviewees did not entirely correspond to the sample in Bamenda. In Switzerland I also came across elderly migrants, and also a section of my interviewees were Francophones. The level of openness towards me and my research questions depended very much on the migrant's juridical situation and also life experience in the country. However, a difficulty regarding fieldwork methods was – compared to Cameroon – that in interaction with Cameroonian migrants in Switzerland, I had to almost entirely depend on interviews. I could hardly observe people using media technologies. Most migrants used internet in their homes. Furthermore, regarding mobile phones, in most cases – apart from coincidental calls during our interviews, and a few times when spending time with migrants – I could not elaborately observe their use of mobile phones. Whereas in Cameroon I was socializing with people on a daily basis, in the context of my own society I had to confine times of social

---

[516] See descriptions of interactions in cyber cafés, chapter 3.

[517] Regarding these groups, I would not tell everybody, that I interacted with them on a regular basis. Some of my other friends would have taken this as a morally doubtful practice, to interact with "criminals".

interaction with migrants according to mine and their everyday routines and duties. Meetings were not always easy to coordinate. However, about one third of my 17 interviewees I met several times. Twice, I also participated in meetings and social gatherings of an Anglophone association in their "headquarters" in Biel, and I followed the association's activities through two of their members. I was mainly interested in examining the migrant's perspectives on mediated social relations with friends and relatives back in Cameroon, in order to have a complementary perspective.

Opportunities and challenges in fieldwork

I suppose that doing fieldwork, trying to adapt to a new setting, and simultaneously pursuing a research topic and intent, is always a challenge. Personal feelings, emotions and reflections are indivisibly bound to experiences, and these are bound to the outcome of research. It of course depends much on the topic of research one is engaged in, and also on one's personal integrity and ability to go along with the challenges that might arise. From my point of view, even though difficulties might be normal, a general sense of "feeling well" in one's field site, where one is working and living for a certain time is crucial. I think I would not have been able to conduct fieldwork under conditions of continuous tensions, difficult social relations, and disturbances. I was in this sense very lucky, since the fieldwork experience was overall a pleasant experience.

Drawbacks – Being an outsider, length of field-stays, background knowledge and language

The "drawbacks" of being an outsider could at times also be seen as a chance and opportunity to relate to issues, which were in danger of being seen as taken for granted by insiders, and certain opportunities which were given regarding access to the field, which I have already highlighted. However, one central issue was that I was always recognizable as outsider, and even though I sometimes had the impression that I could be present in certain spaces in an inconspicuous way, I am aware that I can never indulge in the illusion that my presence did not influence the situation.

The length of field stays is also an issue. Although it seems to be favourable to be in the field repeatedly, and in relatively short intervals, and the fact that social connections continued meanwhile, it is also a fact that my field stays were rather short. I chose to spend quite short intervals in the field due to personal reasons. After my short stay in 2003 for a fieldwork internship of six weeks, I came back for my PhD in 2008, for two and a half months. I came again only six months after I had left, in 2009 for five months, and another time after eleven months, in 2010/11 for three and half months. The last stay was then again eight months later, for five weeks at the end of 2011. The positive aspect of the short intervals was that I had enough time to create some distance, analyse my results from the fieldwork and then test them once more when I returned, but my return was soon enough that I did not "lose connection" meanwhile. Emotionally also, the situation did not become too tense, as colleagues who had stayed for longer terms had told me that they sometimes felt exhausted. Since I tried to plan my stays well in advance, I was able to work in a focussed way and I could do much of the work in short periods of time. I tried to deal with the issue, that I might lack background knowledge due to the limited time I have spent in the setting. I tried to prepare myself by reading, and also by discussing with Cameroonians in Switzerland, migrants and PhD candidates alike, which was helpful to gain insights apart from the stays in the field.

Concerning my accommodation, I lived in different places, but I always rented a room and I did not live in direct co-habitation with others. This was because I always used to leave my room early in the morning only to come back late in the evenings. When I lived temporarily with families – for example when on a visit, or in Yaoundé – I experienced it to be difficult to adopt my "working schedule" in conjunction with family life. To be free to move and not being obliged to announce my return was important for me. The place where I stayed most of the time was convenient for me: I lived in a family's compound, but had my separated room and had the key for the main gate. Although I also spent time with my landlord's family, they did not have the expectation that I would often have dinner with them, and they understood my need to be independent. The other side of the coin would be that I did not live and experience people's daily family life.

What I would consider as a slight drawback concerns my limited ability to speak Pidgin fluently. In an urban context and mostly when interacting with young and more or less well educated people, I was most often addressed in English at first, and I also lead my interviews in English. However, my attempts to converse in Pidgin were generally considered with appreciation. I was able to follow what people were talking in conversations, and I tried to gain knowledge about certain notions in Pidgin, which seemed important for my research, especially in the view that they are not always easy to translate into English and have several layers of meaning.

Positive feelings – between scientific work and personal experience

The peculiar issue about fieldwork, as I also realized for myself, is that one cannot differentiate one's identity as researcher and as "a person", nor is it possible to separate work and non-work. It is about "being there". Overall, I had a positive feeling of being in the field. This has mainly to do with my overall very positive experiences with people, I felt accepted, integrated, and welcomed.

I realized, that when in Switzerland, and spending much time in front of my screen involved in "book work", I felt sometimes disturbed by people's calls and emails, asking for a favour, for connecting them with NGO's and girlfriends, and also relatively often asking for money and gifts which I should bring along on my return, and so on, whereas while being in Cameroon, I could more easily deal with these issues. When in Switzerland I was tied up in work, occupations, duties and my everyday life and concerns, but whilst in the field, I had no other purpose – of course keeping to my schedule – but to engage, interact, trying to understand what was going on, and thus I was more open, adaptive and non-biased in the situation of field work, which anyway took me out of my everyday life and habits. Maybe also because I was not too long on site at one time, I could concentrate on people and "the feeling of being there" and hardly felt signs of being exhausted or things becoming too much. Being there was a continuous balance of trying to reach my aims and deal with issues by being patient – such as when an appointment did not work out or when I felt frustrated having not achieved my aim for the day, or when asking myself if I would ever obtain what I needed regarding specific information. In order to keep a positive attitude and not waste my – short and precious – time in the field, I always had what I called a plan A, plan B, and even plan C. My topic of research and the fact, that I could always converse with various people and make observations in many places in town, was very favourable. Being in the field was also a balance between the inclination of wanting to maintain good relationships with people, and considering my "psychological limits". It helped that people were most often direct in their statements, a fact which helped me to be concise myself about what I liked and disliked.

Furthermore, when I was back in Switzerland, while analysing my collected data and writing, I most often also enjoyed having the opportunity to stay in close contact with – at least some – good friends in Cameroon, through calls, emails, and chat.

Conclusion: Interconnections of my personal experience with the research topic

As I have already stated, I have found it advantageous to relate to fieldwork methods and personal experiences in the later part of this book, because the topics addressed would relate directly to the issues addressed throughout my writing. When following the order of the chapters, I think this comes out clearly. I have related to the issue of migration as well as imaginaries of the West induced by my own presence as a "white man". Furthermore, people's relating to issues in the field of New Media and their use, was interrelated with this sense of "reaching out" to such realms of potentiality. Issues of link-up and claiming relationship concerned me as well, regarding my personal experiences of people asking for my "contact coordinates", and their pursuing a relationship with me, on site, but also when I had returned to Switzerland. Integrated here were also expectations and claims for support, and assuming the costs for communication. Notions of friendship and emotional closeness were also an issue here. I was also able to personally experience the difficulties of being seen as a resource, and could thus have similar experiences to bushfallers[518]. In this sense, I encountered liveness as a sensory experience, in communication acts through New Media, as well as liveness as effort and work in the

---

[518] As well as different at the same time.

maintaining of a "good" relationship during my absence – or its contrary, when communication "faded out" with time and we lost connection. Liveness as a potential were then the characteristics that were attributed to my appearance.

Thus, the topic of my research was connected with my personal experiences in interesting ways – I was therefore – as a person and a researcher – not excluded from the issues I was addressing in my fieldwork and writing. I regard these circumstances as very fruitful, and at the same time also challenging. I cannot deny that this personal interrelating with the topic of my research has influenced my approach and emphasis in my research to a large extent. In the coming chapter, I will come now to conclusions related to the addressed topics and bring together my findings.

8

Liveness, mobility and New Media use: between dislocation and feelings of closeness

In this book I have described how young New Media users deal with space - with distance and closeness. Thereby, I would see the configuration and practices of New Media of communication and information as the internet and mobile phones as both prerequisites and outcome of transnational connections and relations. They lie at the heart of transformations of sociality, which are negotiated along the lines of ideals of sociality, prevalent conditions of New Media use, and the characteristics of different media's specific mediality.

In this conclusion chapter I will examine to what extent my research concerning the interrelations between the fields of New Media use, migration and social transformation in this specific context could point to "something new" and something particular, and in what ways it can be seen as complementing existing findings from other research in these fields. I hope that my research and writing could render some insights into the dynamics of translocal – and transnational – practices of connecting, and motivations of media users from a particular local perspective, not primarily of the ones who actually migrate, but the ones who stay[519]. Furthermore, I hope to have made a contribution to the understanding of how New Media use and mobility practices influence sociality – face-to-face as well as mediated sociality – under given circumstances.

In this chapter I want to relate to the guiding questions, which I have addressed in my research and throughout this writing. I will revise them once again regarding the outcome of my research and their discussion along this writing, and make some continuative interpretations. In order to do this, I will come back to the agentic dimensions of iterational habituation, projective imagining and practical evaluation, which I will reflect in regard to first, the dimensions of mobilities, and second, the dimensions of liveness. Furthermore I will

---

[519] Of course, however, in relation to migrant's perspectives, in order to examine the mutual set up of expectations and social and moral negotiations.

relate my findings to discussions about processes of globalization, such as the appropriation of technologies to local conditions, transforming notions of space and time, and discourses on the novelty of such transformations. To conclude, I will reflect on the relevance of my research and writing, avenues for further research, and sum up my findings in view of what I could possibly contribute in this field of research.

Mobilities, insecurities, and the dimensions of agency

In this book I have illustrated being mobile – or immobile – as containing different dimensions. It is generally related to how people deal with space, distance and closeness, in a physical, virtual and social sense. I intend to relate the different dimensions of mobility to the different dimensions of agency, habitual, projective and practical-evaluative.

In the sense of a physical mobility, as I have tried to show, practices of mobility evolve dominantly along the lines of the meaning and evaluation of physical mobility, in a habitual dimension of agency. New Media can be used to organize migration ventures, and they play a role in the coordination of spatial movements in general. In the sense of a social mobility, New Media are adopted to follow pathways of success to fulfil dreams of a "good life", for both, migrants and non-migrants, by link-up and reaching out to opportunities and social others. In particular by non-migrants New Media are also adopted in order to pursue alternative life projects, apart from migration. A virtual mobility is then related to New Media conditions, in particular in the specific context of their use in Cameroon. There they play a constitutive role in dealing with imaginaries, which are adopted evaluatively according to situational contexts. In this sense, all of the dimensions of mobility are embedded within the flow of time and thus oriented to the past, future and the present alike: However, I will respectively illustrate a specific orientation, which seems to dominate each form of mobility. In order to do that, I will pick out a few illustrative examples, without claiming to render a full overview or an inclusionary analysis of the intersections of agency and mobilities here.

Physical mobility: a habitual dimension of agency, and a prioritized set of action

A habitual dimension is important in youth's practices of striving for success. It is related to the capacity to reactivate lines of sight, which have proven to be fruitful ever since which shape the flow of efforts[520]. These efforts are directed towards moving out beyond the locality to somewhere, where opportunities are assumed to be better. Therefore, "the agentic dimension lies in how actors selectively recognize, locate, and implement such schemas in their on-going and situated transactions" (Emirbayer & Mische 1998:975). I have emphasized in this book that neither being mobile in the sense of physical mobility, nor the importance rendered to transnational mobility and migration ventures, coupled with strong societal imaginations of a "better place" to be, are new in the Cameroonian context. In this sense, migration as an outline for young people's life projects is deeply engrained in the local society's set of practices and strategies. Likewise is the imagination, from the perspective of those who remain in Cameroon, that physical absence of social others possibly points to a great potential, which could not be assumed in co-presence. Schütz (1967) speaks here of systems of relevance, schemas of action which are given selective attention, developed through personal histories and collective experiences in the past. In the Cameroonian context, physical mobility is often prioritized over other strategies, which also points to physical mobility as a projective undertaking here. This is also related to a maintenance of expectation, that certain actions lead to certain ends: In this sense, as Schütz relates to, these patterns of expectations render stability and continuity to action. This is related to by Garfinkel's Etcetera clause, that one can repeat actions and outcomes by assuming constant contexts of actions, as well as trusting that others act in predictable ways (Schütz 1967, Emirbayer & Mische 1998:980, Miztal 1996, Berger 1978). Such schemas of action are often taken for granted, but also contain a great range of ambiguity and openness. Scholars of the Phenomenological

---

[520] The dimension of physical mobility is related to guiding question 1: What role New Media and their opportunities for liveness play in relation to practices and imaginaries of mobility. It contains the two sub-questions, how New Media are used intentionally for purposes of mobility, and what local narratives and imaginations relate to mobility and how New Media contribute to them.

strain of thinking speak of a pre-reflective intentionality, or a pre-reflexive life-world, which is located in social action and also the body (compare to Merleau Ponty 1966, Schütz 1967, Berger & Luckmann 1980).

When relating to physical mobility in this book, I have emphasized that such practices range from actual moving – as a necessity or opportunity - up to its imagination. I have also related to the role New Media possibly play by contributing to the reproduction of the imaginaries of migration as *the* way to achieve success: in order to understand the importance given to physical mobility and migration ventures, it is essential to adopt a local perspective, in the realm where potential migrants originate. However, migrants too – despite their attempts to re-negotiate social ties and duties – stick strongly to the notions of success in the Cameroonian context in their orientations, and thus play their part in reproducing "imaginaries and imageries of greener pastures". These narratives of success are intertwined with idealized versions of mobility leading to success and the sharing of success[521]. Likewise, with the perceived better connections "to the world", opportunities for migration seem to be heightened on the one hand, and on the other hand, the difficulties of transnational migration only again fuel the great evaluation of such an achievement.

Thereby, migration is never an easy path. Only few profited from the opportunity to migrate abroad in colonial times, and migration to the coast for labour in the plantations was a great risk to migrant's health and life. Nevertheless, migration seems to have figured as a prioritized set of actions also in these times[522]. Nowadays, on the level of transnational physical mobility, its high evaluation in the local context stands in contradiction to the circumstances and difficulties regarding migration ventures, which are imposed on those who most intensely desire to migrate, as many youth in the Cameroonian context do[523].

---

[521] It thus also points to an ideal of sociality. See later when examining the different dimensions of liveness and of agency, in the next subchapter.

[522] Compare to chapter 1.

[523] Even though, for many, issues of mobility can become a veritable "obsession" and this "ambiance" is influential in the local context, we should not forget about those who do not intend to migrate and try to make their life in other ways. This is especially the case when people have better prospects on a local level, and often this is again

Political concern about „unwanted" migration increased in the 1960's in Britain, in the 1970's in Western Europe, Australia, and in North America (Münz 1997:221,22). Since the 1990s, immigration has been treated as a „problem" that increasingly affects policies in many Western countries, and influences the effort of immigration control. Western nation-states such as "social welfare states" use political "filters" in order to condition access to functional systems such as economy, legislation, or education (Bommes 1999:29-31,37). The denomination "Fortress Europe" relates to an increasing internationalism within Europe and xenophobia towards strangers, and a rise of right wing politics (Cesarani & Fulbrook 1996:3,4).

Nevertheless, with the availability of New Media, physical mobility continues to figure as a priority in systems of relevances, and even gains importance, in the sense that New Media may provide means to enhance physical mobility, but above all they may lead to a broad range of imaginations about physical mobility and migration[524]. In this sense, the use of New Media interrelates with an apparent easier access to opportunities, which can also result in great disappointment and a feeling of immobility. Physical mobility has always been seen as an attempt to physically "follow" the lines of opportunities, be it on a local, regional or transnational level or scope of movement. Here, New Media have possibly influenced relations to physical mobility.

Interdependencies between physical mobility – migration and travelling - and New Media

I assume that I have been able to show in this book, how New Media contribute to dreaming and motivations for migration through conveyed imageries and transnational social ties, contributing to mental imaginaries. However, New Media undoubtedly also have an impact on migration ventures in the sense of enhancing physical movement. It involves filling out forms for universities abroad and attaining admission, playing the DV lottery or following up visa procedures, but also social connections, which enhance the organization of migration ventures in the sense of rendering financial, organizational or

---

related for example to good and viable connections with people abroad. I have related to this in chapter 2.

[524] Here, the agentic dimensions of projectivity and evaluation come in.

informational support[525]. We could thus assume that New Media of information and communication have not replaced, but rather enhanced transnational mobility - or above all dreams of mobility - and migration ventures. Youth are in general very mobile, even when they lack the means for mobility. To be mobile is also a necessity in order to follow up options, which are not attainable in the local sphere. We have to differentiate here physical mobility in terms of migrating and settling somewhere – even when only temporarily, and in the sense of travelling in a short termed sense. Although I have stated that transnational mobility has become increasingly difficult, we should not forget that there are differences regarding opportunities for travelling, for example regarding a range of countries, which are "easier to travel to". Cameroon has a treaty with countries of the ECOWAS zone[526], and Cameroonians can easily travel to their neighbouring countries, such as Nigeria, Equatorial Guinea, and Gabon, which are for some relatively popular to travel to for business or educational purposes. Furthermore, temporarily migrating on a national level, such as to Douala or Yaoundé, is common. Many young people leave Bamenda and move to other cities in order to study, for business or employment. Opportunities for physical mobility thus depend also on the destination and purpose. Thus a further question is how New Media contribute to physical mobility and travelling on a national, regional or local level, over shorter distances, where these media can serve as means for social networking and coordination of movements. Although I have not addressed this issue in more detail in my book, here, I intend to point out a few interesting aspects of the interrelations of physical mobility and New Media use.

At times travelling cannot be avoided. In the sense as I have said, people are involuntarily immobile, they are sometimes involuntarily mobile as well. Travelling and physical visits are necessary in many instances. Reasons for travelling, which are prevalent and non-negotiable, are for example attempts to find work, applying for a job,

---

[525] As I have described above all in chapter 4.

[526] Within the West African region, „The formation of the economic community of the West African states in 1975 expanded migration opportunities for West Africa since it encompassed a free movement of persons, of residence and of establishment." (Bakewell & De Haas 2007:104). However, the protocol was not applied in that sense. Cameroon has agreements with the ECOWAS zone rather in the field of trade conventions.

setting up a business, or sitting an exam, implying that one needs to "go there in person". In the Cameroonian context, many administrational and professional issues and opportunities – up to paying bills – are related to physical presence. Travelling is thus sometimes a necessity, also in the view that cities lie considerable geographical distances from each other, and Cameroon's highly centralized political and administrative system. On all levels of spatial distance, also in presence availability, personal visits and face-to-face conversation are most important. This is mostly personal and issue related. For some topics, it is usual and at times even demanded that they are discussed face-to-face. Such issues can be family issues, or related to business and support. A personal visit is strongly tied to showing respect and also the seriousness of a matter. Personal presence in various social events is considered as highly important. In such cases, the lack of the opportunity to see somebody face-to-face, when the geographical distance is too great[527], is strongly felt, and sometimes perceived as having a negative impact on the respective social relation.

However, travel can also be perceived as prestigious and "fun". Many young people in Bamenda uttered that when they had the means, they liked travelling. I heard of groups of youth – in this case scammers, who had made a considerable gain – who were inviting friends to Limbe "to have a good time at the beach and in the nightclubs there" (Manfred. Field notes 2010/11, 04.12.10), or an interviewee narrated about his viable bushfaller friend who had started a career as a footballer in a Swiss football club, and when he came to Cameroon for holidays, he invited friends to Douala. "We were staying in a very good hotel, we were eating nice food and drinking much alcohol. In the evenings we went to the nightclub. We had such a good time. This guy really has money: he was spending without even asking. Of course he also paid our trip to Douala and back" (Titus. Field notes 2010/11, 30.10.10). I have already pointed to that travelling can be seen as status enhancing: leisure related travelling, and of course especially international travel[528].

---

[527] The shorter the geographical distance, the more important a face-to-face encounter would figure in evaluations of people. A great geographic distance can excuse not visiting or not being present. Compare to chapter 4, page 223.

[528] See figure F78, a publicity of a travel agency, although only few can afford a flight, not to speak of the considerable hindrances for travelling, due to strict visa policies. Compare to chapter 2.

However, travel – also within national borders – has always been limited due to financial and sometimes time constraints. Here, an interviewee described the effect of mobile phone technology particularly in regard to avoiding travelling: "I was doing call box in Buea when the mobile phone just came up, around 2002 or 2003. There was one Pa (elderly man)... he used to sometimes come early in the morning. When he was doing a call he used to speak for long, sometimes up to ten minutes. At that time people were making very short calls because it was expensive. Then when he finished he even added money on top and said: Thank you my pikin (child). With this expenditure I would have gone to Bamenda and come back. So the phone has helped. Now I have already gone to Bamenda and back because I got every information that I need" (Patricia. Interviews 2009/R). Particularly related to a close spatial zone, as within the city of Bamenda, including outskirts and villages nearby, being reachable back and forth within a few hours, media of communication do obviously partly contribute to replacement of mobility. Alternatively, some issues can be discussed by phone, instead of travelling to the village[529]. The transport infrastructure also and conditions of roads is often insufficient. To reach a remote village in the rainy season is an extremely tedious, time consuming, and dangerous journey, and prices for transport rise with the worsening of the conditions of the road. At times places are better connected through virtual communication technologies than through physical flows of people and goods, roads and transport[530].

F78: Publicity for a travel agency in Buea
F79: Bus travel agency in Dschang, preparing for the journey to Bamenda

---

[529] Some people lamented that visiting has become scarce, because not visiting could be excused by making calls. However, connecting to people in the village communication technologies is sometimes difficult.

[530] Compare to figure F79, preparing for a bus journey.

In particular on a local level, New Media of communication help to organize sociality. Whereas before the advent of mobile phones, spontaneity in socializing on a local level instead consisted in a flexibility of temporal, but less spatial coordinates - since people are connected via mobile phones, both temporal and also spatial coordinates have become flexible to handle. It is related to that previously it was common that when wanting to see somebody, at times people had to wait for the person in question, or waited even in vain. "Earlier you went to meet somebody at Savannah street (in Bamenda), you went there may be in vain. Today you call him, text him: u de house? A dey come" (Isidore. Interviews 2010/11). The spontaneity of socializing, but also an expectation of achieving a certain control over the spatial and temporal configurations of sociality[531], is highly valued by mobile phone users. In this sense, New Media of communication can circumvent unnecessary movements: it can partly substitute travelling, especially on a local and regional level. Moreover, by using New Media, travel can be more easily organized, and reasons for travelling can be combined, for example with visits. I do not think that the phone replaces visits, but rather the phone enhances connections. In this sense it is also possible that travelling takes place due to the mobile phone, concerning contacts and opportunities (Ling & Haddon 2003). Likewise communication media have enhanced the chance to find accommodation: staying with somebody in Yaoundé or Douala, where a great deal of travelling is directed, is rarely a problem. Furthermore, New Media of communication have transformed individual's social networks. Due to mediating parts of social networks and sociality, and facilitated organization of face-to-face encounters, social networks in presence availability have increased in size. Adding to social ties in presence availability are those maintained over distance: social networks have become more far flung.

I would thus not agree that New Media communication and social interaction would replace physical mobility, neither on a local nor a translocal level, but they interrelate in complex ways[532]. Importantly,

---

[531] Some also lamented that people had in general become unreliable when a meeting was arranged, since they could use their mobile phones in order to postpone the meeting or apologize when they were late or not coming at all.

[532] This in turn points to a practical-evaluative dimension of agency, which is involved here.

physical mobility is continuously given importance, although transforming in character. Both – physical mobility and New Media technologies – enhance the possibilities for sociality, sociability and social networking practices, which do in turn contribute to further mobility.

Social mobility: a projective dimension of agency, social distance and hierarchies of power

Social actors do not only follow habitual paths of past routines, but they also deliberately search for, shape and orientate themselves to future possibilities. Emirbayer & Mische (1998:984) argue "that an imaginative engagement of the future is also a crucial component of the effort of human actors." Such engagements range from clearly purposeful aims and goals, up to rather "fuzzy" dreams and desires[533]. According to Emirbayer and Mische, the agency in this dimension lies in its hypothetical attitude toward possible futures. "Projectivity is thus located in a critical mediating juncture between the iterational and practical-evaluative aspects of agency." (1998:984). It contains a certain level of reflexivity and also habitual pursuing of past lines of actions. Schütz (1967) speaks here of life projects as a fundamental category of action, Giddens (1984) of intentional practice, and Baumann (2008) of anxiety, or Appadurai (1996) encompasses the orientation of possible futures to a "larger range of possible lives". Among others, Husserl relates the projective dimension of agency to the openness and interpretative approach to life, in ranges of alternative aims. According to Emirbayer & Mische (1998:989), the strongest habitual relation is given by anticipatory work, which follows the lines of past experiences and stock of knowledge, concerning an expected outcome. Another part consists of the narrative construction of typical trajectories in ranges of possible collective repertoires. These repertoires provide maps of action, but also allow experimentation with new resolutions according to emerging problems. Related to mobility, the projective dimension of agency is intertwined with social mobility, the capacity to imagine and follow up one's personal life project in order to "move forward". Expectations and motivations are tied up with imaginations of a "good life", being able to take over responsibility, acquire status, and partake in

---

[533] Compare to chapter 2.

practices of consumption. Practices of reaching out and link-up by New Media have to be seen as a part of people's coping strategies in a range of diversifying activities to secure livelihood and achieve status, activities on a local level in presence availability, and beyond.

Status is strongly connected with economic success and financial means, travelling and migration, but status is above all about connections to social others, and the negotiations of these social ties as – morally and socially – legitimized connections. I have shown that status involves a sensitive balance between individual and collective interests and needs. For example, social ties with migrants abroad are proportionally highly evaluated as potentially allowing access to imagined opportunities, from the perspective of non-migrants: I have highlighted the opposing views and interests of migrants and non-migrants. The moral legitimacy of status is a highly sensitive issue and an issue of continuous discourses, depending on connections of mutual relations of dependencies, in hierarchies of favouring and profiting[534].

Migrant's social mobility results in a perception of a social distance between migrants and non-migrants. It was expressed by non-migrants, that they felt that migrants distanciated themselves. It can be in communication, but also when migrants come to Cameroon: "When they leave Cameroon, they have their own type of people they meet and interact with. When they come back here they only move in their class of people" (Lizette. Interviews 2010/11). From the perspective of non-migrants, migrants emphasize their superior status in the sense that as providers, they would also be dominating transactions, exchanges and closeness in mediated social ties. They can be seen as imposing their will on non-migrants in communication and regarding support giving. However, the main critique here is provoked by expressions of migrant's apparent mistrust and suspicion towards non-migrants, who are assumed to be idle, lazy, greedy, deceitful, and always responsive. It is thus about non-migrant's articulating dignity and pride, when they emphasize their independence from migrants, or more generally, when they mock migrant's foolishness, their being out of place and snobbishness. "Africans have the image, that when they move abroad, they are big, they don't care any longer. I do not think that one is really

---

[534] This part on social mobility is related to the guiding question 3 of this book, how sociality – and liveness - is negotiated through New Media.

big because one is going abroad. Maybe the person can come back rich, but money is not everything" (Felix. Interviews 2010/11). It thus points to non-migrant's attempts to express moral superiority in view of migrant's selfish lifestyle, and their assumed sense of superiority regarding social status.

From the perspective of migrants, however, they feel deprived from status by only being seen as providers and resources[535]. Migrants need thus to find ways in order to follow up their life projects in strategies which balance their intention and urge to follow up notions of solidarity, as well as finding ways to reduce the pressure and to concentrate on their own social achievement as a priority. In such ways, they work on their economic or material base of investment in their society of origin, in order to eventually secure their influence on a local level despite their physical absence. The social difference implied by their being abroad is used in order to reconnect locally, and let their achievement come to fruition by re-installing themselves "at home". In these ways, they also gain a considerable power in their communities of origin. As described in chapter 6, as a tendency, migrants might reach the peak of their influence, when they themselves have reached a certain age, eleviated themselves from financial debts, and have proven their viability in their society of origin for a certain time. However, generally, young migrants can also slip into more powerful, but unsettled positions. Power hierarchies can be shifted regarding age, gender and position, in this sense they can possibly be reversed, since "being abroad" is seen as a dominant factor and strategic emphasis. In any case negotiations of power hierarchies are intensified. New Media of communication are used to connect and disconnect in order to pursue one's aims and advantage. It seems that power demonstrations even build on the use of New Media of communication: by, for example, reflexively conveying information and silencing other information, opening up or withdrawing in conversation, urging and putting the communication partner under pressure, or by controlling communicative exchange in order to pursue one's aims.

These experiences stand in contrast to assumption of some globalization scholars and ICT "prophets", who praise New Media as

---

[535] I have related extensively to migrants' perspectives, and how they deal with their situation, in chapter 6.

equalizing power relations in the sense of giving wide parts of responsible citizens access to equal chances and equal access to these media (compare to Friedman 2005, McLuhan 1964). Furthermore, characteristics that are often attributed to New Media, as connective, interrelating, unifying media, can similarly be used to differentiate, to disconnect, and particularize (Neubert & Macamo 2008:271,287, compare to McLuhan 1967).

Social mobility: connections to social others, status, and alternative pathways

Status is tied to different dimensions of mobility: generally, it is about a potential to be connected, in which New Media play a crucial role. Being connected to viable and influential social others is an opportunity for youth to move ahead, in particular in a country such as Cameroon, where social networks are crucial and dominated by clientele relations. New Media furthermore offer means to connect – at least imaginatively – to potentials assumed to exist in the world.

On a local level, and also among youth who have not migrated, conspicuous consumption practices are crucial to highlight connectedness and social status[536]. Such practices are related to a public sphere, where imageries and narratives circulate, where social hierarchies, positions and status are negotiated in competitive arenas of mutual perception and acknowledgement – or mockery. In general, in youth social groups, high levels of peer pressure are at play, to compete, take part, and gain the status of a successful individual. Many of these demonstrations of status or performances are related to faraway realms of opportunities in one way or another. Thus, demonstrations of status are firmly rooted in local imaginaries, practices and strategies for status achievement, and they are directed at status achievement in a specific locality, where they gain validity, which has a strong habitual component. Another aspect of such performance is the ability to anticipate success, self-assurance, and being focussed on one's aims. This is related to in numerous narratives in the local context emphasizing the great achievement of those who "have made it". It points to a strong future oriented aspect regarding models of success,

---

[536] See figure F80, mobile phones are items of prestige, indicating one's connectedness. Additionally, see chapter 3.

which at the same time relate to those of the past[537]. An interesting point here is, that models of success do not differ substantially among youth or the elderly and powerful in Cameroonian society. Differentiations are rather made through "more" and "better" concerning markers of success, such as – most importantly – expressed in practices of conspicuous consumption (compare Mbembe 2008). I have shown in this book that taking part in consumption is blurred: there are not necessarily distinctive markers, which render clues about the basis on which success is built. It means that the openness of interpretations is taken as a basis in order to perform connected and successful identities, also by those who lack a distinctive basis on which to achieve success, and by those who are not seen to be entitled to take part, such as youth. In such ways fact and fiction become interchangeable, or, as De Boeck (2003:59) expresses: "A change has occurred in the ways in which the representation and the represented reality relate to each other"[538]. It raises insecurities about the basis of and distinctive pathways to success, and it stimulates a general suspicion that those demonstrating success engage in subversive practices. At the same time, imaginations of basically legitimized pathways – as "genuine and hard work", education or migration - act pertinently in view of these insecurities, although the belief in these pathways dwindles in view of the life realities of most youth in the Cameroonian context (compare Fokwang 2008, Fuh 2010)[539].

---

[537] The narratives on decisiveness of actions stand in contrast to the often lacking knowledge of how to go about such aims. Here the dependency on migrants is great, to inform, organize and often financially support migration.

[538] Thereby, in his description of Kinshasa's "second world", or the world of the occult, De Boeck (2004) points to that "fiction" or the "unreal" ought not to be seen as mere fantasy, but implies a basis for action, which is "real" in the sense of a social reality. Furthermore, signifiers offer diverse interpretations and connections to "reality", so much so that a "reality of the double" dominates in interpretations of phenomena. I cannot go deeper into Boeck's subtle levels of interpretations here.

[539] Youth are lacking capital, for example to start a business, and having a sound education does not mean that young people would be able to pick up a job, when they are lacking financial means and connections. Financial problems and the urge for a quick redistribution of money on a local level is seen as potentially solvable only by migration, offering possibilities to save, to invest, and to come back later, in terms of solving these problems once and for all.

F80: Being connected is itself prestigious, here a mobile phone is proudly presented
F81: Scammers working at home

Confusions regarding pathways to success and the increasing arbitrariness between demonstrations and basis of success have also to be seen in the light of migration ventures becoming more difficult to achieve, and the disappointments of relations to migrants. It is also related to the attitude of many young people who see themselves as being part of an interconnected world, where they ought to play a role. When this access is denied to them, they would deliberately search for another way. This is how they might come about illegal practices such as scamming[540]. By scammers, these practices are most often only seen as a temporary means to survive, as they still strive to attain success by other, morally consolidated ways. Furthermore, scamming is seen as a youth's activity for gaining recognition among peers, and not suitable for gaining reputation as a responsible person in a wider societal context. As one interviewee said: "Just imagine, when I am a grandfather one day, how can I still be scamming! What could I tell my grandchildren? I am going to the cyber café to earn my life… no way. As I am getting older I must stop it." (Bryant 2009/FS). It is also considered as a way to emancipate oneself from the family, and eventually from unreliable migrants, and in this sense circumvent power hierarchies (compare Frei 2012). Furthermore, despite the local narratives that "scamming money can never do something good", some scammers contribute a considerable part of their siblings' school fees, and support their families. Most of my interviewees who were involved in scamming activities, said that they were generally not asked about the source of money, when morally

---

[540] See figure F81, scammers working at home. More often of course, they work in the public cyber cafés.

legitimized acts of redistribution were concerned. In such ways, sources of wealth and identities of providers have become arbitrary[541].

Hence, regarding pathways for social mobility, local practices seem - at least partly – to oppose an ideal view - of Western agents of development policies for example – that New Media ought to play a dominant role regarding economic purposes and business in order to enhance people's livelihood. My findings show, that practices for following life projects are strongly oriented towards profiting through social networks by posing claims on moral duties[542]. Instead of focussing on local economic innovations, youth focus to a large extent on status creation or on migration ventures, and economic innovations can figure in an illegal sphere. Furthermore, as I have shown, different trajectories for life projects can be transformed in their meaning and evaluation.

Virtual mobility: an evaluative dimension of agency, New Media infrastructure and a digital divide

The last dimension of agency relates to what "responds to the demands and contingencies of the present." (Emirbayer & Mische 1998:994), which is included in reflective and interpretative work on the part of social actors, in the sense of judgements according to situational conditions. Furthermore, Emirbayer & Mische (1998:994) state: "By increasing their capacity for practical evaluation, actors strengthen their ability to exercise agency in a mediating fashion, enabling them (at least potentially) to pursue their projects in ways that may challenge and transform the situational contexts of action themselves (…)." Here, the authors emphasize the interrelations between agency and structure[543]. So far, it points to an evaluation of situational conditions for actions, mediated through according practices. Encompassed within this mediation work is the characterization of unresolved problems, which is most strongly related to the past in apprehension of the present moment. It is also related to what Emirbayer & Mische (1998:998) call a

---

[541] Generally, the pathway to success through scamming is seen as highly problematic; it is often also related and compared to the discourses on bushfalling. Compare to chapter 3.

[542] However, such social networking can undoubtedly have an – indirect – impact on opportunities and livelihood.

[543] I will come back to the interrelations between agency and structure later in this chapter.

deliberation of issues at hand in the light of a broader field of possibilities and choices, which has here a strong connotation of future trajectories. Consequently, it also relates to a decision and execution of an action in view of the achieved overall appraisal of the situation. All situations are in turn defined in relational and temporal terms. Additionally, notions of tactics and manoeuvre, and making a difference though one's actions come in here (Emirbayer & Mische 1998:999, Simone 2005, Simone & Abouhani 2007, Mead 1932, Giddens 1984). Virtual mobility is in such ways related to the capacity to use New Media and adopt them according to aims, needs, the specific circumstances and conditions of their use, in an evaluative judgement of the users.

When I want to speak about a digital divide here, it is closely related to the relation between the agency of users and structure or temporal-relational contexts of action[544]. I have related to local conditions as a base to which media habits and uses are adapted to and built on. In a local context, felt disadvantages concern infrastructure, access to, and performance of technology, and thus the conditions for virtual mobility or New Media use. Even in regard to pre-given structures the negotiability of circumstances is given, the estimations of New Media users also depend highly on the evaluation of their ability, to act towards such conditions by mediating them through practices which are oriented to deal with them most appropriately under given circumstances: such are concerning material conditions as well as subjective evaluations. Although I have also highlighted what I have called "failures of technology" in an urban area, as well as to the most usual difficulties of young New Media users of staying connected, generally, the expectations towards the accessibility and also functionality of New Media in urban areas are high, as are the expectations of connectedness. It seems that urbanites have stronger notions of what it means to be connected, than rural people, who are so to say used to a "certain disconnection". Tazanu (2012:170) relates in his work to an interesting detail, that as a tendency, his effort to call relatives in the village was evaluated positively, whereas in an urban area, it was necessary to call more often in order to satisfy urbanites' exigencies for connectedness

---

[544] This relates to the guiding question 2, how New Media practices shape and influence social spaces on a local level, and vice versa. Especially, it relates to the 1st sub-question, how New Media's materiality contributes to their use.

with migrants abroad. In my book I have mostly dealt with practices of New Media use under conditions of a general accessibility of New Media, as well I have presupposed basic capacities of New Media users. In this subchapter I will relate to situations where connectivity through New Media is most limited, such as is the case in rural areas and concerning the lack of most basic exigencies for New Media use.

In certain places accessibility to New Media is limited in the Cameroonian context: beyond urban areas, the internet is simply not available. However, mobile phone communication is – increasingly - well established also in rural areas: Having network or not in a village refers to shifting levels of "bush". In rural areas power is also often not available. When I visited Moussa in a remote village for a few days, we organized a generator in order to watch films. This news spread around the area and people came from afar in order to use this opportunity to charge their mobile phones. When intending to make or receive a call, a hill needs to be climbed[545]. Incoming calls need to be prearranged: if expecting a call one needs to be located in a confined area and wait to receive it, or beep in order to be called. This means that calls do not have the aura of spontaneous communication under such conditions. In the village it is also difficult to purchase credit. There are no call boxes, and one can only obtain credit from others transferring it to one's phone. When a mobile phone has a problem, there are no repairing opportunities[546]. After visiting another friend in a village, when travelling back to town together, he took a large number of other people's mobile phones along which needed to be repaired.

---

[545] See figure F82. See also figure F83, a specific location at the wall, where network is available, and where the phone can be hung on the wall, in order that the inhabitants of the compound would be able to possibly receive calls.

[546] Or if there is a repairer located in the village, he would lack the required spare parts.

F82, 83: In the rural areas, connection is possible in certain places, where one has to go to when a call has been arranged. Here, on a hill (Esu, NW), and in the compound only at a particular spot behind the kitchen wall (Fontem SW).

However, apart from infrastructure, the main reasons for a feeling of disconnectedness are apparent in the difficulties of staying connected, such as "keeping the mobile phone in use", "fuelling the phone with credit", or lacking internet literacy, drawbacks, which limit people regarding their New Media use[547]. In this book I have hardly related to the issue of illiteracy, since I was doing my research among – more or less – literate young people in an urban area. Mobile phone technology is often praised as circumventing the inability to read and write, or knowledge of a written language. In fact, it enables non-literate people – also those with limited technology knowhow – to communicate. However, illiteracy involves limitations regarding dealing with calls and additional communication features of the mobile phone. For example text messaging is limited to literates, but also some of my – basically literate – interviewees uttered that they hardly wrote text messages because they were not good in expressing themselves in writing. Non-literate people would not be able to store contacts, or to read names in the phone's address book, when intending to make calls, or they cannot read the name of the caller displayed when a call comes in, as I have observed[548]. In this sense, literacy has an influence on people's abilities to handle calls and calling, which need in such cases to be bridged by the help of literate others in presence availability.

---

[547] I have described such drawbacks elaborately in chapter 4, and related to the notion of "keeping the mobile phone in use", in chapter 3.

[548] I witnessed a friend's grandmother in the village making a call through the friend's mobile phone, on the occasion of her visit. She needed help to make a call.

In any case, evaluations about local New Media conditions in terms of infrastructure and abilities of New Media users – such as literacy and financial means – are integrated in overall evaluations of mediated communication performances, locally, but also regarding transnational mediated communication. I have mentioned that failures of technology can be taken as excuses, thereby building on imaginaries of media's conditions in the environment of the communication partner. In the same sense as constant and easy availability is construed towards migrants abroad, migrants equally use the weaknesses of media's infrastructure and functioning in Cameroon as excuses, such as that they tried to call but "could not get through". Or non-migrants claim on migrants' taking over the costs for calls, which must be less expensive for them abroad, or they insist on being called because the internet lines would be weak. The ultimate characteristics of such explanations are evaluations of each other's social positions, thus, technological failures can serve to camouflage failures in sociality and discrepancies in regard to social duties. In such ways, infrastructural conditions underlying New Media use can also contribute to insecurities regarding connecting and connectedness. These examples point to both, a creative ability of New Media users to deal with – real and imagined, technological and other – conditions, but also, they point to the factuality of limitations of the functionality of New Media in the local context. The so-called "digital divide" is an issue here. However, it cannot be simply explained by lacking access to New Media, and it is not automatically removed when guaranteeing this access, but has several aspects[549]. As I have described in this book, New Media can also be seen as opposing "a Western notion" and pre-assumption of what these media could be and do. A digital divide is thus felt in differing ways, in a broad sense as a lacking level of participation in opportunities of a "good life".

Virtual mobility: reproduction of imaginaries, and their according interpretation

When talking about virtual mobility, I would also highlight the possibility of mobility through media, while staying in one place, instead of physically moving. Earlier in this chapter, I related to the

---

[549] Compare also to the next subchapter, concerning the agentic dimensions and the different dimensions of liveness.

414

interrelations between physical mobility, travelling or migration in different scopes of spatial reach. I have also shown how New Media and social mobility interrelate. Here I want to again come back to the issue of imagination. Although, as I have explained in this book, imaginations, for example of "white man's kontri", of the opportunities of a migration venture, and so on, are to a large extent guided and influenced by long standing narratives and images of success in the local context, New Media and imagination have a great evaluative force, and they are used in order to reflexively deal with these imaginaries. New Media can be seen as a means for active media users in order to take part in the production and reproduction of imageries and imaginaries in practices of imagination, and practices of New Media use can be seen as coping strategies, in order to strive for a "good life", and to deal with given conditions, which points out to both other dimensions of agency, habitual and projective. Such imaginaries are also dealt with in a framework of interpretation and judgement according to the arising circumstances and situations.

These imaginaries and imageries deriving from the virtual realm of media do have very tangible impacts on everyday life conditions, on motives, actions, and policies, and thus, they are also consciously used and deployed. Furthermore, the immigration authorities of the European Union have recognized the power of local imageries, and try to integrate this in some attempts to control or reject emigration from Cameroon, since Cameroonians usually have only minimal chances of obtaining a visa or a residence permit in European countries[550], but Cameroonians and Nigerians make up a great number of African asylum seekers. I relate here to a public awareness film issued by the IOM sponsored by the European Union and as well the Swiss Ministry of Migration BFM, broadcasted in Cameroon and Nigeria. It shows a young man calling his father "at home"[551]. The environment of the father is neat and decent, whereas the environment of the migrant is dark, cold, and rainy. He is phoning from a phone booth, huddling in his anorak, telling his father on the phone that everything is fine, that he had

---

[550] Apart from when they obtain a residence permit for education or when obtaining a visa upon marriage.

[551] See the link to the spot on YouTube, http://www.youtube.com/watch?v=AJa8k1FDPeI.

found work and a place to stay. While saying this, images of him running and hiding from the police, sleeping under a bridge in the cold, and begging on the streets, are faded in. The film wants to convey what is already clear to non-migrants in Cameroon – that life can in fact be difficult "over there", and also that "migrants very often do not tell the truth" about their lives. Furthermore, the film intends to deconstruct the high status migration induces in the local context. Here, however, there are sufficient positive examples of bushfallers returning, displaying their wealth in Cameroon, and also, a failure is instead attributed to individuals, leaving the imagined opportunities of "a better life in the West" intact. A further interpretation is that Western policy makers try to infiltrate such negative images in order to prevent Africans from migrating. In this sense, evaluations are made along the lines of local narratives and life stories of migrants, the conditions in which the film is seen, and the assumed intentions of the producers of the film.

Imageries and imaginaries could also purposefully be employed and worked on in transnational communication, in order to reach a certain aim or make a certain impression towards the communication partner. This points to the contribution of migrants to the reproduction of narratives and imaginations about "the West" in the Cameroonian context, which is related to the conflicting priorities of interpersonal communication and strong societal imaginaries, and to what extent mutual communicative exchange contributes to invalidating these imaginaries or not. On the one hand, migrants can stick to the high status attributions, which are rendered to them through migration, out of pride, as some non-migrants put it, or they do not communicate their life circumstances out of shame and fear of being seen as failure. On the other hand, migrants are prone to being seen as traitors by non-migrants when they do not stick to the image, that they ought to perform a role as providers, and non-migrants might not be ready to believe migrants when they try to convey the reality of their life circumstances abroad. I have tried to show in this book[552] that practices of New Media use, regarding information and communication, do not necessarily lead to a re-evaluation of youth's sketches of a good life. The internet especially, turns out to be a media which does not necessarily contribute to clarification and knowledge, but it can be adopted to follow up youth's

---

[552] Especially in chapters 2 and 6.

imaginaries and aims. The same can be true in the field of interpersonal mediated communication, which often contributes to the reproduction of strong societal imaginaries.

Such diverse interpretations occur in a framework of evaluative judgements in a world where pathways for a better future are seen as fraught with immense opportunities but also inconceivable dangers. Competing and contradicting images and imaginaries can raise insecurities, which can possibly to a certain extent be absorbed by individuals' intentionally relating to strong convictions. An evaluative dimension of agency comes to the fore, when New Media users are reflectively dealing with prevalent conditions and opportunities for imagining.

Liveness, transforming sociality, and the dimensions of agency

Relating to the different dimensions of liveness[553], liveness can be seen as a sensory experience, as potential and as work. I suppose that mediated connectedness and the opportunities for liveness must have specific impacts on sociality. This is so regarding liveness as a sensory experience, which provides a sense of closeness and grasping the communication partner in mediated social interaction. This sense is related to an ideal version of face-to-face sociality, deriving from past social experiences. Regarding a sense of projectivity and the basic possibility to be connected, it is the dimension of liveness of a potential, which points to high expectations in regard to the availability of social others through New Media of communication. Then, regarding liveness as work, it is about maintaining social ties over distance and time, by "working on the gap", including negotiations about conduct and situating or locating the communication partner, by adopting different tools of New Media in present interaction, in the practical-evaluative dimension of agency. Put simply, we could see liveness as a specific mode of sociality (compare Zhao 2003, Couldry 2004, Auslander

---

[553] Which I have introduced in chapter 1.

1999)[554], transcending from a local – face-to-face – to a mediated sociality[555].

Liveness as sensory experience: a habitual dimension of agency, and a notion of an ideal sociality

Relating to a habitual dimension of agency, Emirbayer & Mische (1998:980) state that actors locate typifications of past and present contexts of action "in relation to other persons, contexts, or events within matrices composed of socially recognized categories of identity and value." These "matrices" contain "nuanced lines of inclusion and exclusion, acceptability and non-acceptability, within crosscutting contexts of action." Even though these matrices are taken for granted, they also entail a reflexive evaluation where they fit and thus create stability, and where they need to be reworked. This evaluation, situated between taking reference to prevalent modes of sociality in regard to their evaluation in new frameworks, act pertinently related to liveness as a sensory experience, when taken to the level of mediated sociality. In general, liveness as a sensory experience relates to the capacity of media users, to experience the social other mediated through diverse senses. Media sociality is from this point of view an extension to liveness in face-to-face encounters, which seems to be taken as a base of evaluation, when mediated social interaction and communication is described and evaluated in a habitual dimension of agency. This reference contributes to a perceived quality of mediated social interaction, and has an influence on qualities of social ties[556].

Face-to-face interaction, physical propinquity and presence availability (Giddens 1984) still form the core of social life and are emotionally central, but mediated social interaction is increasingly adding, intersecting, and playing a more important role. According to the evaluations of the interviewees, social ties, which draw on place and sharing a location, built in co-presence and thus based on trust and

---

[554] The perception of liveness is not only bound to social others, but also to media contents and imagination.

[555] As I have already pointed out in the previous subchapter concerning mobilities, I will emphasize one specific and dominating dimension of agency in each dimension of liveness, by adopting a few examples. The dimensions of liveness will most decisively relate to the guiding question 3, since I relate to them as modes of sociality.

[556] I have related to this specifically in chapter 5 in this book.

familiarity, are central and also provided the base on which future mediated social ties are built on. In communication performance, physical presence is seen as best to grasp each other, since social and emotional cues are visible and perceivable, social conduct can be observed, cross-connections contribute to the fulfilment of social exigencies, and the interaction takes place in a specific social context, under conditions of a shared background and knowledge and in a presupposed context of mutual social control. Obviously, this is a highly idealized version of face-to-face sociality, regarding the sensory grasping and interpretation of other's motivations, and physical propinquity providing a sense of mutual control and closeness.

An evaluation of actual face-to-face interactions points clearly to such an ideal shift of such a mode of direct sociality: There are certain peculiarities of face-to-face social interaction in the Cameroonian context – and in particular in Bamenda[557]. On a local level, spending time together and being with each other, is as a tendency seen as more important than deep rooted communicative exchange. I am speaking here of repeated daily encounters of those in physical propinquity. This can be the case even more decisively concerning a rural, village based everyday sociality. Förster (2010) describes that verbal articulation is not considered as central to social relationships in the village, but in physical co-presence silence is also common. As I have described in this book, being together is related to in expressions such as "showing one's face", "showing concern", or "doing something for each other"[558]. It was deplored by young New Media users, that in mediated sociality "you can just talk. All is talking" (Anastasia. Interview 2010/11), or every exchange was reduced to verbal expression. It leaves out other highly evaluated sensory experiences, seeing a person, being together in mutual adjustment, sharing issues concerning everyday life. Simultaneity and spontaneity is an issue here. The ideal mediated sociality is defined

---

[557] As I have already mentioned earlier in this book, Bamenda is an urban place, but also has many characteristics of rather rural areas, people know each other in the quarters, there are strong ethnic ties and connections to the villages of origin, a low economic performance – people have a lot of time to spend with each other - and a limited central area in town where the possibility to meet each other is high. In this sense, Bamenda can be considered as a rather provincial town, which also influences modes of sociality accordingly.

[558] Compare to chapter 4.

according to possibilities for simultaneous interaction, and constant and spontaneous availability, such as on a local level a person can drop in at the compound of friends and relatives at any time. Spending time together also relates to relations of mutual knowing, and a sharing of a base of knowledge concerning each other's relations, duties and rights. Here, however, slippages come in, as well as in the realm of face-to-face sociality. I have mentioned in my book, that mistrust towards others is sometimes great, even among the most close social ties of family and friends[559]. It transpires that not only migrants turn out to be treacherous when leaving to go abroad, but that deceit is deeply engrained in the social fabric of the local society. Mistrust is mainly related to an insecurity about an evaluation of temporal-relational positions of others, or: who is doing what with whom. Although face-to-face sociality and presence availability is seen as favourable to assess others in such terms, total social control is rather a myth, but points out to the importance of such evaluations of others in this society (Simone 2005, Fokwang 2008, Tazanu 2012). I have also pointed to that New Media of communication raises the fear of concealment and treacherous activities out of public control, also under conditions of presence availability[560]. This fear is increased when social others move out of presence availability. When social ties – for example with those who have migrated – merely rely on mediated communicative interaction, shifts in sensory experience are evaluated differently. According to media's specific mediality, by simultaneous communicative exchange, technological features can serve to strive for a simulation of co-presence, which is of course only an approximation. Media confine sensory experience of the communication partner to certain senses, the conditions of mediating are seen as imperfect. Lacking social and emotional cues in mediated communicative interaction, and of assessments of other's motivations, culminate in the fear that migrants will slip beyond control[561]. As a consequence, it is the morality of an ideal sociality, which is more strongly emphasized. At the same time, experiences and narratives about migrant's disconnecting and lacking responsiveness are as prevalent as

---

[559] Compare to chapter 6.

[560] Compare to chapter 3.

[561] This assessment happens for example along the lines of presumptions about other's potential availability and responsiveness through New Media, see next subchapter.

the narratives about migration and the potentiality of "white man's kontri". In this sense, an ideal face-to-face sociality is super elevated in regard to slippages, which are likely – and often presumed - to occur in mediated communication.

The local situation of New Media practices, technological environments and conditions for mediated social interaction seems to underpin such a view and evaluation[562]. This concerns the local face-to-face context and sociality in presence availability, but as well the realm of media. As most New Media use is effected in public spaces, these practices contribute to the sensory experiences in these spaces, for example the integration of phone calls and talks with physically absent others into face-to-face sociality: I have described in this book, how communication with absent others has become a means of status demonstration[563]. However, the publicness of mediated social interaction is also seen as a risk. Narratives of internet scams, unauthorized entering of others' mailboxes, installed spyware and a general lack of privacy in cyber cafés[564], eavesdropping on phone calls, catching text messages directed to others, and so on, are wide-spread in the local context. For these reasons many people can restrict their mediated social interactions from the public: in particular mobile phones allow a certain withdrawal and concealment of mediated social interaction, or some people strive to have private internet access, due to such reasons. Preventing others in co-presence from witnessing mediated social activities and communication is, I suppose, rather an effect due to fears of deceit, than due to desires for privacy in the sense of designating distinguished "private" activities[565]. In this sense, an orientation to social interaction with absent others can enforce a feeling that media users can circumvent local social ties and duties, and that they can be potentially accessing resources through mediated social interaction without others knowing. New Media and mediated sociality raise insecurities about who is doing what with whom. These ideas point

---

[562] This section refers to the guiding question 2, how media's materiality impact on local social spaces, and vice-versa.

[563] See figure F84.

[564] See figure F85, a situation of communication in a cyber café. It can be due to various reasons, why these people share a computer. In this case it was lacking internet literacy of the female user, who needed a friend's help.

[565] Compare to chapter 3.

to discrepancies in a notion of an ideal sociality in co-presence, where suspicion is a prevalent feeling, however, anxieties might be enforced due to imaginaries of great opportunities when reaching out beyond the locality. In this sense, in the context of expectation maintenance, references are implicitly oriented toward an ideal face-to-face sociality, but also seamlessly adding to previous but enforced anxieties.

Liveness as sensory experience, a sense of place, and a shared space for togetherness

Apart from physical appearance, and thus emotional and social cues, it is the absence of a shared physical base for togetherness, for shared knowledge and a sense of grasping a persons motivations, relations and activities, and eventually locating each other due to differentiating life-worlds of the communication partners, which often "complicates" mediated communication. At the same time, this lack is super-imposed by according imaginations and presumptions.

In simultaneous mediated communication, it is about creating a virtual time and space for sociality under condition of physical absence, or creating communality in media. It is something, which is most often not verbally related to, which slips into the communicative exchange as another dimension of the social interaction, contributing to create a "feel" of media, as Tazanu (2012:64) has called it, and of the communication partner. It relates to the maintenance of expectations, for example concerning the imagination of the communication partner's environment and media conditions (Emirbayer & Mische 1998:981). In mediated communication, one knows implicitly that the other person is right now too sitting in front of a screen, or speaking through the phone, which is creating a kind of connection in an imagined sense. Wilding (2006) and Licoppe (2009) have called it connected presence or connected relationships[566]. Miller and Slater (2001:4) relate to the term "virtuality", that "the term suggests that media can provide both means of interaction and modes of representation that add up to "spaces" or "places" that participants can treat as if they were real": It is referred to as "meeting in chat", as if "chat" is a location to meet and socialize. However, I think this serves for coordination since internet communication is most often place based - since the mobile phone is

---

[566] Compare to chapter 5.

not, this coordination is not necessary[567]. I would thus not speak of a place for mutual communication, but of an imagined space[568] for togetherness, mutually built up in communicative exchange, and when simultaneous, a time of intimacy and mutual attention (Goffman 1972,1983). It is a virtual and placeless world constructed in social interaction.

Thereby, the role of place, or a physical material location for social interaction plays an interesting role, which highlights ambiguous tendencies[569]. Although some young New Media users, who are in contact with migrants abroad, try to deliberately create a sense of migrant's environment in view of the difficulties of creating a conclusive impression of the communication partner in mediated social interaction, for many the imagination of migrant's environments is not considered as a priority. The sense of togetherness, in the sense of expressing loyalty is considered far more important, and this loyalty needs again and again to be reconfirmed by calls and other mediated social interaction. In return, it is precisely this difference regarding the communication partner's life-worlds, which provokes misunderstandings and tensions in transnational social ties[570]. Achieving a "feel" of the location of the communication partners abroad in a material or geographical sense, gains importance in two instances: first it is related to an underlying desire to understand migrants conditions of daily life, when intending to work on mutual closeness, and secondly it concerns the feeling that migrants slip out of control. Regarding generating a sense of place in order to enhance closeness, a feeling of place can support mediated communication, and

---

[567] With exceptions, since environmental influences and constraints – or time limits - can disturb a phone conversation, hence, calls are sometimes also spatially coordinated.

[568] A space, which is not a place, but of course influenced by places. The locations of the communication partners, as a base for the use of communication technologies and interrelated conditions, local face-to-face social networks and imagination - or memory - of each other's location, influence mediated communication to a large extent. See chapter 5, or later in this chapter.

[569] Here specifically guiding question 2 comes in.

[570] This was apparently recognized by some of my interviewees, who stated that they needed to obtain a sense of their migrant friends' environments, in order to be able to understand their life conditions abroad. It of course also depends strongly on the closeness and intensity of the maintained contact, whether extra clues gain importance in mediated communicative interaction. Compare to chapter 5.

is in itself a characteristic of it, since New Media users are situated in a specific locality[571]. While communicating, media users are part of their immediate environment, and they imagine – most often – the places where their communication partners are located - which can be related to what Moores (2004) calls "doubling of place", that communication partners are in more than one place at the time, or as Fortunati (2002:515,521) has pointed to people involved in mediated communication are only "half present" in the local sphere[572]. I suppose that the less cues are conveyed from the communication partner – for example in a non-simultaneous email - the more detached from a real place is a communication act experienced. The more cues are conveyed from the communication partner's environment while communicating – as for example when calling from or to Cameroon, noises in the street can be heard – the stronger is the communicative interaction related to a specific place – depending on the communication partner's ability to imagine the place[573].

F84: Using the mobile phone with ear plugs, Yaoundé
F85: Sharing a computer in a public cyber café, Bamenda

Regarding the growing importance of place when the fear of a loss of control over migrants is concerned, this relation to place is dominated by implicit assumptions about places, for example concerning the imagination of great opportunities abroad. The evaluation of

---

[571] I will relate again to the importance of a feeling of place for generating a feeling of closeness, later on in this subchapter, concerning the dimension of liveness as work.

[572] Compare to chapter 4.

[573] In some cases, non-migrants said that they wondered why it was always so quiet, when migrants called, in the sense that they did not obtain many cues to the migrant's environment. Compare to chapter 6.

transnational social ties refers to a large extent to such presumptions. This comes to the fore when comparing transnational social ties to those mediated within national boundaries. Even though presence availability is not given when somebody moves for example from Bamenda to Douala, and numerous stories are told relating to the possibility of such a move creating social and emotional distance and even losing touch, many people make a decisive difference between these social ties within the country, compared to those abroad. A potential presence availability seems to be assumed vis-à-vis those within the country, although people might only see each other once a year on Christmas, and do not even call each other often. The main issue here is the assumed potential availability, and the great shift to becoming "out of reach" when somebody leaves the country. Occurring frictions in relationships with migrants, which imply for non-migrants, that their access to these opportunities through them is endangered, are accordingly interpreted as the failures of migrants, who have assumed a selfish and greedy lifestyle, due to the influence of their environment on them. Regarding liveness as sensory experience, I have highlighted New Media user's orientation to habitual pathways, when their handling of transformed social situations – here mediated – is concerned.

Liveness as potential: a projective dimension of agency and expectations of being connected

Concerning liveness as a potential, an imaginative engagement with the future is distinguished in different lines of interests, in a projective dimension of agency. This comes to the fore in particular related to differing interests of migrants and non-migrants. Thereby, a strong habitual relation is given by anticipating outcomes concerning mediated social interaction in these transnational social ties according to practices of New Media uses, and narrative construction of typical trajectories concerning repertoires of collective sets of action (compare Emirbayer & Mische 1998:989). In order to subsume the potentiality of New Media for media users, I would break it down to the potential for connectedness. The availability and accessibility of New Media of communication and information enhance a feeling for the potentiality of this connectedness through the opportunities for liveness, which they offer. It likewise enforces the urge to maintain connections to others by

New Media on the basis of a felt morality. These repertoires provide maps of action for life projects and imaginative scenarios, but also follow up reflective choices and inventive manipulation, related to expected outcomes of trajectories and repertoires of action oriented to the future (Emirbayer & Mische 1998:988,990). Before the advent of New Media, being absent or being abroad meant that people hardly had any opportunity to stay in close contact. Now as this has changed, negotiations occur along the line of the expectation to be connected despite physical absence.

On the basis of a felt moral condition to be and stay connected to those abroad, evaluations of migrant's New Media uses relate to an evaluation of an ideal connectedness, which derives from a notion of an idealized face-to-face sociality. Earlier in this chapter I have described the importance of the reference to a notion of an ideal sociality or liveness in face-to-face contexts and encounters, which is super elevated in narrative constructions, as promising total social control and an overview of other's social relations and activities. This idealized notion of an ideal connectedness has gained a new meaning in the context of transnational sociality. It is here, where connectedness in the sense of a seamless sociality and solidarity gains a heightened importance, despite physical dislocation, in view of imagined life circumstances and media environments abroad and the desire to control access to such opportunities through migrants. In this sense, migrant's unconstrained accessibility to New Media of communication is presumed, likewise is their economic viability for taking over the expenditures for communication, and also implied by it, their ability to support – they are evaluated according to their assumed duty to keep social ties alive. These assumptions figure strongly in the background for the evaluation of migrant's media habits and conduct in media use[574]. From the perspective of non-migrants, the feeling of urge is especially important when the impression arises that migrants withdraw from this moral condition: when social interaction is solely mediated, lacking physical presence, social cues and differentiating life-worlds make it more

---

[574] Note that a tendency to a practical-evaluative dimension of agency is included here.

difficult to access others, and also it is easier to avoid being accessed[575]. I have related to the narrative that migrants tend to dissociate and disengage from the maintaining of such an ideal sociality in mediated social interaction as a matter of fact.

The moral urge to stay connected underpins the notion that migrants abroad ought to be always accessible and reachable, which likewise comes to the fore in New Media habits. Being accessible and reachable – on phone, or online – involves continuous evaluations of the quality of the respective mediated social tie. In Cameroon the mobile phone is constantly switched on, as "one should never miss a call", and there are only few social spaces and situations where – receiving or effecting – calls is perceived as unacceptable, and also here it is a matter of negotiation and explanation. The acceptance of calls being a part of face-to-face sociality is great, being connected to others is seen as a prerequisite for success and in itself prestigious. Although there are explanations given in circulating narratives as to why migrants abroad would not always be able to answer calls, or why they switch their phone off or put it on silent, because they are busy, these narratives also contain a feeling of bitterness and doubt if such conditions in migrant's daily lives would not just serve as excuses[576]. As I have already pointed out, on the part of migrants, such inconsistencies in communicative performance can also be partly explained by failures of technologies, such as "the lines are not passing", the communication partner is not online, or an email was sent but not received. Such excuses are to an extent accepted, since they are common incidents, which cannot be verified, but they always contain the doubt that they only camouflage a non-interest in communicating. Another dimension of evaluations of migrants' conduct in mediated sociality is the degree to which they seem to take initiative for communication. Non-migrants usually complain that migrants are not taking the initiative for communication. This evaluation is based on the assessment, that migrants are ultimately to be held responsible for maintaining connections: they are held liable to their

---

[575] Compare to explanations, why migrants tend to dissociate and distance themselves from non-migrant's attempts of getting in contact with them, also later in the subchapter concerning the dimension of liveness as work.

[576] Not communicating – in the sense of not being reachable and responsive – under the view of supposedly being easily connected, says much about the qualities of a social tie.

associated moral duty, to show generosity, which is extended to mediated communicative interaction. Furthermore, on the premise of the assumed tendency, that migrants emphasize social distance, as well as their tendency to use New Media as a means to disconnect rather than connect, they are seen as morally doubtable individuals. Migrant's withdrawing makes non-migrants to be rendered to a powerless position, since they feel dependent on migrant's goodwill to stay in contact with them.

The experience of disappointment in regard to transnational social ties has a direct impact on the use of New Media. A general practice by non-migrants is what I have called link-up communication, which serves as an inexpensive means to remind migrants of their duties, which they seem to circumvent. I have mentioned in this book, that link-up practices of New Media use are enhanced by conditions in the Cameroonian context, which are most importantly lacking financial means, and lacking computer and internet literacy. Link-up practices however seem to be more inclusively explained by integrating explanations, which are based on disappointments about sociality experienced in mediated social interaction with migrants. At times, non-migrants are not able, but also not willing, to invest more in transnational ties, since their experiences do not match their expectations. Link-up communication is in turn negatively evaluated on the part of the migrants: Ways of contacting and conduct in mediated communication are taken as a base for the evaluation and consideration of social ties[577]. In this sense, such habits of communication can have further distanciating effects in transnational social ties, and New Media's assumed potential for connectedness is not followed up because migrants abroad do not perform according to the moral integrity, which is assumed – as a potential – by non-migrants.

Liveness as potential, arising frictions, or transforming meaning of relatedness

As I have described, evaluations of migrants' conduct in transnational social interaction is based on the assumed potential for an ideal sociality over physical distance. It is related to New Media users'

---

[577] Compare to the next subchapter, related to an evaluative dimension of agency, regarding liveness as work.

potential capacity to "smoothly" follow up their life projects through social others, and especially, through those who have moved abroad. I have already mentioned in this chapter, that related to mediated social ties maintained on a national level – accordingly out of presence availability – being in close contact is not attributed the same high importance as related to transnational social ties. It is interesting, that it is in particular on a transnational level, where a "perfect sociality" is imagined and expected on the base of the availability of New Media of communication.

I have mentioned that the high evaluation of transnational ties may transform the base for sociality. This is so regarding a shifting base of claiming relationship and consideration of the importance of a social tie, and shifting hierarchies and attributed responsibilities. Regarding a shifting base of claiming relationship, I have described in this book, how somebody who has moved abroad raises attention and consideration. It could be a reason to try to reconnect to an old classmate, or to emphasize a social tie to a distant cousin, with whom one was not in close contact while living in Cameroon. The heightened importance rendered to transnational social ties is reflected for example in how people have ordered social contacts in ranges of importance in the address books of their phones, or in their messenger chat address lists, where, apart from peers, relatives, close family members or friends, migrants often figure in an extra category under "important persons", or in an extra section of "bushfallers". Regarding shifting hierarchies, a younger sibling abroad can be held responsible for supporting the parents instead of the eldest brother in Cameroon. Or, as one interviewee expressed: "(…) It only depends on if somebody is making money. (…), even when the mother is abroad. Here the children would take care of her, if she is abroad, everybody would depend on her and complain" (Valeria. Interviews 2010/11). In this sense, being abroad is valued more important than age, gender, and also the relation towards a migrant. Although these shifts can be explained by an attribution of great potential regarding transnational social ties, which is based on super-imposed imaginaries, another part of the explanation seems to derive from physical dislocation and specific medialities in New Media. From the perspective of non-migrants, New Media of communication can be usefully employed in order to follow up personal social networks, without others in presence availability necessarily knowing. This concerns social ties in presence availability, up to national and transnational social ties: who is doing what with whom is less transparent to others. It also implies a shift regarding "rights" to claim relationship or to ask for support, towards certain others. Provided that one has a contact number or an email address, demands can be conveyed, circumventing others' judgement whether one is entitled to do so, due to a lack of cross connections and shared realm of face-to-face sociality. This points to shifting ways of how to effect such communicative acts or demands through New Media of communication. In face-to-face sociality, a demand might encompass careful evaluation and a conveying of the demand in a personal encounter. At least it would be an elaborate phone call on a local level, when a meeting is not possible. Here the impact of geographical distance comes in: when conveying a demand in New Media of communication, especially on a transnational level, communicative performance is often reduced to sending an email, making a short call, or possibly just link up through a beep or a call-me-back text message. Such communicative performance needs again to be related to an assessment of great potential: when somebody is abroad, it raises expectations, which are often valued as higher than social relatedness or notions of conduct towards a socially higher-ranking person. Furthermore, although migrants rise in social position, they can also take over ambivalent positions, such as a younger sibling studying abroad can be seen as socially subordinate, and not largely respected, but is likewise asked for support.

It is seen to a certain extent to be unavoidable, that migrants with time shift their attention and responsiveness, when they give more importance to their environment and friends in their country of residence, instead of their kin in Cameroon, as it is lamented. However, in view of migrants, who – due to the high expectations and claims posed upon them - reduce their social ties to the closest core of family and friends, and who evaluate even those ties according to performances such as politeness, understanding, and reasonability of demands, and it is obvious that disappointment will be felt by those non-migrants, who are not included in the small circle of favoured social ties[578]. Accordingly, from the view of non-migrants, relations to very close relatives were often described as good, and understanding of their situation abroad was expressed. Furthermore, regarding migrant's supportiveness, differentiated evaluations were given under consideration of their life circumstances. It is the evaluation of the relations to more distant relatives and friends, which brings disappointment to the fore. Since these ties are for most young New Media users the ones in question, narratives about the disloyal and treacherous migrants are dominant.

When distanciation is felt in transnational ties, it is tied to feelings of disappointment, not in the potential ability of New Media technologies, but towards specific and general social others, here the migrants, whereas, as one interviewee expressed: "When the communication partner is good, time and space cannot not limit (...)" (Gladys. Interviews 2010/11/EC). In this sense, the potential of New Media to bridge spatial and social distance, is super elevated in regard to transnational social ties, and the non-fulfilment of the respective expectations by migrants is likely to be explained by their personal failure. Liveness as a potential is thus seen as a critical trajectory.

Liveness as work: an evaluative dimension of agency, and tensions arising in mediated social ties

In liveness as work and maintaining a virtual presence, a reflective and interpretative work on the part of social actors comes to the fore. Actions are performed in accordance with evaluative judgements from the part of the actors in relation to the situational conditions in which they act. These situational conditions are emergent events or temporal and relational contexts. Actors use their agency in a mediating fashion, between the contexts and conditions in which they act, and deliberately influence those conditions. Thereby, deliberation, in the sense of a reflexive judgement of action to apply in specific instances in the light of a broader view of aims, and decision making, in the sense of a resolution towards a specific action, refer to a general "incompleteness of situations", such as that deliberate judgement and decision also have an

---

[578] This relates as well to shifts in the evaluation of mediated social ties, and their re-evaluation, regarding their emotional and social closeness and relatedness, which can imply shifts. See the next subchapter.

arbitrary character allowing for alternative actions (Emirbayer & Mische 1998:998,999). In this sense, in the course of this book, I have on the one hand repeatedly related to tendencies, which correspond to prevalent normative narratives and general assessments of "being connected" in the local context, and on the other hand to the actor's practical handling of situations, which often also contains alternative solutions towards emergent social situations[579]. In the following two subchapters concerning liveness as work, I will relate first to prevalent tendencies in the light of normative evaluations of social situations, and secondly to a deliberation of such presumptions and taking adapted action. Liveness as work relates to a continuous effort concerning a morally acceptable maintaining of a social tie over distance and time. It can be related to a continuous strife to secure liveness in mediated communication or to avoid too much liveness, when preferable in specific social situations. Work on mediated social ties encompasses references to a stock of knowledge, which draws on social experience in the past, on respective media experience, and work in the sense of adapting to specific situations in relation to specific social ties. This work thus relates to learning processes, to habits, and to a variability of social situations[580].

Concerning the reference to prior experiences, I have mentioned the prevalent idea that migrants have the tendency to change when they are abroad, in the sense that they detach themselves from those at home. It leads to the impression that migrants emphasize social distance, and thus negate the base of social relatedness or friendship, which has been built under conditions of presence availability, on relatedness or on equality. Regarding mediated communicative exchange, it implies that non-migrants are often not ready to invest in transnational social ties, as the outcome of such effort would not correspond to their expectations, and in order to remind migrants of their duties, they turn to mere link-up communication. Such forms of communication are also employed in the

---

[579] Thereby, I have tried to see social networking practices and social networks as a continuum, from a local level of face-to-face interaction and presence availability up to mediated social relations on a translocal level. In the same sense, I have also tried not to see the use of different media of communication as separated from each other: it relates to context and the social tie, what means of communication are preferred in a specific situation.

[580] In this section the guiding question 3 is most obvious and prevalent.

view that migrants abroad are seen as viable and bound to show concern and responsibility. Non-migrants adopt media habits according to such imaginations and given conditions and abilities, which are a part of the link-up forms of communication, such as setting priorities for communication, going to the cyber café at regular intervals, or reserving money to call specific people. An overall assessment relates to maintaining a wide network of social ties, setting priorities in regard to social ties under consideration of the according social capital involved, as well as finding different ways of how to link up with those, who they consider as viable contacts.

From the perspective of migrants, they also learn from inconsistencies in mediated communication, which are here mainly related to a sense of exploitation and distorted views from non-migrants, claiming on their responsiveness and supportiveness. No matter what relation they have to specific social others back in Cameroon, the tendency of people to change towards somebody who has migrated abroad, leads migrants to re-evaluate social ties, based on their evaluations of social relatedness and emotionally rewarding social ties. Disappointment is also involved here, when for example a good friend seems not to be truly interested in a migrant's fate, but keeps on asking for money. Regarding maintaining liveness in the sense of a virtual presence, migrants experience the same tension in their efforts. They are judged on what they are able to give back to the community by tangible and visible achievements, and on the basis of these more claims are made. In mediated communicative interaction, migrants shift their attentiveness regarding beeps and demands for support, and their consideration of social ties shift towards a core of close social ties in which they invest. Their media habits balance between a desired level of closeness, feelings of duty and controlling of unwanted contacting and claims. It can encompass not answering calls, keeping separate phones, not reacting to beeping, and many more practices.

As I have described, between migrants and non-migrants practices of staying in contact – and underlying interests to do so – seem to oppose each other, which can be explained by shifts between imaginations and related expectations, attributed duties and rights, as well as media's mediality, which increase the potential for tension and friction in transnational social ties. Differential perspectives are assessed according to conduct in mediated communication[581]. Both migrants and non-migrants work on social ties according to their assumed potentiality, which is evaluated in relatedness, closeness, and duties. As a tendency, whereas migrants expect and insist on investments into communication also from the part of non-migrants, and therefore on more reciprocity, non-migrants have the tendency to orient themselves to stable relations according to presumed social and economic positions. I have mentioned, that the stronger a social tie according to relatedness or social and/or emotional closeness, the more likely rather one-sided flows of attention arise, and are expected to arise. Claiming support from those who are in a better position is seen as a given by non-migrants, whereas migrants feel intimidated by too many claims from too many people. If non-migrants only invest in social ties with migrants, when a demand for support is involved, it is likewise experienced as impolite, when such a demand is not embedded in smooth communicative exchange, whereas from the view of non-migrants, limitations in using New Media of communication in the context of Cameroon as well as a defined responsibility from the part of the migrants, is taken as a basis on which to justify their communicative conduct. Whereas non-migrants claim on regular calls, migrants feel overburdened to stay in contact with a great number of people and to deal with the respective expenditures. In such ways, notions of a morally confined sense of conduct tend to shift away from a mutually shared and confirmed basis of reference and they constantly undermine each other.

Working on mediated social ties: differing evaluations of closeness and differentiating life-worlds

By deliberately "working on the gap", or negotiating manoeuvring capacities on both sides in transnational social relationships, "incomplete situations" and the arbitrary character of actions, which are directed towards dealing with emerging events and situations, it may allow solutions which please both parties involved. Although general assumptions, prior experiences, and New Media habits of migrants

---

[581] The fact that non-migrants views on the duties of migrants abroad are largely super-imposed by strong imaginaries, then points to a habitual dimension of agency, which also strongly guides these practices.

434

respectively non-migrants towards their communication partners are of great importance, relations to specific social others are continuously revised, negotiated, and articulated in mediated social exchange. Thus the "manoeuvring capability" of New Media users refers to a framework of tendencies and evaluations, as well as to individual social ties. Such negotiating and articulating is perceived, addressed and effected in mediated communicative interaction.

Evaluations of social ties depend on their perceived validity of social and emotional closeness. These qualities and their evaluations are also refer to the different categories of social ties, from parents, siblings, and close friends up to weaker connections of mates and acquaintances. Furthermore, the evaluation of these social relations differs according to the differences of age, social status and achievement, influencing mutual relations of duties and rights, and the conditions for closeness. Whereas social closeness is based on social relatedness, and also has a stance of moral duty, emotional closeness relates to social ties based on free will, and the work and effort of maintaining them. Being separated over long periods of time can be seen as potentially endangering closeness and loyalty in social ties. In view of physical absence and spatial dislocation, mere social relatedness is often not perceived as sufficient any longer as a basis for "good" and apt social relationships[582]. Considerations of social ties shift to emotionally more rewarding social ties in mediated social interaction. In particular migrants often value emotional closeness as important: it relates to their evaluation of "home" and close social ties under sometimes difficult conditions abroad. Migrants often related with excitement and expressions of attachment to "ma people back home". Regarding non-migrants I was often surprised, that migrants' absence was often related to in a rather unemotional way. The fact that they "went out to do something for us" or "to become somebody", as it was expressed, was apparently more highly evaluated. As one non-migrant stated: "Things can change. (...) At times relationships can break off. But with family, even if you can only talk on phone, things will never change (...)" (Barbara. Interviews 2009/R). Hence, the conditions or the basis on which social ties ought to be maintained and kept alive, depend

---

[582] See chapter 4, where I have defined the notion of a "good relationship", or "good communicative performance".

435

on evaluations, which can differ among migrants and non-migrants[583]. Although migrants generally stated that family ties were considered as the most important, concerning more distant relatives, claims were no longer taken as necessarily valid, and friends also had to "prove their loyalty". This assessment points to the tendency that emotional closeness gains a more important stance compared to social closeness or relatedness, from the perspective of migrants[584]. A crucial basis for such an assessment of social ties is conduct in mediated communication, which is effected in long-term work, within frameworks of mutually accepted positive conduct in communication performance. What Giddens (1984) calls "tact", in the sense of knowing how to behave towards a certain person, in certain social situations, has then the strongest relation to an evaluative dimension of agency[585]. As an interviewee expressed: "Devote your time, energy and money for the person, be there for her. Phone calls, sharing of pictures and videos, getting information about the person's life. If you cannot meet through media then the relationship may weaken. You have to meet continuously" (Kingsley. Interviews 2010/11/EC)[586].

Another part of what I call "work" impacts on imaginations of each other's life-worlds. This is of particular importance for migrants: It is not easy to work on imaginations, which often superimpose mediated social interaction. Imaginations consist, from the perspective of non-migrants, of imaginaries of the location of the migrant, his environment, daily life, and social networks - possibly also regarding the memories of known social others who also reside abroad. From the migrant's perspective, they are made up of memories of "home", conditions and social ties. On both parts imaginaries are only fragmentary. Non-migrant's imageries of the migrant's environment and life conditions are super-imposed by strong imaginaries and narratives in the Cameroonian context. Additionally, migrants often only selectively convey information about

---

[583] Compare different evaluations of social and emotional closeness in chapter 5.

[584] And possibly also from the perspective of non-migrants, when a serious concern of wanting to work on a social tie despite physical separation and dislocation, is involved as a motivation.

[585] Compare to chapter 5, where I have elaborately described different understandings of conduct or tact.

[586] See figures F86 and F87, which show the importance of New Media opportunities in the Cameroonian context.

their lives. From the perspective of migrants, although they share the knowledge with non-migrants about environment, daily life and social others, non-migrants might have their reasons to purposefully withhold certain information[587]. Furthermore, the longer migrants are absent, the stronger the possibility that their imaginaries and memories do not correspond to the life reality and conditions "back home" any longer. In my book I have also related to examples of social ties, where the "localizing" of the communication plays an important role, which is effected in New Media communicative interaction. It can contain asking questions, giving explanations or sending pictures, in order to enhance understanding of the life conditions in another location, with the underlying desire to enhance "the grasp" on the communication partner, as a condition for feelings of mutual closeness and understanding. It thereby depends on the willingness of communication partners to balance inconsistencies and to work on the "gaps", in order to positively validate transnational social ties.

To sum up: Taking liveness as a lens is an attempt to bring the realms of migration, New Media use and imagination together: ultimately, it is about shifts in modes of sociality, and evaluations of such modes of sociality in a transnational context. The different agentic dimensions have served to access different nuances of young New Media user's capacity to deal with specific temporal-relational contexts. Liveness is gradual, and it is thus a kind of measurement and base of evaluation, whether and to what extent liveness is achieved and developed in social interaction. It can serve to work on arising insecurities about others' relations, activities and motivations in a mediated realm of sociality, or by avoiding liveness, distance can be deliberately created. Thus, liveness is both potential and practice which can differ from each other. Realized liveness is always mutual, and it is accompanied by the negotiation of qualities in social relationships. I have emphasized that mediated liveness is just an approximation, it is not a substitution for physical co-presence, as expressed by one interviewee: "Would you tell somebody face-to-face that you miss the person? No. You can only say that when you are physically separated and only talking to the person" (Francis. Interviews 2010/11). Due to intensified processes of negotiation within mediated social ties – also

---

[587] Compare to chapter 6.

impacting on a local level – mediated sociality has become an issue of concern.

F86: Diverse offers for SIM cards, call boxes and computers, Bona Moussadi, Yaoundé

F87: Signboard of a cyber café, Bambili, outskirts of Bamenda

Dealing with space, time, distance and closeness - reflections on "globalization processes" and interrelations of agency and structure

What I have been looking at in the course of this book is best summed up by how people – New Media users in this case – organize and negotiate mobility, media activities and imaginations in spatial and temporal terms. Or in other words, how people manage distance – spatial, social, and emotional – by their practices of imagination and communication. When referring to issues such as mobility, migration and use of New Media, I am generally speaking about movements, shifts in perspectives and structural and social transformations. Such processes and transformations can be paraphrased by the notion of "globalization". We have to ask how "globalization processes" consolidate their impacts on a local level and how they impact on interpersonal levels of social interaction, and how actors relate and deal with "global impacts" on a local level and in sociality.

What is implied here is an examination of how agency relates to structure (compare Emirbayer & Mische 1998, Giddens 1984, Bourdieu 1977). Emirbayer & Mische (1998:1004) point to a double constitution of agency and structure: "temporal-relational contexts support particular agentic orientations, which in turn constitute different structuring relationships of actors toward their environments. It is the constitution of such orientations within particular structural contexts that gives form to effort and allows actors to assume greater or lesser degrees of transformative leverage in relation to the structuring contexts of action". The intertwinement of structure and agency needs to be seen in a temporal framework; the same as references of actors towards structure shift, structure is also continuously transforming. In this process, action is never completely determined by structure, and likewise action can never be completely freed from structure and conditions, and their interrelation is continuously negotiated in unfolding situations (Emirbayer & Mische 1998:1004). Under "structure" I subsume here the conditions, which have been relevant for this writing, which encompass different dimensions and transactions. The first dimension relates to social norms, including ideals and normative assessments, here mainly concerning sociality and communication. The second dimension contains what I want to call structural life-world conditions, which encompass given structures – enabling and constraining – which impact on mobility and New Media use, such as policies, education, economy, and (New Media-) infrastructure. These dimensions encompass transactions from local to transnational levels, from face-to-face to mediated sociality, as well as from personal to transpersonal – individual to collective – sets of action.

The notion of globalization

"Globalization" is a bundle of different and diverging theories, ideas and perspectives, in different scientific traditions, including different discourses ranging from the macro level of international relations, global trade, politics and media, to micro levels that are more related to an individual's or (trans-) local communities' practices of exchange and interaction, impacts on everyday life on a local level, and perceived transformations that are related to "globalizing practices" in one way or the other. The notion of "globalization" is a blurring expression, used in both scientific and popular discourses and often with various and differing connotations. It can be used as referring to descriptions of processes and transformations, as well as to explanations for the same, and furthermore, the understanding of the notion "globalization", can signify processes of transformation caused by human beings as well as normative and moral statements of people about such processes and transformations (compare Dürrschmidt 2002: 12). Regarding the various and also contrasting discourses there are several definitions of "globalization", depending on the perspective, approach and emphasis of the authors in different disciplines and periods of discourses, and the debates they mainly engage in. Famous debates regarding "globalization" are, first, the homogenization or heterogenization debates, whether "culture" tends to standardize or differentiate regarding processes of globalization (Ritzer & Tomlinson, Robertson 1992, Appadurai 1996, Lash & Urry, Hannerz 1996, Beck 1997), second, the debates about impulse, or stimulation, which deal with the causes or the force that induces "globalization processes" (Wallerstein, Giddens 1995, Appadurai 1996, Harvey 1990, Robertson 1992, a.o.). Herein included are discourses on the global economic system, global New Media systems and debates concerning transformed relations of space and time[588]. Third are the so-called continuity- or discontinuity debates (Giddens 1995, Lash & Urry, Robertson 1992), whether such processes can be seen as leading to completely new developments or not. On the following pages I will briefly address the first set of debates regarding "negotiation of culture" and a creative adopting of technologies. I will relate to the second set of debates regarding transforming relations of space and time, and finally, to the third set of debates regarding discourses of novelty. I will try to integrate the dimensions of structure - norms of sociality and life-world conditions - into the discussion concerning globalization as a set of social transformations.

Homo- versus heterogenization, or reflexively appropriating New Media technologies

In order to find a useful framework, which could serve as a background to understand the practices of mobility and processes of social and cultural transformations that go along with these, we could take a closer look at the homogenization-heterogenization discourse. Its first aspect is the dichotomy of increasing homogenization of local "culture" which is related to "the internationalization of culture" or commoditization and global consumerism. Adversely, its second aspect is the "diversification" of local cultures through the influence of a "global culture", which is often related to resistance and negotiating

---

[588] They could be related to the second as well as the third set of discourses.

440

"local culture" in diverse manners in order to "resist" global cultural influences, and which is consequently expressed in the term "heterogenization of culture". The "outcome" of the relation of both, global and local identities, coming along with the processes of negotiation, appropriation, embracement, accentuation and opposition to both, local and global culture, is in scientific and popular discourse often related to as "hybridization", "syncretism", or "glocalization" (Robertson 1992) and "creolization" (Hannerz 1992, 1996), the world and localizations described as "the global ecumene" (Hannerz 1996) or "collage world" (Harvey 1990) or processes of exchange as "global flows" (Appadurai 1996). The idea of "homogenization" versus "heterogenization" or diversification of culture(s) comes to the fore when, as for example in the case of digital and electronic media technologies, "Western technology" as a medium of "Western culture" is adapted to a specific context and conditions. Here, neither the discourses about homogenization – of "culture" – nor about heterogenization fully capture the understanding of such processes of social and cultural transformations. The notion of "homogenization of culture" underestimates the importance of local forms of reinterpretation and the emphasizing of cultural elements of local "tradition", whereas "heterogenization" does not recognize the importance of strong allusions and valuation and permeation of "global culture" in the local context. Nevertheless, the mentioning of such concepts could point to processual, situational, and ambiguous tendencies, which are described as "negotiations" or "negotiating" social and cultural issues.

In the writings of Hahn (2004), Hahn & Ludovic (2008), Spittler (2002), Van Binsbergen & Geschiere (2005), and others, the concept of appropriation can be adopted for understanding the different patterns of negotiating the symbolic meanings of „Western culture" in a certain local context. Similarly Hannerz (1996) describes processes of global culture being negotiated, appropriated, rejected, transferred in its meaning and related to new contexts in "dimensions of reflexivity" (compare Beck 1994, Lash & Urry 1994, Förster 2004, 2006). Giddens (1991) describes "reflexivity" as a reflexive appropriation of knowledge on an individual and societal level, which forms an important part of (re-

)production of society and identity[589]. Here it is necessary to differentiate between nuances of appropriation processes: on the one hand, for example "Western cultural representations" could become incorporated by taking over a local meaning, and would not be differentiated or recognized as formerly external. On the other hand, local appropriation is not always complete or thorough, local processes of meaning making could be dominated by external meanings (Förster 2005a:39). In any case, appropriation points to an integration of practices, generating local meanings, juxtaposing interpretations and forms of use, between local conditions and agentic capacities of actors. Technology in general, and media technologies in particular, have always been seen as a main facilitator of transnational connections and flows: I will relate here to their local integration in terms of media contents, meaning, uses, infrastructure and technology.

Regarding interrelations between technologies and technology users, discourses range between technological deterministic views and views emphasizing agency. Media technology had often been looked upon as determining the audience. However, the "conventional" audience concepts have transformed towards the emphasis of reflexive audiences in the course of the 1980's (Spitulnik 1993,2002, Wilson 2009, Bermejo 2007). Even though we cannot deny that there are "dominant" media contents, it is not an "overwhelming" of audiences by media imageries, but rather a reflexive examination and integration in local processes of meaning making (compare Bucholtz 2002:19). Media contents undoubtedly have an influence on cultural and social transformations in a local context, thereby, these impacts are not to be separated from overall influences, processes of transformations are complex aggregations of dispositions – located internally and externally to society - which can be activated by media technologies, leading to a variety of – often non-anticipated – transformative processes. As a matter of fact, New Media – and media in general, importantly TV – convey diverse images of sociality, forms of social organization, lifestyles and attitudes. The same is conveyed in interpersonal communication

---

[589] Giddens emphasizes, that although such practices of reflexivity are inherent to societal organization, they gained importance considering the various global flows, exchange and interconnections (Giddens 1991).

circuits: new online forms of sociality arise, and emotional closeness is fostered despite physical dislocation.

On the level of their local meaning, New Media technologies have been integrated into local systems of sociality. This is why their importance is so highly valued in the local context. New Media as tools for mediated sociality have gained specific meanings in this context, which is related to their specific medialities. New Media sociality is performed according to local normative evaluations, interpretations, resources and practices, within a new framework for sociality. Thereby, New Media technologies also raise ambivalent notions, assessed in normative evaluations of their potential, which can be positive or negative, according to users' intentions. An assessment of New Media technologies as external comes into play when relating to technology, not to their use. It was expressed that "Black man use white man's technology for their own purpose", or as it was expressed by scammers, that they would "beat the whites with their own weapons".

Regarding the appropriation of technological features, on the one hand, technological tools have an impact or shape user's practices respectively, in the sense that the integration of certain technological tools within a repertoire of technology use is limited to a certain technological framework. On the other hand, media technologies can be picked up or not, and adopted for alternative uses: the more media users are technologically versed, the more they are able to purposefully use these technologies in order to reach their aims, and to relocate technological "boundaries". It can thus rather be seen as an interplay of technological frameworks and creative adoption (compare Lavie, Narayan & Rosaldo 1993). In particular, however, it is an interplay with negotiations of a new sociality in New Media, as I have related to. Technological features are employed according to emergent needs for sociality, and differentiated according to New Media's specific medialities, which are preferred, or profited from, left out, reworked, or even invented, and which can be used according to different and unfolding mediated social situations.

F88: Chinese mobile phones can also be repaired
F89: Diversification of businesses: cyber café, call box, documentation, and selling fruits, Yaoundé

On a level of media infrastructure or technology, due to local conditions, local adaptations are made which could differ from global mainstream uses. For example, Chinese phones can be repaired although they have been produced as disposable goods[590]. Climbing hills in order to connect to a network in villages broadens the possibility of profiting from telecommunication technologies despite severe limitations. Prevalent constraints of New Media users, such as limited financial means, are dealt with in specific media practices as short calls or beeping – connecting without spending. Email sending services or depending on other's help, circumvent media user's non-literacy. The internet is successfully used in crime related practices for new sources of income, and New Media related services are combined in locally meaningful business combinations[591]. In return, New Media are used less for certain other purposes, as for example, the internet is more important for social interaction than searching for information, or text messaging is not as important as beeping.

Space and Time – a differentiation of concepts

Concepts of mobility and practices of New Media use are obviously strongly related to different views of space. It is understood in more recent migration and mobility studies that space cannot be thought as a „container" in which actions take place, without being conditioned by it, or conceptualized independently of subjective experience (Müller-

---

[590] See figure F88.
[591] See figure F89.

Richter 2001:16). Various globalization scholars, social scientists and philosophers have dealt with concepts of space and time related to the discussion of globalization phenomena and based on, broadly speaking, the idea of the "production" of space by social agents (compare to Giddens 1990, Robertson 1992, Hannerz 1996, Harvey 1990, Lash & Urry 1994, Appadurai 1996, Lefebvre 2000 (1974), a.o.). Simmel (1990/1907) was one of the first who described „space" as social space, constituted by social action and agency of actors (in: Pries 2002:22). Bourdieu (1977,1993) offers a typology of three different relations which constitute "space": First is the physical composition of material space as the result of material practices, second are representations of perceived space as the result of perceiving practices - systems of order and valuation - and third, the representational space, as the result of ideological-imaginative practices, which lead to a constitution of space with or without connection to a physical space, or an imagined space[592] (Ossenbrügge 2004:22,23).

For Giddens (1990) the separation of the concepts of time and space allows putting social practices into focus. Giddens understands "disembedding" as lifting social relationships out of local contexts of interaction and by the notion "reembedding" he relates to their space-time transcending reorganization (Giddens 1990:79,80)[593]. Disembedding and reembedding occurs through practices of New Media use, which transcend space by creating a unity of time in simultaneous social interaction. Different media technologies have different consequences or offer different options for the control of time and space (Scannell 2007:127, Ling & Campbell 2009). The bridging of a non-unity of place is the underlying key notion of media, which is "mediation" (compare Mazzarella 2004). "Mediation" points to an indirect reception and perception of the „content" or message of interaction, and furthermore to a double perception of „worlds" where media of communication are involved. This also involves the

---

[592] Which could be, for example, the materiality of New Media sites in the city space, the evaluation of such social spaces, some of them pointing to "modernity" and prestige, and the representational space of practices of imagination.

[593] Comparable notions regarding transforming relations of time and space is Harvey's „time-space compression"; related to the processes of increasing interconnections one could also compare to Hannerz' notion of „the global ecumene" and „creolizing world", or Harvey's „collage world".

imagination of the communication partner's life-worlds or their localization during mediated communicative interaction. In such ways, New Media serve to bridge social situations, which limit sensory perception of the communication partner's life-world and social and emotional cues, by "manipulating" temporal and spatial relations.

To enhance the description of transformation processes within spatial and temporal dimensions we can relate to "social spaces": Bourdieu describes spaces as frames for social practices. He thereby focuses instead on processes and social relations as constitutive elements of social fields, rather than culture, identity or function (compare Glick-Schiller, Caglar & Guldbrandsen 2006)[594]. In a phenomenological view it is about how people experience and act in spatial and temporal dimensions (compare Schütz 1967, Scholl & Tholen 1996:16). New Media contribute to the shaping of social spaces' material attributes, to a mental dimension of imaginaries adding to social spaces, and also to the organizing of social practices, dimensions, which are closely interrelated. Thus, everyday interaction with others creates social spaces, which can be approached with regard to two dimensions: the opportunities for interactions given by a certain space, which have been designed by members of a society, and the intentional acts of agents (Förster 1999). Given structures and conditions, which impact on New Media use – such as for example material media infrastructure, financial means, media habits and conventions of sociality[595] – and on mobility – such as policies and other obstacles limiting free movement, in particular on a transnational level – certainly influence the manoeuvring capacities of agents, and contribute to shaping their relations to temporal and spatial configurations.

Discourses on novelty - or do New Media transform sociality?

The question which lies at the core of the discourse on novelty, is whether developments attributed to globalization processes are to be considered as new phenomena of what could then be called a "new

---

[594] Similar are Hannerz' „habitats of meaning", Appadurai's „scapes", Harvey's „social and temporal practices", and, furthermore, Giddens has dealt with concepts of locales, regions or regionalization as referring to „the zoning of time-space in relation to routinized social practices" (Giddens 1985, in: Pries 2002:28).

[595] Such as following up social networks in order to obtain information, or linking up practices related to expectations towards migrants.

global era", or these processes could be seen instead as a continuation of previous developments in an intensified manner. Förster (2005:41-43) has related to these discussions by distinguishing three dimensions of examination of the question of novelty, the overarching dimension of society and societal change, the dimension of social action, and the dimension of individual experience, perception and orientation for action.

The issue of novelty comes into focus where New Media are concerned. In the attribute of the "new" of "New Media", their "globalizing attributes" are accentuated, and already point to an assumptive novelty. Transforming perceptions and relations of space and time sometimes lead to an all-encompassing idea of a "new global era" with a novelty in a qualitative sense (Förster 2005:33-36, 42), and also related to the discourses on a "periodization" of globalization processes (compare to Watson 1998, Lash & Urry 1994) in order to classify these globalization processes chronologically. This idea of a "new global era" or "new global age" can be considered as problematic and inherently includes discussions on trajectories of developments, which undoubtedly have a history reaching back in time. Furthermore, in the local Cameroonian context, New Media and mobilities do not just occur out of a "blank space". Interpersonal social relationships have been mediated before, and various media such as print, radio and TV were prior to the internet and mobile phones. Also regarding their influence on practices of imagination, the internet – and also the mobile phone with its opportunity to stay in communicative contact with people not physically present on site, as well as media contents – add to media which have been present previously, such as most importantly TV, radio, and letters or messengers. In the same sense people have experiences with media of communication such as the phone, which had been the landline phone for decades, and even computers[596]. In this sense, introduction or rather the gaining ground of technologies, as in this case New Media, is a gradual process. Similarly, migration practices have always existed, but have become more prevalent due to their global scale, and also their medialization. As Vertovec (2007:151) expresses: "The widening of networks, more activities across distances, and

---

[596] If they have not a personal computer experience themselves, they would have at least heard about these technologies.

447

speedier communications reflect important forms of transnationalism in themselves. However, they do not necessarily lead to long-lasting, structural changes in global or local societies. Migrants have historically maintained long-distance social networks, and the fact that messages or visits take a shorter time does not always lead to significant alterations in structure, purpose or practices within the network." In a similar sense, Held and McGrew (2002:69ff) object to seeing globalization as a new phenomenon, and plead for examination of its historical forms and the "spatio-temporal and organizational attributes of global interconnectedness in discrete historical epochs". In this context, the discussion on „novelty" points to what role New Media use and migration practices play within the complexities of social and cultural change in a world of complex interrelations. One of the central underlying questions of my book is thus whether New Media use and mobility lead to a qualitative transformation of sociality, or if the transformations are only of a quantitative nature.

What might be uncontested is the fact that through these processes that we call "processes of globalization", the scope and intensity of "global cultural flows" – to use Appadurai's (1996) notion – that influence the immediate "locality", have apparently increased. Globalization can be seen as "(…) a process (or set of processes) which embodies a transformation in the spatial organization of social relations and transactions – assessed in terms of their extensiveness, intensity, velocity and impact – generating transcontinental or interregional flows and networks of activity, interaction, and the exercise of power" (Held & McGrew, Goldblatt & Perraton 2002:68). It leads to the intersection and increasing interconnection of frameworks of action over spatial distances, and therefore the dissolving of clearly distinguishable and localized personal and societal contexts (Förster 2005:51). With the advent of New Media mediated social interaction has intensified. In the local context, before the advent of the mobile phone and internet, people were hardly connected to each other beyond a level of presence availability. The reach of sociality and the range of social networks increase through New Media technologies, thus link-up has gained a new importance. In interconnections in various societal fields, New Media are thus a crucial "force".

Practices of social interaction need to adapt to transformed conditions. Mediated liveness is clearly distinguished from liveness opportunities in face-to-face sociality, but normatively, mediated liveness is assessed in regard to a valuation of an ideal notion of sociality, which refers to face-to-face social interaction, presence availability, and a comprehensive "grasp" of the other. Slippages, which occur through physical dislocation in mediated social interaction, are negotiated and super-elevated along the lines of an ideal sociality in terms of instant availability, controllability and solidarity. Thereby, negotiations of sociality take place on the basis of assumptions of a great potentiality of the life-worlds of communication partners, a consideration of local media environments and conditions of New Media use, and collective normative references. Social action in the sense of a basic orientation to an ideal of sociality is thereby not substantially transforming (compare Förster 2005), but needs to be seen within a new framework of "globalizing tendencies".

On the level of a New Media user's perception, opportunities for sociality and social networking are boosted: New Media are apt means of easily link up with social others, and allow access to a wide range of social ties and choice regarding intentions of maintaining social ties over distance and on a local level. Opportunities for instant mediated sociality have become part of people's everyday practices of sociality. High expectations towards the potentiality of New Media in interrelation with imaginations of great potentials beyond the local sphere are combined with an ideal of sociality, which cannot be matched in such constellations. In this sense, mediated communication is prone to failure, which must lead to intense negotiations on sociality. In this sense it can be the perceived gap between the actual and what is assumed to be potentially possible, which has come to the fore as a new quality of life experience. It is an experience where exclusions are also felt more vigorously, when structural conditions bounce back on agency of individuals and groups, but at the same time so many references point to possible pathways.

How can this research and thesis contribute to a better understanding of the field?

There are several issues, which I could not – sufficiently – include in this book, such as transportation, local visual representations in a range of media, online activities in different fields, online social networks, sensory experience in New Media use, gender aspects, and fieldwork methods. These "gaps" point to the importance to address such issues in research. To conclude I will sum up my findings and highlight the specific contribution of this book in this field of research.

Suggestions for further research, and issues not – sufficiently – addressed in this book

Concerning the issue of mobility, there are further interesting interrelations of forms of mobilities, which could be addressed in research. I have tried to focus on the issues of mobility and New Media use in this book, however sometimes without discussing their relations to other "globalization tendencies". Similarly, there are other technologies, which interrelate with mobility and New Media, which I could not address in this book. Here I want to specifically mention transportation and related infrastructure, public transportation systems such as taxis and motor-bikes, streets and their maintenance, relating to mobility itself as a practice, such as travelling and the bridging of space in a physical sense including a bodily sensation of it (compare Nkwi 2009 for the Cameroonian setting, Lyons & Urry 2005). In my book I have only marginally addressed the question of whether New Media use can substitute or enhance physical mobility (compare Burrell 2009)[597], because it was leading beyond the scope of my work, and I have referred more strongly to virtual travelling and practices of imagination. However, it would be an interesting complementary issue in order to better understand the "managing" of time and space in practices of media use, since physical mobility's impact on the organization of the social world must be considered as crucial.

---

[597] I have addressed this topic in this conclusion chapter. Furthermore, I have pointed to physical movement especially in the sense of its evaluation regarding its adding to status and as life projects for youth, in chapter 2.

Furthermore, New Media, such as the internet and mobile phones are part of an entire array of media, which convey imageries of faraway worlds in the local context, and thus contribute to global imaginaries. It is thereby difficult to attribute practices of imagination to specific media and to sort out the particular role of the internet and mobile phones. I have also pointed out here that the visual – in conveyed images in media – contributes importantly to societal imaginaries. Regarding an array of diverse media in Bamenda[598], I suppose that it would be interesting to examine visual culture in the city space, in order to produce a compilation of imageries conveyed through diverse media, such as TV, publicity, signboards, fashion styles, performances, and so on, and how these point to imaginaries and how they are embedded in practices (compare Förster 2005).

A further topic which would be interesting to address, is media's contents related to different fields of interest of New Media users, and respective activities. Here I think especially of activities in the field of politics. As we have seen in the most recent developments in the "North African revolutions", social networks on the internet can play an important role in organizing, informing and motivating people for a common aim, in this case political resistance. Regarding Cameroon, I hardly met people however who were politically active using the tool of the internet in online and social networks at the time when doing my research. Among migrants, apart from one person, who said that he was active in the opposition party SCNC, and another who was active in politically relevant chat rooms, the majority were not active on the internet regarding using this media as a virtual platform for exchange, also including diaspora or pan-African organizations online (compare Bräuchler 2003, Burrell & Anderson 2008, Cunningham & Sinclair 2000). In the Cameroonian context, for most youth with whom I have interacted, political issues seemed not to be important. Many stated explicitly, that they were not interested in politics, statements, which however point to peculiar forms of resistance, to political disinterest (compare Fokwang 2008). These forms of political resistance, in the view of alternative ways to success – which can be migration, scamming, or reaching out by New Media in general - under the conditions of

---

[598] Compare to chapter 3, how imaginaries are represented in imageries and practices in the city-scape.

prevalent inequalities of power, clientele networks and corruption, point to interesting avenues for further research[599].

Of great importance for young New Media users in Bamenda are social network sites. A few people were involved in online social networks on topic related sites, such as in the case of a young man in relation to his occupation with the UN, another regarding topics concerning his NGO, and two young men on sites for "IT cracks". Apart from those few, most were involved in online social networks, which were not topic related, on friendship and moreover dating sites. I have mentioned that in particular using Facebook – or at least having a Facebook account - has become common (Danah 2008, Silver 2000)[600]. Most interviewees related to online social interaction as "fun", but they did not consider such contacts as morally valuable. However, others stated that they pursued long-term and "serious" online friendships. Another interesting example are the conversations with the "dupes" or potential fraud victims by the scammers, conversations, which point to local interpretations of "cues of trustworthiness", which need to be adopted in order to convince and to successfully finalize "the business". Hence, such online social ties and interaction would be interesting to examine in particular related to social and emotional cues, the building of trust, and what role a commonly shared life-world plays in social interaction. In such research, also the conversation contents would gain more importance. Related to communication among media users connected through New Media, be it the internet and/or mobile phones, I have not gone into the contents of conversation[601] extensively, using methods such as conversation analysis.

Related to the issue of addressing communication contents are sensory experiences in New Media communication, which I have only

---

[599] Another field would be online contents related to a religious and spiritual field. Here I have mentioned the interest of some youth in some sites on the internet with a religious content. I have however not met anybody who was active in online churches - only one young woman said that she asked for prayers when in difficulties, on the site of a Nigerian church institution. In particular the Pentecostal churches are relatively active on the internet.

[600] Compare specifically to chapter 5. I have collected sufficient material to write about online relationships, however it is a topic, which I only marginally related to in this book.

[601] I have addressed it to a certain extent – related to some key issues – in chapter 5.

partly addressed. I have discussed the materialization of media technologies in the local sphere, and I have described media users' adopting of different technological features, such as chat, email, phone calls and beeping, and accordingly the users' evaluations of these technological features regarding creating a feeling of closeness towards the communication partners. In turn I have addressed users' "bodily experience" of these media only to a lesser extent. Furthermore, I have only marginally addressed the users' sense of styles in written language - such as in chat - or the ways of conversing in short phone calls of 59 seconds for example, related to the question of how the transformation of language use could influence perceived qualities of social ties (compare Crystal 2001). What would be also interesting here is the issue of what role the visual plays in internet communication and information retrieval. I would argue that the visual is important in adding to verbal interaction, such as by the use of webcams, sending pictures, links, films, videos, or adding emoticons.

Concerning an emphasis on gender related to New Media use, I have probably had the tendency to emphasize a male perspective[602]. I have tried to explain the reasons for this imbalance, related to media users in public cyber cafés, to the scamming issue as a male business, up to the fact that finding interviewees was easier for me among male youth. A "male bias" also relates to literature – for example about youth – which as a tendency adopt a rather male perspective. Regarding observations in youth social spaces, I have mainly emphasized a "male performance", which is more conspicuous and better observable in the public sphere[603]. I have briefly sorted out a few aspects of "typically female" uses of New Media such as regarding calling, beeping and in general habits of internet use[604] (Boneva & Kraut 2002, Johnson-Hanks 2007), and I have also briefly related to a female perspective on migration, respectively what differences come to the fore regarding expectations towards female migrants[605]. It would however be

---

[602] I have however tried to highlight female participations, for example in New Media related business branches, and the tendency of young women linking up often with strong intentions of economic support through men.

[603] Here I have shown "female's contribution" to particularly these demonstrations of male virility particularly. See especially chapters 1, chapter 3, and chapter 4.

[604] See chapter 5.

[605] See chapter 1, and chapter 6.

interesting to concentrate on female views and practices, in the field of the intersection of migration and New Media use, in further research.

I have already pointed out that I have experienced New Media as a research topic as closely tied to my own physical presence in the site of my fieldwork. The concentration on interrelations of New Media use with mobility and related imaginations of a "better world" were directly influenced by my personal experiences in the local context (compare Archambault 2009, Robben & Sluka 2006). Such reflections tied to cross-cultural research applied to such border transcending topics, are worth following up in studies concerning reflective methodological tools for fieldwork in Social Anthropology.

I hope that in this subchapter I could highlight interesting topics for further research, opening up new fields, or adding to the topics addressed in this book.

F90: Signboards for DV lottery, passport photographs and international phone calls in Yaoundé
F91: Wireless connection at Dreamland Restaurant, Commercial Avenue, Bamenda
F92: Imaginary of reaching out to the world - in a cyber café, Bamenda

Concluding résumé

Since the turn of the new millennium, opportunities for instant mediated sociality have been appropriated by young urban New Media users in the range of their everyday practices of sociality. In particular regarding New Media, the internet and mobile phones, I suppose that the intensity of this social experience, users' agency, and also the evaluation of connectedness as a potential, make a difference here. New Media and their opportunities to create liveness highlight an ideal notion of sociality, now to be fulfilled under conditions of physical separation:

normatively, mediated liveness is assessed in regard to an evaluation of an ideal notion of sociality, which refers to instantaneity, control and a comprehensive "grasp" of the other. Slippages, which are likely to be experienced in mediated social interaction refer then to high expectations towards the potentiality of New Media in interrelation with imaginations of great opportunities: combined with notions of an ideal sociality, which are difficult to match in such constellations, mediated communication is prone to fail. The "feel of gaps" becomes a central feeling and experience in many youth's lives: The gap between the actual and the potential comes to the fore regarding the felt disadvantageous everyday life conditions in the local context, which is perceived as detrimental to youth's notion of "a good life", as well as social tensions and distanciation, which may arise in mediated social ties. Crucial in these negotiation processes are the differing cultural and social settings, in which communication partners are located, and respectively the contesting imaginations of them.

Feelings of immobility and disconnectedness are thus based on premises of the potential of being mobile and connected. In view of increased expectations regarding liveness, also a sense of uncertainty and anxiety, and an awareness of distance, separation, and exclusion arises more vigorously among youth in the local context.

Outlook and thanks

Again and most specifically, I want to express my gratitude to my friends in Cameroon, and Cameroonian friends in Switzerland. I want to thank them for their willingness to spend their time with me, answer my questions, and give me all possible support, which was needed to make this work possible. The topic of this research was extremely fascinating throughout and has not lost its attraction over time. Hopefully, in future, I will be able to integrate my acquired knowledge and experiences into my professional life, and possibly be able to further contribute to the field and extend my insights. Even though my research and the work on my book will be concluded now, I hope I will be able to visit Cameroon again in near future. Meanwhile, I will try not to lose "connection". Although, as I have tried to show in this book, being physically absent and the demands of everyday life in differing life-worlds could "complicate" liveness, at least the means to work on social ties are given, since people are only a phone-call, a beep, a text message, an email, or a chat conversation away.

References

Abu-Lughod, Lila (2002) 'Screen Egyptian Melodrama – Technology of the Modern Subject?', in Ginsburg, Faye D, Abu-Lughod, Lila & Larkin, Brian (eds) *Media Worlds. Anthropology on New Terrain*, Berkeley, Los Angeles, London: University of California Press, pp. 115-133

Adepoju, Aderanti (1995) 'Migration in Africa. An overview', in Aina, Tade Akin & Baker, Jonathan (eds) *The migration experience in Africa*, Uppsala: Nordic African Institute, pp. 87-108.

Adey, P. (2010) 'Aerial Life: Spaces, mobilities, affects', Oxford: Wiley-Blackwell Publishers.

Agar, J. (2003) 'Constant touch: a global history of the mobile phone', UK: Icon books.

Aina, Tade Akin & Baker, Jonathan (1995) 'Introduction', in Aina, Tade Akin & Baker, Jonathan (eds) *The migration experience in Africa*, Uppsala: Nordic African Institute, pp. 11-28.

Albrow, Martin (1997) 'The global age: State and society beyond modernity', Stanford: Stanford University Press.

Allan, G. A. (1979) 'A sociology of friendship and kinship', London: Allen and Unwin.

Alpes, Jill (2011) 'Bushfalling: How young Cameroonians dare to migrate', Doctoral thesis, University of Amsterdam, online.

Anderson, Benedict (1991), 'Imagined Communities: Reflections on the Origins and Spread of Nationalism', London: Verso.

Appadurai, Arjun (1996) 'Modernity at large. Cultural dimensions of globalization', in Gaonkar, Dilip & Lee (eds), *Public Worlds*, Volume I.

--- (1995) 'The production of locality', in Fardon, A (ed), Counterwork, London: Routledge.

--- (2009) 'Disjuncture and difference in the global cultural economy', in Lievrouw, Leah A. & Livingstone, Sonia (eds) *New Media, Visions, histories, mediation*, Volume I, London: SAGE Publications.

Appadurai, A. & Beckenridge, Carol (1989) 'On moving targets', in Public culture, Vol. 2. I-IV.

Archambault, Julie Soleil (2009) 'Being cool or being good: researching mobile phones in Mozambique', Anthropology Matters Journal 2009, Vol. 11 (2).

Ardener, Edwin, Ardener, Shirley & Warmington, W.A. (1960) 'Plantation and village in the Cameroons. Some economic and social studies', London: Oxford University Press.

Ardevol, Elisenda (2005) 'Dream Gallery: online dating as a commodity', online.

Argenti, Nicolas (2007) 'The intestines of the state. Youth, violence, and belated histories in the Cameroon Grassfields', Chicago & London: University of Chicago Press.

Armbrust, Walter (1996) 'Mass Culture and Modernism in Egypt', New York: Cambridge University Press.

Atekmangoh, Christina (2011) 'Expectations abound: family obligations and remittance flow amongst Cameroonian "bushfallers" in Sweden. A gender insight', SIMT24, Masters Thesis (2 years) in Development Studies, Spring term 2011, Lund University.

Auhagen, An Elisabeth (2002) 'Freundschaft und Globalisierung', in Hantel-Quitmann, Wolfgang & Kastner, Peter (eds) *Die Globalisierung der Intimität. Die Zukunft intimer Beziehungen im Zeitalter der Globalisierung*, Giessen: Psychosozial Verlag, pp. 87-116.

Auslander, Philip (1999) 'Liveness. Performance in a mediatized culture', 2nd Edition, London & New York: Routledge.

Awambeng, Christopher M. (1991) 'Evolution and growth of urban centres in the North-West Province (Cameroon) Case studies (Bamenda, Kumbo, Mbengwi, Nkambe, Wum)', Bern: Peter Lang Verlag.

Bakewell, Oliver & De Haas, Hein (2007) 'African migrations: continuities, discontinuities and recent transformations', in Chabal, Patrick, Engel, Ulf & de Haan, Leo (eds), *African Alternatives*, Leiden: Brill, pp. 95-118.

Banégas, Richard & Warnier, Jean-Pierre (2001) 'Nouvelles figures de la réussite et du pouvoir', Politique Africaine 82 (2001), pp. 5-21.

Barber, B. (1983) 'The logic and limit of Trust', New Brunswick, NJ: Rutgers University Press.

Barnes, J.A. (1972) 'Social Networks', Reading, MA: Addison-Wesley.

Baros, Wassilios (2001) 'Familien in der Migration. Eine qualitative Analyse zum Beziehungsgefüge zwischen griechischen Adoleszenten und ihren Eltern im Migrationskontext', Frankfurt a.M.: Peter Lang Verlag.

Baudrillard, Jean (1995 (1981)) 'Simulacra and Simulation', Ann Arbor: University of Michigan Press.

Baumann, Zygmunt (2008) 'Flüchtige Zeiten. Leben in der Ungewissheit', Hamburg: Hamburger Edition.

Baumann, Margret & Gold, Helmut (2000) 'Mensch Telefon. Aspekte telefonischer Kommunikation', Heidelberg: Kataloge der Museumsstiftung Post und Telekommunikation.

Bayart, Jean-François (1999) 'The social capital of the felonious state or the ruses of political intelligence', in Bayart, Jean-Francois (ed), *The criminalization of the state in Africa*. Oxford: James Currey and Bloomington.

--- (1979 (1985)) 'L'état au Cameroun', Paris: Presses de la Fondation Nationale des Science Politiques.

Beck, Ulrich (1992) 'Risk society', London: Sage.

--- (1994) 'The reinvention of politics: towards a theory of reflexive modernization', in Beck, Ulrich, Giddens, Anthony & Lash, Scott (eds) *Reflexive Modernization, Politics, Tradition and Aesthetics in the modern social order*, Cambridge: Polity Press.

Beck, Rose Marie & Wittmann, Frank (2004) 'African media cultures – transdisciplinary perspectives', Köln: Köppe Verlag.

Beisswenger, Michael (2002) 'Getippte „Gespräche" und ihre trägermediale Bedingtheit. Zum Einfluss technischer und prozeduraler Faktoren auf die kommunikative Grundhaltung beim Chatten', in Schröder, Ingo W. & Voell, Stephane (eds) *Moderne Oralität, Ethnologische Perspektiven auf die plurimediale Gegenwart*, Marburg: Curupira.

Bell, D. & Newby, H. (1976) 'Community, Communion, class and community action', in Herbert, D. & Johnson, R. (eds) *Social areas in the city*, Vol. II, London: John Wiley and Sons.

Bell, David & Kennedy, Barbara M. (2000) 'The cybercultures reader', New York: Routledge.

Berger, Peter L. & Luckmann, Thomas (1980) 'Die gesellschaftliche

Konstruktion von Wirklichkeit. Eine Theorie der Wissenssoziologie', Frankfurt a. M.

Berger, P.L. (1978) 'The problem of multiple realities: Alfred Schütz and Robert Musil', in Luckmann, T. (ed) *Phenomenology and sociology*, Harmondswoth: Penguin Books, pp. 343-367.

Bernal, Victoria (2004) 'Eritrea on-line. Diaspora, Cyberspace, and the public sphere, American Ethnologist, Vol. 32, N° 4., pp. 660-675.

Berry, Keith & Clair, Robin (2011) 'Contestation and opportunity in reflexivity: An introduction', Cultural studies-critical methodologies, 11.2, pp. 95-97.

Berutti, Gilda (2008) 'Urban public spaces in the augmented city', in Eckhardt, Frank (ed) *Media and Urban Space. Understanding, investigating and approaching Mediacity*, Berlin: Frank& Timme Verlag.

Bian, Yanjie (1999) 'Getting a job through a web of Guanxi in China', in Wellman, Barry & Berkowitz, S. D. (eds) *Social structures: a network approach*, Cambridge: Cambridge University Press, pp. 255-278.

Blotevogel, H.H. (1995) 'Raum', in ARL (ed) Handwörterbuch der Raumordnung, Hannover, pp. 733-740.

Bommes, Michael (1999) 'Migration und nationaler Wohlfahrtsstaat. Ein differenzierungstheoretischer Entwurf', Wiesbaden: Westdeutscher Verlag.

Boneva, Bonka & Kraut, Robert (2002) 'Email, Gender, and personal relationships', in Wellman, Barry & Haythornthwaite, Caroline (eds) *The internet in everyday life*, Oxford: Blackwell Publishing, pp. 372-403.

Bourdieu, Pierre (1993) 'The field of cultural production', New York: Columbia University Press.

--- (1998) 'Practical reason: A theory of action' Cambridge: Polity Press.

--- (1977) 'Outline of a theory of practice', Cambridge: Cambridge University Press.

--- (1986) 'Forms of capital', in J. C. Richards (ed) Handbook of Theory and Research for the Sociology of Education, New York: Greenwood Press.

Bourdon, Jerôme (2000) 'Live television is still alive: on television as an unfulfilled promise', Media Culture Society, Nr. 22:531, online.

Bräuchler, Brigit (2003) 'Cyberidenities at war: religion, identity, and the internet in the Moluccan conflict', in Indonesia 75 (April 2003), pp. 123-151.

Braune, Inge (2008) 'Aneignung des Globalen. Internet-Alltag in der arabischen Welt. Eine Fallstudie in Marokko', Bielefeld: Transcript Verlag.

Brettell, Caroline, B. (2008) 'Theorizing Migration in Anthropology. The social construction of networks, identities, communities, and globalspaces', in Brettell, Caroline, B. & Hollifield, James F. (eds) Migration Theory. Talking across disciplines, 2nd edition, New York, London: Routledge, pp. 113-160.

Brinkman, Inge, De Brujin, Mirjam & Bilal, Hisham (2009) 'The mobile phone in Khartoum', in De Brujin, Mirjam, Nyamnjoh, Francis & Brinkman, Inge (eds) Mobile phones: the new talking drums of everyday Africa, Leiden: Langaa & African Study Centre.

Brown, B., Green, N., & Harper, R. (eds) (2001) 'Wireless world: social, cultural and interactional aspects of wireless technology', Springer Verlag.

Bucholtz, Mary (2002) 'Youth and Cultural Practice', Annual Review of Anthropology, Vol. 31 (2002), pp. 525-552.

Bukow, Wolf-Dietrich (2000) 'Die Familie im Spannungsfeld globaler

Mobilität', in Buchkremer, Hansjosef, Bukow, Hans-Dietrich & Emmerich, Michaela (eds) *Die Familie im Spannungsfeld globaler Mobilität. Zur Konstruktion ethnischer Minderheiten im Kontext der Familie,* Hemsbach: Leske + Budrich, Opladen.

Burrell, Jenna & Anderson, Ken (2008) 'I have great desires to look beyond my world: trajectories of information and communication technology use among Ghanaians living abroad', New Media Society 2008, 10, 203, Los Angeles, New Delhi and Singapore: SAGE Publications.

Burrell, Jenna (2009) 'Could Connectivity replace mobility? An analysis of Internet café use patterns in Accra, Ghana', in De Brujin, Mirjam, Nyamnjoh, Francis & Brinkman, Inge (eds) *Mobile Phones. The New Talking Drums of Everyday Africa.* Cameroon, The Netherlands: Langaa & African Studies Centre.

Burt, Richard (2002) 'Slammin' Shakespeare in Acc(id)ents Yet Unknown: Liveness, Cinem(edi)a, and Racial Disintegration', Shakespeare Quarterly, Vol. 53, N° 2, Screen Shakespeare (Summer 2002) pp. 201-226.

Calhoun, Craig (1992) 'The infrastructure of modernity: indirect social relationships, information technology, and social integration', in Haferkamp, H. & Smelser, N.J. (eds) *Social change and modernity.* Berkeley: University of California Press.

---. 1991. Indirect relationships and imagined communities: large scale social integration and the transformation of everyday life. In: Bourdieu, Pierre & Coleman, James S.. (eds.) *Social Theory for a changing society,* New York: Russell Sage Foundation, pp. 95-117.

Caron, André H. & Caronia, Letizia (2007) 'Moving Cultures. Mobile communication in everyday life', London: McGill-Queen's University Press.

Castells, Manuel (2007) 'Communication, power and counter-power in the network society', International Journal of Communication 1(1), pp. 238-66.

--- (2001) 'The internet galaxy. Reflections on internet, business, and society', Oxford: University Press.

--- (1996) 'The information age: Economy, Society and culture', Vol. I, in *The rise of the network society*, Oxford: Blackwell.

Castells, Manuel, Fernandez-Ardevol, Mireia, Linchuan Qui, Jack & Sey, Araba (2007) 'Mobile communication and society: a global perspective', MIT Press.

Castles, Stephen (2007) 'The factors that make and unmake migration policies', in Portes, Alejandro & DeWind, Josh (eds) *Rethinking Migration. New Theoretical and Emperical Perspectives*, New York, Oxford: Berghahn Books, pp. 29-61.

Cesarani, David & Fulbrook, Mary (1996) 'Introduction', in Cesarani, David & Fulbrook, Mary (eds) *Citizenship, nationality and migration in Europe*, London & New York: Routledge, pp. 1-16.

Chilver, Elizabeth M. (1967) 'Paramountcy and protection in the Cameroons: the Bali and the Germans, 1889-1913 / Elizabeth M. Chilver', in Gifford, Prosser & Louis, Roger (eds) *Britain and Germany in Africa. Imperial rivalry and colonial rule*, New Haven: Yale University Press, pp. 479-511.

Christensen, Pia, Hockey, Jenny & James, Allison (2001) 'Talk, silence and the material world: patterns of indirect communication among agricultural families in northern England', in Hendry, Joy & Watson, C.W. (eds) *An Anthropology of indirect communication*, London & New York: Routledge.

Clifford, James (1994) 'Diasporas', Cultural Anthropology 9(3), pp. 302-338.

Coleman, J.S. (1990) 'Foundations of Social Theory', Cambridge, Mass.: The Belknap Press of Harvard University Press.

Comaroff, Jean & Comaroff, John L. (1999) 'Occult economies and the violence of abstraction: notes from the South African Postcolony', American Ethnologist, Vol. 26, n° 2, May 1999, pp. 279-303.

Couldry, Nick (2004) 'Liveness, "Reality", and the mediated habitus from Television to the Mobile Phone', The Communication Review, Vol. 7, Issue 4, Oct. 2004, London: Department of Media and Communications. London School of Economics and Political Science, pp. 353-361.

--- (2003) 'Theorising media as practice', Social Semiotics, Dec. 2003, London: London School of Economics and Political Science.

--- (2009) 'Does „the media" have a future?', European Journal of Communication, Vol 24 (4), pp. 437-449.

Cresswell, Tim (2002) 'Introduction: Theorizing place', in Verstraete, Ginette & Cresswell, Tim (eds) *Mobilizing place, placing mobility. The politics of representation in a globalized world*, Amsterdam: Colophon.

--- (2006) 'On the Move: Mobility in the Modern Western World', New York: Routledge.

Crow, G. and Allan, G. (1994) 'Community Life. An introduction to local social relations', Hemel Hempstead: Harvester Wheatsheaf.

Cruise O'Brien, Donal B. (2003) 'Symbolic confrontations. Muslims imagining the state in Africa', London: Hurst & Company.

Crystal, David (2001) 'Language and the internet', Cambridge: Cambridge University Press.

Cunningham, Stuart & Sinclair, John (2000) 'Floating lives: The Media and Asian Diasporas', St. Lucia, Queensland: University of

Queensland Press.

Danah, Boyd (2008) 'Taken out of context: American teen sociality in networked publics', PhD Dissertation. University of California-Berkeley. School of Information.

Darley, Andrew (2000) 'Visual Digital Culture. Surface play and spectacle in new media genres', London & New York: Routledge.

De Boeck, Filip (2008) 'Kinshasa. Tales of the „invisible city" & the second world', in Geschiere, Peter, Meyer, Birgit & Pels, Peter (eds) *Readings in Modernity in Africa*. Bloomington & Indianapolis: Indiana University Press, pp. 124-135.

De Boeck, Filip & Plissart, Marie-Francoise (2004) 'Kinshasa. Tales of the Invisible City', Ludion.

De Brujin, Mirjam (2007) 'Mobility and Society in the Sahel: An exploration of mobile margins and global governance', in Hahn, Hans Peter & Klute, Georg (eds) *Cultures of migration. African perspectives*, Berlin: LIT Verlag, pp. 109-128.

De Bruijn, Mirjam, Nyamnjoh, Frances & Brinkman, Inge (2009) 'Mobile Phones: The new talking drums of everyday Africa', Cameroon, Leiden: Langaa & African Studies Centre.

De Brujin, Mirjam, Van Dijk, Rijk & Foeken, Dick (eds) (2001) 'Mobile Africa. Changing patterns of movement in Africa and beyond', Brill.

De Bruijn, Mirjam, Van Dijk, Rijk & Gewald, Jan-Bart (2007) 'Social and historical trajectories of agency in Africa', in Chabal, Patrick, Engel, Ulf & de Haan, Leo (eds) *African Alternatives*, Leiden: Brill, pp. 9-20.

Den Ouden, Jan H. B. (1987) 'In search of personal mobility: Changing interpersonal relations in two Bamiléké chiefdoms, Cameroon', Africa, 57 (1) 1987.

Dewisch, René (1996) "'Pillaging Jesus": healing churches and the villagization of Kinshasa', Africa, Vol. 66, N° 1, 1996, pp. 555-586.

Diefenbach, Heike & Nauck, Bernhard (1997) 'Transnationale Migration', in Pries, Ludger (ed) *Soziale Welt*, Sonderband 12, Baden-Baden: Nomos Verlagsgesellschaft, pp. 277-292.

Diouf, Mamadou (2003) 'Engaging Postcolonial culture: African youth and public space', African Studies Review, Vol. 46, No 2, (Sept. 2003), pp. 1-12.

Dobler, Gregor (2007) 'Solidarity, Xenophobia and the regulation of Chinese business in Oshikango, Namibia', in Alden, Chris, Large, Daniel & Soares de Oliveira, Ricardo (eds) *China returns to Africa*, London: Hurst.

Donner, Jonathan (2007) 'The rules of beeping: Exchanging messages via intentional "missed calls" on mobile phones', Journal of computer-mediated communication, 13(1), article 1.

Döring, Nicola (1999) 'Sozialpsychologie des Internet. Die Bedeutung des Internet für Kommunikationsprozesse, Identitäten, soziale Beziehungen und Gruppen', Göttingen: Hogrefe Verlag für Psychologie.

Dreher, Jochen (2008) 'Protosoziologie der Freundschaft. Zur Parallelaktion von phäomenologischer und sozialwissenschaftlicher Forschung', in Raab, Jürgen, Pfadenhauser, Michaela, Stegmaier, Peter, Dreher, Jochen & Schnettler, Berndt (eds) *Phänomenologie und Soziologie. Theoretische Positionen, aktuelle Problemfelder und empirische Umsetzungen*, Wiesbaden: VS Verlage für Sozialwissenschaften, pp. 295-306.

Durkheim, Emile (1953) 'Individual and Collective Representations', Sociology and Philosophy, London: Cohen and West, pp. 1-34.

--- (1964) 'The division of labour in society', New York: Free Press.

Dürrschmidt, Jörg (2002) 'Globalisierung', Bielefeld: Transcript Verlag.

Egloff, René (Forthcoming) 'Photographie in Bamenda. Eine Ethnographische Untersuchung in einer Kamerunischen Stadt', Universität Basel, unpublished Dissertation.

Eisenstadt, Smuel (1954) 'The Absorption of Immigrants. A Comparative Study', London: Routledge.

Emirbayer, Mustafa, Mische, Ann (1998) 'What is agency?', American Journal of Sociology, Vol. 103, n° 4 (January 1998), Chicago: University of Chicago Press.

Endress, Martin & Srubar, Ilja (2003) 'Alfred Schütz. Theorie der Lebenswelt 1. Die pragmatische Schichtung der Lebenswelt', Alfred Schütz Werkausgabe, Band V.1, Konstanz: UVK Verlag.

Espinosa, Kristin & Massey, Douglas (1997) 'Undocumented migration and the quantity and quality of social capital', in Pries, Ludger (ed) *Transnationale Migration*, Soziale Welt, Sonderband 12, Baden-Baden: Nomos Verlagsgesellschaft, pp. 141-162.

Evina, Roger Charles (2009) 'Migration au Cameroun Profil National 2009, Geneva: International Organization for Migration (IOM), online.

Eyoh, Dickson (1998) 'Through the prism of a local tragedy: Political liberalization, regionalism and elite struggles for power in Cameroon', in Murray, Last et al. (eds) Africa, Journal of the International African Institute, Vol. 68, n° 3, 1994, London: IAI, pp. 338-359.

Faist, Thomas (1997) 'Migration und der Transfer sozialen Kapitals oder: Warum gibt es relativ wenige internationale Migranten?', in Pries, Ludger (ed) *Transnationale Migration*, Soziale Welt, Sonderband 12, Baden-Baden: Nomos Verlagsgesellschaft, pp. 47-62.

Fardon, Richard (1996) 'The Person, Ethnicity and the Problem of "identity" in West Africa', in Fowler, Ian & Zeitlyn, David (eds) *African Crossroads. Intersections between History and Anthropology in Cameroon*, Oxford: Berghahn Books, pp. 17-42.

--- (2000) 'African Broadcast Cultures: radio in transition', Oxford: James Currey.

Fischer, Claude S. (1992) 'America Calling. A Social History of the Telephone to 1940', California: University of California Press.

Ferguson, James G. (2008) 'Global Disconnect: Abjection & the Aftermath of Modernism', in Geschiere, Peter, Meyer, Birgit & Pels, Peter (eds) *Readings in Modernity in Africa*, London: The Int. African Institute, pp. 8-16.

Feuer, Jane (1983) 'The Concept of "Live Television": Ontology as Ideology', in E.A. Kaplan (ed) *Regarding Television*, Los Angeles, CA: American Film Institute/University Publications of America, pp. 12–22.

Fleischer, Annett (2006) 'Family, obligations, and migration. The role of kinship in Cameroon', MPIDR working paper WP 2006-047, November 2006: Max Planck Institute for Demographic Research.

Fluitman, Fred & Momo, Joseph Jean Marie (2001) 'Skills and work in the informal sector. Evidence from Yaoundé, Cameroon', Int. Training Centre of ILO, Turin 2001, online.

Fokwang, Jude Thaddeus (2008) 'Being young in Old Town: Youth subjectivities and associational life in Old Town', Dissertation, Graduate Dep. of Anthropology, University of Toronto.

Förster, Till (1999) 'Raum und Öffentlichkeit in einer dörflichen Gesellschaft Westafrikas', Iwalewa-Forum 1-2,99, pp. 49-74.

--- (2004) 'Pratiques de la mondialisation. Une perspective ethnologique', in Laurent Monnier/Yvan Droz (eds), Côté Jardin, côté cour: Anthropologie de la maison Africaine, Paris: Presses Universitaire de France, pp. 209–226.

--- (2005) 'Layers of Awareness: Intermediality and changing practices of visual arts in Northern Côte d'Ivoire and Cameroon', African Arts 38.4: 32–37, pp. 92–93.

--- (2005a) 'Globalisierung au seiner Handlungsperspektive', in Loimeier, Neubert & Weissköppel (eds) *Globalisierung im lokalen Kontext,* Münster: Lit-Verlag.

--- (2009) 'Greener Pastures. Afrikanische Europabilder vom besseren Leben', in Kreis, Georg (ed) *Bilder Europas,* Basel: Schwabe Verlag, pp. 59-78.

--- (2010) 'Neue Medien – neue Wege. Imagination und das Leben der Bilder in Afrika', Leviathan. Berliner Zeitschrift für Sozialwissenschaft, 38.3.

Fortunati, Leopoldina (2002) 'The mobile phone: towards new categories and social relations', Information, Communication & Society, N°4/5.2002, Routledge, pp. 513-528.

Frei, Bettina (2003) 'Internet use in urban Cameroon (Bamenda) September to November 2003', Un- veröffentlichte Seminararbeit, University of Basel.

--- (2005) 'New dimensions of space and time. Change of social relationships and identity by the internet in Cameroon', Unveröffentlichte Lizentiatsarbeit, University of Basel.

--- (2012) '"I go chop your dollars". Scamming practices and notions of moralities among youth in Bamenda', in H.-P. Hahn & Kastner, Kristin (eds) *Urban life-worlds in motion. African Perpectives,* Bielefeld: Transcript Verlag.

Frei, Bettina & Tazanu, Primus (2011) 'Mit der Heimat verbunden: Per Mobiltelefon realisieren Migranten Entwicklungsprojekte', Afrika Bulletin, Nr. 142, Ausg. Mai/Juni 2011, pp. 8-9.

Friedman, Thomas (2005) 'The world is flat: a brief history of the twenty-first century', NY: Strauss Giroux.

Fuh, Divine (2010) 'Competing for attention: Masculinity and prestige amongst male youth in Old Town, Bamenda', Unpublished dissertation, University of Basel.

Fulbrook, Mary & Cesarani, David (1996) 'Conclusion', in Cesarani, David & Fulbrook, Mary (eds) *Citizenship, nationality and migration in Europe*, London & New York: Routledge, pp. 209-217.

Geary, Christraud M. (1996) 'Political Dress: German-style military attire and colonial politics in Bamun', in Fowler, Ian & Zeitlyn, David (eds) *African Crossroads. Intersections between history and anthropology in Cameroon*, Oxford: Berghahn Books, pp. 165-192.

Geschiere, Peter (1997) 'The modernity of witchcraft. Politics and the occult in postcolonial Africa', Charlottesville & London: University Press of Virginia.

Geschiere, Peter & Fisiy, Cyprian (1994) 'Domesticating personal violence: witchcraft, courts and confessions in Cameroon', in Murray, Last et al. (eds) Africa. Journal of the International African Institute, Vol. 64, n° 3, 1994, London: IAI, pp. 323-341.

Geschiere, Peter & Gugler, Josef (1998) 'Introduction: the urban-rural connection – changing issues of belonging and identification', in Murray, Last et al. (eds) Africa. Journal of the International African Institute, Vol. 68, n° 3, 1994, London, pp. 309-319.

Geschiere, Peter, Meyer, Birgit & Pels, Peter (2008) 'Introduction', in Geschiere, Peter, Meyer, Birgit & Pels, Peter (eds) *Readings in*

Modernity in Africa, London: The International African Institute, pp. 1-7.

Geurts, Kathryn Linn (2002) 'Culture and the Senses. Bodily Ways of Knowing in an African Community', Berkeley: University of California Press.

Gillwald, Alison, Milek, Anne & Stork, Christoph (2010) 'Gender assessment of ICT access and usage', Africa, Vol. 1, 2010, policy paper 5, online.

Giddens, Anthony (1976) 'New Rules of Sociological Method', Basic Books, Hutchinson: New York.

--- (1994) 'The transformation of intimacy. Sexuality, love and eroticism in modern societies', Cambridge: Polity Press.

--- (1990) 'The consequences of modernity', Cambridge: Polity Press.

--- (1991) 'Modernity and Self-identity. Self & society in the late modern age', Cambridge: Polity Press.

--- (1992) 'The transformation of intimacy', Cambridge: Polity Press.

--- (1984) 'The constitution of society. Outline of the theory of structuration', Cambridge: Polity Press.

Glazer, Nathan & Moynihan, Daniel Patrick (1979) 'Beyond the melting pot', London: M.I.T. Press.

Glickman, Harvey (2005) 'The Nigerian "419" Advance fee Scams: Prank or Peril?', Canadian Journal of African Studies/Revue Canadienne d'Études Africaines, 39, 3, pp. 460-89.

Glick-Schiller, Nina, Basch, Linda & Szanton-Blanc, Cristina (1997) 'From Immigrant to Transmigrant: Theorizing Transnational Migration', in Pries, Ludger (ed) *Transnationale Migration*, Soziale Welt.

Sonderband 12, Baden-Baden: Nomos Verlagsgesellschaft, pp. 121-140.

--- (1992) 'Towards a Transnational Perspective on Migration: Race, Class, Ethnicity, and Nationalism Reconsidered', New York: New York Academy of Science.

Glick-Schiller, Nina & Fouron, Georges (1998) 'Transnational lives and national identities: the identity politics of Haitian immigrants', in Smith, Michael Peter & Guarnizo, Luis Eduardo (eds) *Transnationalism from below*, New Jersey: State University of New Jersey, pp. 130-164.

Goffman, Erving (1963) 'Behaviour in public places: notes on the social organization of gatherings', NY: Free Press.

--- (1972) 'Interaction rituals: Essays on face to face behaviour', Harmondsworth: Penguin Books.

--- (1973) 'Interaktion: Spass am Spiel Rollendistanz', München: R. Piper & Co. Verlag.

--- (1983) 'The interaction order', American Sociological Review, 48.1., pp. 1-17.

Goheen, Miriam (1996) 'Men own the fields, women own the crops. Gender and power in the Cameroon Grassfields', Madison: University of Wisconsin Press.

Goldlust, John (2001) 'Globalizing Community: Jews in Space', in Garbett, Kingsley & Murray, David (eds.) Social Analysis, Issue 45.

Goldring, Luin (1998) 'The power of status in transnational social fields', in Smith, Michael Peter & Guarnizo, Luis Eduardo (eds) *Transnationalism from below*, New Jersey: State University of New Jersey, pp. 165-195.

--- (1997) 'Power and status in transnational social spaces', in Pries, Ludger (ed) *Transnationale Migration*, Soziale Welt, Sonderband 12, Baden-Baden: Nomos Verlagsgesellschaft, pp. 179-196.

Grätz, Tilo (ed) (2010) 'Mobility, transnationalism, and contemporary African societies', Newcastle: Camebridge Scholars Publishing.

Gross, Neil & Simmons, Solon (2002) 'Intimacy as a Double-Edged Phenomenon? An Empirical Test of Giddens', Social Forces, Vol. 81, No 2, (Dec. 2002), pp. 531-555.

Guarnizo, Luis Eduardo & Smith, Michael Peter (1998) 'The locations of transnationalism', in Smith, Michael Peter & Guarnizo, Luis Eduardo (eds) *Transnationalism from below*, New Jersey: State University of New Jersey, pp. 3-34.

--- (1997) 'Power and status in transnational social spaces', in Pries, Ludger (ed) *Transnationale Migration*, Soziale Welt, Sonderband 12, Baden-Baden: Nomos Verlagsgesellschaft, pp. 179-196.

Gupta, Akhil & Ferguson, James (1992) 'Beyond Culture: Space, identity and the politics of difference', Cultural Anthropology, Vol. 7, n° 1, pp. 6-23.

Gugler, Josef (2002) 'The son of the hawk does not remain abroad: the urban-rural connection in Africa', African Studies Review, Vol. 45, 1, Apr. 2002, pp. 21-41.

Habermas, Jürgen (1990) 'Strukturwandel der Öffentlichkeit. Untersuchungen zu einer Kategorie der bürgerlichen Gesellschaft', Frankfurt/Main: Suhrkamp.

Hahn, Hans Peter (2004) 'Global Goods and the Process of Appropriation', in Probst, Peter und Spittler, Gerd (eds) *Between Resistance and Expansion. Dimensions of Local Vitality in Africa*, Münster: LIT-Verlag, pp. 211-230.

--- (2007) 'Migration as Discursive Space – Negotiations of Leaving and Returning in the Kasena Homeland (Burkina Faso)', in Hahn, Hans Peter & Klute, Georg (eds) *Cultures of Migration. African Perspectives*, Berlin: LIT Verlag, pp. 149-174.

Hahn, Hans Peter & Klute, Georg (2007) 'Cultures of Migration: Introduction', in Hahn, Hans Peter & Klute, Georg (eds) *Cultures of migration. African perspectives*, Berlin: LIT Verlag, pp. 9-30.

Hahn, Hans Peter & Kibora, Ludovic (2008) 'The domestication of the mobile phone: oral society and new ICT in Burkina Faso', Journal of Modern African Studies 46, I (2008), Cambridge Univ. Press, pp. 87-109.

Hafkin, Nancy J. & Huyer, Sophia (2008) 'Women and gender in ICT statistics and indicators for development', Research articles, MIT Press. http://itidjournal.org/itid/article/viewFile/254/124.

Hakken, David (1999) 'Cyborgs@cyberspace?: An ethnographer looks to the future', New York: Routledge.

Hannerz, Ulf (2006) 'Studying down, up, sideways, through, backwards, forwards, away and home: reflections on the field worries of an expansive discipline', in Coleman, Simon & Collins, Peter (eds) *Locating the field. Space, place and context in Anthropology*, Oxford: Berg, pp. 23-42.

--- (2003) 'Being there... and there... and there! Reflections on multi-sited ethnography', Ethnography. June 2003, Vol. 4, n° 2, Sage online publications, pp. 201-216.

--- (1996) 'Transnational Connections. Culture, people, places', London & New York: Routledge.

Hannken, Helga (2003) 'Internationale Migration von und nach Afrika. Der weite Weg zurück nach Eritrea', in Jensen, Jürgen (ed) *Immigration – Emigration – Remigration*, Reihe: Interethnische

Beziehungen und Kulturwandel. Ethnologische Beiträge zu soziokultureller Dynamik, Hamburg: LIT-Verlag.

Hansen, Ketil Fred (2003) 'The politics of personal relations: beyond neopatrimonial practices in Northern Cameroon', Africa, Vol. 73, N° 2, (2003), pp. 202-225.

Hantel-Quitmann, Wolfgang (2002) 'Die Globalisierung der Intimität. Die Zukunft intimer Beziehungen im Zeitalter der Globalisierung', in Hantel-Quitmann, Wolfgang & Kastner, Peter (eds) *Die Globalisierung der Intimität. Die Zukunft intimer Beziehungen im Zeitalter der Globalisierung*, Giessen: Psychosozial Verlag.

Hantel-Quitmann, Wolfgang & Kastner, Peter (eds) (2002) 'Die Globalisierung der Intimität. Die Zukunft intimer Beziehungen im Zeitalter der Globalisierung', Giessen: Psychosozial Verlag.

Haraway, Donna (1991) 'A Cyborg Manifesto: Science, Technology, and Socialist-Feminism in the Late Twentieth Century', Simians, Cyborgs and Women: The Reinvention of Nature, NY: Routledge, pp. 149-181.

Hardin, Russell (1993) 'The Street-Level Epistemology of Trust', Politics & Society 21 (4), pp. 505-529.

Harvey, David (1990) 'The condition of postmodernity', Cambridge: Blackwell Publishers Inc..

Hawisher, Gail E. & Selfe, Cynthia L. (2000) *Global Literacies and the World-Wide-Web*, NY: Routledge.

Hjarvard, Stig (2002) 'Mediated encounters. An essay on the role of communication media in the creation of trust in the "global metropolis', in Stald, Gitte & Tufte, Thomas (eds) *Global encounters. Media and cultural transformation*, Luton: University of Luton Press.

Hoggett, P. (1997) 'Contested communities', in P. Hoggett (ed)

Contested Communities. Experiences, struggles, policies, Bristol: Policy Press.

Homfeldt, Hans Günther, Schweppe, Cornelia & Schröer, Wolfgang (2006) 'Transnationalität, soziale Unterstützung, agency', Nordhausen: Traugott Bautz.

Horst, Heather & Miller, Daniel (2005) 'From Kinship to Link-up. Cell phones and social networking in Jamaica', Current Anthropology, Vol. 46, N° 5, December 2005, Wenner-Gren Foundation for Anthropological Research.

Horst, Heather A & Miller, Daniel (2006) 'The cell phone. An Anthropology of communication', Oxford: Berg.

Hoskins, A. (2001) 'Mediating Time: The Temporal Mix of Television', Time Society, 10(2/3), pp. 213–233.

Howells, Richard (2003) 'Visual Culture', Cambridge: Blackwell Publishing Company.

Humphreys, Lee (2005) 'Cell phones in public: social interactions in a wireless era', New Media & Society, London: SAGE Publications.

Hyden, Göran; Leslie, Michael; Ogundimu, Folu (eds) (2003) 'Media and Democracy in Africa', The Nordic Africa Institute.

Ifeka, Caroline (1992) 'The mystical and political powers of Queen Mothers, Kings and Commoners in Nso', Cameroon', in Ardener, Shirley (ed) *Persons and Powers of women in diverse cultures*, New York, Oxford: Berg, pp. 135-158.

Ignacio, Emily Noelle (2005) 'Building Diaspora: Filipino Community Formation on the Internet: Filipino cultural community formation on the internet', Rutgers University Press.

Ito, Mizuko, Okabe, Daisuke, Matsuda, Misa (2005) 'Personal, portable, pedestrian: mobile phones in Japanese life', Cambridge: MIT Press.

Iveson, Kurt (2007) 'Publics and the City', Oxford: Blackwell Publishers Ltd..

Jackson, Peter, Crang, Philip & Dwyer, Claire (2004) 'Introduction. The spaces of transnationality', in Jackson, Peter, Crang, Philip & Dwyer, Claire (eds) *Transnational Spaces*, pp. 1-20.

Jensen, Jürgen (2004) 'Plural societies and transnational social spaces – modern African complexities', in Ossenbrügger, Jürgen & Reh, Mechthild (eds) *Social spaces of African societies. Applications and critique of concepts about "transnational social spaces*, Münster: LIT Verlag, pp. 35-76.

Johnson-Hanks, Jennifer (2007) 'Women on the Market: Marriage, Consumption and the Internet in Urban Cameroon', American Ethnologist, 34(4), pp. 642-658.

Jua, Nantang Ben (2003) 'Differential responses to Disappearing Pathways: Redefining Possibility among Cameroon youths', African Studies Review, 46 (2), pp. 12-36.

Jules-Rosette, Benetta (1990) 'Terminal signs: computers and social change in Africa', Berlin, New York: Mouton de Gruyter.

Jurriëns, Edwin (2004) 'Cultural Travel and Migrancy. The artistic representation of globalization in the electronic media of West Java', Leiden: KITLV Press.

Kaberry, Phyllis M. (1952) 'Women of the Grassfileds. A study of the economic position of women in Bamenda, British Cameroons', London: Her Majesty's Stationery Office.

Katz, J.E. (ed) (2003) 'Machines That Become Us: The Social Context of Personal Communication Technology', New Brunswick, NJ:

Transaction Publishers.

Katz, Janes & Aakhus, Mark (2002) 'Perpetual Contact. Mobile Communication, Private Talk, Public Performance', Cambridge: University Press.

Katz, James & Rice, R.E. (2002) 'Social consequencesof internet use: Access, involvement and interaction', Cambridge: MIT Press.

Kennedy, Paul & Roudometof, Victor (2002) 'Transnationalism in a global age', in Kennedy, Paul & Roudometof, Victor (eds) *Communities across borders. New immigrants and transnational cultures*, London & New York: Routledge, pp. 1-26.

Kinge, Appolinaire (2004) 'The Cameroun experience of the informal sector', 7th meeting of the expert group on informal sector statistics (Delhi Group), Delhi 2004, online.

Klein, Jennie (2000) 'Review: Real Events', Journal of Performance and Art, Vol 22, n° 1 (Jan. 2000).

Knibbe, Kim (2009) ',,We did not come here as tenants, but as landlords": Nigerian Pentecostals and the power of maps', African Diaspora 2 (2009), pp. 133-158.

Kofler, Angelika (2002) 'Migration, Emotion, Identities. The subject meaning of difference', Braumüller Verlag.

Kohnert, Dirk (2007) 'On the renaissance of African modes of thought – The example of the belief in magic and witchcraft', in Schmidt, Burghart, and Rolf Schulte (eds) *Witchcraft in Modern Africa: Witches, witch-hunts and magical imaginaries*, Hamburg: Dokumentation and Buch (DOBU), pp. 39-61.

Kollock, Peter & Smith, Marc (1999) 'Communities in cyberspace', London: Routledge.

Kriem, Maya S. (2009) 'Mobile telephony in Morocco: a changing sociality', Media Culture Society 2009, 31, 617, Sage, online.

Krotz, Friedrich (2001) 'Die Mediatisierung kommunikativen Handelns. Der Wandel von Alltag und sozalen Beziehungen, Kultur und Gesellschaft durch die Medien', Wiesbaden: Westdeutscher Verlag.

Künzler, Daniel (2007) 'Who wants to be a millionaire? Global capitalism and fraud in Nigeria', University of Zurich. 22.1.2007, online.

--- (2009) 'The Figure of Success as Content and Consequence of the Video Film Industries in South-Eastern Nigeria and Ghana', Revised version of a paper presented at the International Conference "Nollywood and Beyond: Transnational Dimensions of the African Video Industry" at the Johannes Gutenberg University, Mainz, May 15, 2009.

Kusenbach, Margarethe (2008) 'Mitgehen als Methode. Der "Go-Along" in der phänomenologischen Forschungspraxis', in Raab, Jürgen, Pfadenhauser, Michaela, Stegmaier, Peter, Dreher, Jochen & Schnettler, Berndt (eds) *Phänomenologie und Soziologie. Theoretische Positionen, aktuelle Problemfelder und empirische Umsetzungen*, Wiesbaden: VS Verlage für Sozialwissenschaften, pp. 349-358.

Landzelius, Kyra (2006) 'Native on the Net. Indigenous and Diasporic Peoples in the Virtual Age', London & New York: Rouledge.

Larkin, Brian (2008) 'Degraded images, distorted sounds: Nigerian video and the infrastructure of piracy', in Geschiere, Peter, Meyer, Birgit & Pels, Peter (eds) *Readings in Modernity in Africa*, Bloomington & Indianapolis: Indiana University Press, pp. 146-154.

--- (1997) 'Indian films and Nigerian lovers: Media and the creation of parallel modernities', Africa 67 , no. 3, pp. 406-440.

Larsen, Jonas, Urry, John & Axhausen, Kay (2006) 'Mobilities, Networks, Geographies', Hampshire: Ashgate Publishing Limited.

Lawie, Smadar, Narayan, Kirin & Rosaldo, Renato (eds) (1993) 'Creativity / Anthropology', Ithaca & London: Cornell University Press.

Lawson, Mark (2001) 'Where were you when Nasty Nick was expelled?', Guardian, 18 August, 2-3.

Leader, Kathryn (2010) 'Closed-Circuit Television Testimony: Liveness and Truth-telling', Law Text Culture, 14(1), 2010, pp. 312-336.

Lee, Barett A. & Campbell, Karen E. (1999) 'Neighbour networks of Black and White Americans', in Wellman, Barry (ed) *Networks in the Global Village. Life in Contemporary Communities*, Colorado: Westview Press, pp. 119-146.

Lee, Sarah (1999) 'Private uses in public spaces: a study of an internet café', New Media Society 1999. 1:331, London: Sage publications.

Lefebvre, Henri (2000 (1974)) 'The Production of Space', in Dünne, Jörg & Günzel, Stephan (eds) *Raumtheorie. Grundlagentexte aus Philosophie und Kulturwissenschaften*, München: Suhrkamp.

Ley, David & Waters, Johanna (2004) 'Transnational migration and the geographical imperative', in Jackson, Peter, Crang, Philip & Dwyer, Claire (eds) *Transnational Spaces*, pp. 104-121.

Levitt, Peggy & Glick-Schiller, Nina (2007) 'Conceptualizing Simultaneity: a transnational social field perspective on society', in Portes, Alejandro & DeWind, Josh (eds) *Rethinking Migration. New Theoretical and Emperical Perspectives*, New York, Oxford: Berghahn Books, pp. 181-218.

Levitt, P. (1998) 'Social remittances: Migration driven local-level forms of cultural diffusion', International Migration Review, Vol. 32, N°4, pp. 926-948.

Licoppe, Christian (2009) '"Connected" presence: the emergence of a new repertoire for managing social relationships in a changing communication technoscape', in Lievrouw, Leah A. & Livingstone, Sonia (eds) *New Media*. Volume III, Practices: Interaction, Identity, Culture, London: SAGE, pp. 69-94.

Lindquist, Johan A. (2009) 'The anxieties of mobility. Migration and tourism in the Indonesian borderlands', Honululu: University of Hawai'i Press.

Ling, R. (2004) 'The mobile Connection: The cell phone's impact on Society', S Francisco: Morgan Kaufmann.

Ling, R. and Haddon, L. (2003) 'Mobile Telephony, Mobility and the Co-ordination of Everyday Life', in Katz, J. (ed) *Machines that Become Us: The Social Context of Personal Communication Technology*, Transaction Publishers, New Brunswick, New Jersey, pp. 245-66.

Ling R., & Campbell, S.W. (2009) 'Introduction: The Reconstruction of Space and Time through Mobile Communication Practices', in R. Ling, & S. W. Campbell (eds) *The Reconstruction of Space and Time: Mobile Communication Practices*, New Brunswick, NJ: Transaction Publishers, pp. 1-16.

Luhmann, Niklas (1979) 'Trust and power', Chichester: Wiley.
--- (1973) 'Vertrauen. Ein Mechanismus der Reduktion sozialer Komplexität', 2. Erweiterte Auflage. Stuttgart: Ferdinand Enke Verlag.

Lull, James (2000) 'Media, Communication, Culture: a global approach', Cambridge: Polity Press.

Lyons, G. & Urry, J. (2005) 'Travel time use in the information age', Transportation Research, 39(A), pp. 257-276.

Macamo, Elisio & Neubert, Dieter (2008) 'The New and Its Temptations: Products of Modernity and their Impact on Social Change in Africa', in Afe, Adogame, Echtler, Magnus & Vierke, Ulf (eds) *Un- packing the New*, Wien, Berlin: LIT Verlag, pp. 271-304.

Malaquais, Dominique (2001) 'Arts de feyre au Cameroun', Politique Africaine, 82, pp. 101-18.

Mankekar, Purima (1993) 'National texts and gendered lives: an ethnography of television viewers in a North Indian city', American Ethnologist 20.3, American Anthropological Association, pp. 543-563.

Mannheim, Karl (1936) 'Ideology and Utopia. An Introduction to the Sociology of Knowledge, New York: Harcourt, Brace & World.

Manovich, L. (2004) 'The poetics of augmented space', online www.manovich.net

Markham, Annette N. (1998) 'Life online. Researching real experience in virtual space', Altamira Press.

Martin, Jeannett (2007) 'What's new with the "Been-to's"? Educational migrants, return from Europe and migrant's culture in urban southern Ghana', in Hahn, Hans Peter & Klute, Georg (eds) *Cultures of migration. African perspectives*, Berlin: LIT Verlag, pp. 203-238.

Massey, Doris (1995) 'The Conceptualization of Place. In D. Massey and P. Jess (eds) *A Place in the World? Places, Cultures and Globalization*, Oxford: Oxford University Press/Open University, pp. 45-77.

Massey, D., Espinosa, S., Kristin, E. (1997) 'What's driving Mexico-US migration? A theoretical, empirical and policy analysis', American Journal of Sociology 102, pp. 939-999.

Massey, D., Arango, J., Hugo, G., Kouaouci, A., Pellegrino, A. & Taylor, J.E. (1994) 'An Evaluation of international migration theory: The North American case', The Population and Development Review.

Massey, Douglas S., Arango, Joaquin, Hugo, Graeme, Kouaouci, Ali, Pellegrino, Adela & Taylor, J. Edward (eds) (1998) 'Worlds in Motion. Understanding international migration at the end of the millennium', Oxford: Clarendon Press.

Massey, D., Alarcon, S., Durand, R., Gonzalez, J., Humberto (1987) 'Return to Aztlan: The social process of international migration from Western Mexico', Berkeley: University of California Press.

Mayrhofer, Elke (2003) 'Afrikanische Diaspora. Terminus, Konzept und die Bedeutung von „home"', in Zips, Werner (ed) *Afrikanische Diaspora. Out of Africa – Into new worlds*, Münster: LIT Verlag, pp. 53-74.

Mazzarella, William (2004) 'Culture, Globalization, Mediation', Annual Reviews in Anthropology, 2004. 33, pp. 345-67.

Mbaku, John Mukum (2005) 'Culture and Customs of Cameroon', London: Greenwood Press.

Mbarika, Victor W.A., Mbarika, Irene (2006) 'Africa Calling', IEEE Spectrum, online www.spectrum.ieee.org.

Mbembe, Achille (2008. The New Africans Between Nativism & Cosmopolitanism. in Geschiere, Peter, Meyer, Birgit & Pels, Peter (eds.) *Readings in Modernity in Africa*. London: The Int. African Institute.

Mbembe, A. & Roitman, J. (1996) 'Figures of the subject in times of crisis', in Yaegar, P. (ed) *The geography of identity*, University of Michigan Press, Ann Arbor.

McCarthy, Anna (2001) 'From Screen to Site: Television's Material Culture, and Its Place', MIT Press, October, Vol. 98 (Autumn 2001), pp. 91-111.

McIntyre, Joseph A. (2004) 'Away from home: Hausa speaking refugees in Hamburg', in Ossenbrügge, Jürgen & Reh, Mechthild (eds) *Social spaces of African societies. Applications and critique of concepts about "transnational social spaces*, Münster: LIT Verlag, pp. 147-174.

McIntyre, Ronald & Woodruff Smith, David (1989) 'Theory of intentionality', in Mohanty, J. N. & McKenna, William R. (eds) Husslers Phenomenology: A Textbook, Washington DC: Centre for Advanced Research in Phenomenology and University Press of America, pp. 147-179.

McLuhan, Marshall (1999 (1964)) 'Understanding media. The extensions of man', Cambridge: MIT Press.

--- (1996 (1967)) 'The Medium is the massage. An inventory of effects', in Agel, Jerome (ed), Singapore.

Mead, George Herbert (1932) 'The philosophy of the present', Chicago: University of Chicago Press.

Merleau-Ponty, M. (1966) 'Die Phänomenologie der Wahrnehmung', Berlin: De Gruyter.

Meyrowitz, Joshua (1985) 'No Sense of Place: The Impact of Electronic Media on Social Behavior', New York: Oxford University Press.

Miller, Daniel & Slater, Don (2001) 'The Internet. An ethnographic approach', Oxford: Berg.

--- (2009) 'Being Trini and representing Trinidad', in Lievrouw, Leah A. & Livingstone, Sonia (eds) New Media, Volume III, *Practices: Interaction, Identity, Culture*, London: SAGE.

Misztal, Barbara A. (1996) 'Trust in modern societies', Cambridge: Polity Press.

Mitchell, W.J.T. (1984) 'Image/Imago/Imagination', New Literary History, Vol. 15, n° 3. Spring, pp. 503-537.

Mohr de Collado, Maren (2005) 'Lebensformen zwischen „Hier" und "Dort". Transnationale Migration und Wandel bei den Garinagu in Guatemala und New York', Aachen: Shaker Verlag.

Moores, Shaun (2003) 'Media, Flows and Places', MEDIA@LSE Electronic Working Papers, N° 6, London: School of Economics and Political Science.

--- (2004) 'The doubling of place: electronic media, time-space arrangements and social relationships', in Couldry, Nick & McCarthy, Anna (eds) *Mediaspace. Place, Scale and Culture in a Media Age*, London: Routledge, pp. 21-36.

Morawska, Ewa (2001) 'Structuring Migration: The Case of Polish Income-Seeking Travellers to the West', Theory and Society, 31, pp. 47-80.

Münz, Rainer (1997) 'Woher – wohin? – Massenmigration im Europa des 20. Jahrhunderts', in Pries, Ludger (ed) *Transnationale Migration*, Soziale Welt, Sonderband 12, Baden-Baden: Nomos Verlagsgesellschaft.

Murphy, Patrick D. & Kraidy, Marwan M. (eds) (2003) 'Global Media Studies. Ethnographic Perspectives', New York: Routledge.

Mytton, Graham (1983) 'Mass communication in Africa', London: E. Arnold.

Närman, Anders (1995) 'The dilemmas facing Kenya school leavers. Surviving in the city or a force for local mobilization?', in Aina, Tade

Akin & Baker, Jonathan (eds) *The migration experience in Africa*, Uppsala: Nordic African Institute, pp. 167-180.

Navas Sabater, Juan, Dymond, Andrew & Juntunen, Nina (2002) 'Telecommunications and information services for the poor: toward a strategy for universal access', World Bank Publication.

Niedrig, Heike & Schroeder, Joachim (2004) 'Bildungsperspektiven jugendlicher Transmigranten. Chancen und Barrieren im Bildungswesen aus der Sicht afrikanischer Migrantenjugendlicher in Hamburg', in Ossenbrügger, Jürgen & Reh, Mechthild (eds) *Social spaces of African societies. Applications and critique of concepts about "transnational social spaces"*, Münster: LIT Verlag, pp. 77-110.

Ndjio, Basile (2008) 'Evolués & Feymen', in Geschiere, Peter, Meyer, Birgit & Pels, Peter (eds) *Readings in Modernity in Africa*, London: The International African Institute, pp. 205-214.

Nkwi, Walter Gam (2009) 'From the elitist to the commonality of voice communication: the history of the Telefone in Buea, Cameroon', in De Bruijn, Mirjam, Nyamnjoh, Frances & Brinkman, Inge (eds) *Mobile Phones: The new talking drums of everyday Africa*, Cameroon, Leiden: Langaa & African Studies Centre.

Nyamnjoh, Francis (2005) 'Images of Nyongo amongst Bamenda Grassfielders in Whiteman Kontri', Citizenship Studies, Vol. 9, No. 3: Routledge, pp. 241-269.

--- (2005a) 'Africa's Media, Democracy and the Politics of Belonging', London & New York: Zed Books.

--- (2005b) 'Fishing in troubled waters: disquettes and thiofs in Dakar', Africa 75 (3) 2005, pp. 295-324.

--- (2002) '"A Child is One Person's Only in the Womb": Domestication, Agency and Subjectivity in the Cameroonian Grassfields', in Werbner. R. (ed) *Postcolonial Subjectivities in Africa*,

London: Zed Books.

Nyamnjoh, Francis & Page, Ben (2002) 'Whiteman Kontri and the enduring allure of modernity of the Cameroonian youth', African Affairs (2002) 101, London: Royal African Society, pp. 607-634.

--- (1999) 'Cameroon: A Country United By Ethnic Ambition And Difference', African Affairs, Vol. 98 (390), pp. 101-118.

OECD (2007) 'International Migration Outlook. Annual Report', Sopemi 2007 Edition, OECD 2007.

Ong, Aihwa (1999) 'Flexible citizenship: the cultural logic of transnationality', Durham, NC: Duke Univ. Press.

Ortiz, Laura Velasco (2005) 'Mixtec Transnational Identity', Tucson: The University of Arizona Press.

Page, Ben (2007) 'Slow going: the mortuary, modernity and the hometown association in Bali-Nyonga, Cameroon', Africa n° 77 (3) 2007.

Parsons, T. (1969) 'Politics and social structure', Glencoe: Free Press.

Pelican, Michaela (2010) 'Local perspectives on transnational relations of Cameroonian migrants', in Grätz, Tilo (ed) *Mobility, Transnationalism, and Contemporary African Societies*, Cambridge: Cambridge Scholars Publishing, pp. 178-191.

Pelican, Michaela & Tatah, Peter (2009) 'Migration to the Gulf States and China: local perspectives from Cameroon', African Diaspora 2 (2009), pp. 229-244.

Pertierra, Raul (2005) 'Mobile phones, Identity and discursive intimacy', Human Technology, Vol. 1 April 2005, pp. 23-44.

Pfaff, Julia (2007) 'Finding one's Way through Places – A

Contemporary Trade Journey of Young Zanzibari Traders', in Hahn, Hans Peter & Klute, Georg (eds) *Cultures of Migration. African Perspectives*, Berlin: LIT Verlag, pp. 61-88.

Phelan, Peggy (1993) 'Unmarked: the politics of performance', London & New York: Routledge.

Pink, Sarah (2001) 'Sunglasses, suitcases and other symbols', in Hendry, Joy & Watson, C.W. (eds) *An Anthropology of indirect communication*, London & New York: Routledge.

Plaut, P. (2004) 'Non-commuters: The people who walk to work or work at home', Transportation 31, pp. 229 – 255.

Popitz, Heinrich (2006) 'Soziale Normen', in Pohlmann, Friedrich & Essbach, Wolfgang (eds), Suhrkamp Taschenbuch.

Portes, Alejandro (1997) 'Immigration Theory for a new century: Some problems and opportunities', International Migration Reviews 31, pp. 799-825.

Pradelles de Latour, Charles-Henry (1994) 'Marriage payments, debt and fatherhood among the Bangoua: A Lacanian analysis of a kinship system', Africa, Vol. 64, n° 3, 1994, London: IAI, pp. 21-33.

Pries, Ludger (1997) 'Neue Migration im transnationalen Raum', in Pries, Ludger (ed) *Transnationale Migration*, Soziale Welt, Sonderband 12, Baden-Baden: Nomos Verlagsgesellschaft, pp. 15-46.

Putnam, Robert D. (1993) 'Making Democracy work. Civic Traditions in modern Italy', Princeton: Princeton University Press.

Raab, Jürgen (2008) 'Präsenz und mediale Präsentation. Zum Verhältnis von Körper und technischen Medien aus Perspektive der phänomenologisch orientierten Wissenssoziologie', in Raab, Jürgen, Pfadenhauer, Michaela, Stegmaier, Peter, Dreher, Jochen & Schnettler, Berndt (eds) *Phänomenologie und Soziologie. Theoretische*

Positionen, aktuelle Problemfelder und empirische Umsetzungen, Wiesbaden: Verlag für Sozialwissenschaften.

Rafaeli, Sheizaf (2009) 'Interactivity: From New Media to Communication', in Lievrouw, Leah A. & Livingstone, Sonia (eds) *New Media*, Vol. III, Practices: Interaction, Identity, Culture, London: SAGE.

Rajewsky, Irina O. (2002) 'Intermedialität', Tübingen/Basel: A. Francke Verlag, pp. 22-41.

Rasmussen, Terje (2002) 'The internet as a world medium', in Stald, Gitte & Tufte, Thomas (eds) *Global Encounters – media and cultural transformation*, Luton: University of Luton Press, pp. 85-106.

Rheingold, H. (1993) 'The Virtual Community', New York: Harper Perennial.

Rivers, Deanna Sue (2005) 'Zapotec use of e-commerce: the portrait of Teotitlan del Valle, Mexico', Dissertation, Michigan: Pro Quest, UMI Dissertation Services.

Robben, A.C.G.M. & Sluka, Jeffrey A. (2006) 'Ethnographic fieldwork: An Anthropology', Wiley-Blackwell.

Robertson, Roland (1992) 'Globalization', London and Newbury Park: Sage.

Robins, Kevin (2002) 'Encountering Globalization', in Held, David & McGrew, Anthony (eds) *The global transformation reader. An introduction to the globalization debate*, 2nd edition, Cambridge: Polity Press.

Roitman, Janet (2008) 'A successful life in the illegal realm', in Geschiere, Peter, Meyer, Birgit & Pels, Peter (eds) *Readings in Modernity in Africa*, London: The International African Institute, pp. 214-220.

Rouse, Roger (2004) 'Mexican migration and the social space of postmodernism', in Jackson, Peter, Crang, Philip & Dwyer, Claire (eds) *Transnational Spaces*, pp. 24-37.

--- (1989) 'Mexican Migration to the United States: Family Relations in the Development of a Trans-national Migration Circuit', PhD Diss: Stanford University.

Rowlands, Michael & Warnier, Jean-Pierre (1988) 'Sorcery, power and the modern state in Cameroon', in Finnegan, Ruth (ed) Man, New Series, Vol. 23, N° 1, London: Royal Anthropological Institute, pp. 118-132.

Rudin, Harry (1938) 'Germans in the Cameroons, 1884-1914. A case study in modern imperialism', Yale: Yale University.

Rydin, Ingegerd & Sjöberg, Ulrika (2008) 'Mediated Crossroads. Identity youth culture and ethnicity', Theoretical and Methodological Challenges, Sweden: Nordicom.

Sarbaugh-Thompson, Majorie, Feldman, Martha S. (1998) 'Electronic Mail and Organizational Communication: Does Saying "Hi" Really Matter?', Organization Science, Vol. 9,, N° 6 (Nov. 1996): INFORMS.

Scannell, P. (1996) 'Radio, Television and Modern Life: A Phenomenological Approach', Oxford: Blackwell.

--- (2007) 'Media and communication', London: SAGE Publications.

Schapendonk, Joris (2010) 'Staying put in moving sands. The stepwise migration process of sub-Saharan African migrants heading north', in Engel, U. & Nugent, P. (eds) *Respacing Africa*, Leiden, Boston: Brill.

Schein, Louisa (2002) 'Mapping Hmong Media in Diasporic Space', in Ginsburg, Faye D, Abu-Lughod, Lila & Larkin, Brian (eds) *Media*

Worlds. Anthropology on New Terrain, Berkeley, Los Angeles, London: University of California Press, pp. 229-246.

Schröder, Ingo W. & Voell, Stephane (2002) 'Einleitung: Moderne Oralität. Kommunikationsverhältnisse an der Jahrtausendwende', in Schröder, Ingo W. & Voell, Stephane (eds) *Moderne Oralität. Ethnologische Perspektiven auf die plurimediale Gegenwart*, Bamberg: Curupira, pp. 11-50.

Schütz, Thomas (1967) 'The phenomenology of the social world', Evanston, IL.: Northwestern Univ. Press.

Sciadas, George (2005) 'From the digital divide to digital opportunities. Measuring infostats for development', Canada: Orbicom, online.

Sey, Araba (2011) 'We use it different, different: Making sense of trends in mobile phone usein Ghana', New Media Society 2011, 31.3.

Sheller, M. & Urry, J. (2006) 'Mobile technologies of the city', New York: Routledge.

Silver, David (2000) 'Looking Backwards, Looking Forward: Cyberculture Studies', in Gauntlett, David (ed) *Web.studies: rewiring Media Studies in the Digital Age*, Oxford: Oxford University Press, pp. 19-30.

Simmel, Georg (1999 (1908)) 'Soziologie. Untersuchungen über die Formen der Vergesellschaftung', Gesamtausgabe Bd. 11, Frankfurt a. M.: Suhrkamp.

--- (1971) 'Georg Simmel on individuality and social forms' by Donald, N. (ed), University of Chicago Press.

--- (1950) 'The sociology of Georg Simmel', by Wolff, K.H. (ed), New York: Free Press.

Simone, Abdoumaliq (2005) 'Urban Circulation and the Everyday Politics of African Urban Youth: The Case of Douala, Cameroon', International Journal of Urban and Regional Research, Vol. 29.3, pp. 516-532.

--- (2008) 'On the worlding of African cities', in Geschiere, Peter, Meyer, Birgit & Pels, Peter (eds) *Readings in Modernity in Africa*, Bloomington & Indianapolis: Indiana University Press, pp. 135-146.

Simone, Abdoumaliq & Abouhani, Abdelghani (eds) (2007) 'Urban Africa: Changing contours of survival in the city', London: Zed books.

Slater, Don (2005) 'Comments to Miller and Slater. From Kinship to Link-up', Current Anthropology, Vol. 46, N° 5 (December 2005).

Smart, Alan & Smart, Josephine (1998) 'Transnational social networks and negotiated identities in interactions between Hong Kong and China', in Smith, Michael Peter & Guarnizo, Luis Eduardo (eds) *Transnationalism from below*, New Jersey: State University of New Jersey, pp. 103-129.

Smith, Daniel Jordan (2007) 'A culture of corruption: everyday deception and popular discontent in Nigeria', Princeton and Oxford: Princeton University Press.

Smith, Marc & Kollock, Peter (eds) (1999) 'Communities in Cyberspace', London: Routledge.

Smith, M. K. (2001) '"Community" in the Encyclopedia of informal education: online.

Smith, Michael Peter & Guarnizo, Luis Eduardo (eds) (1998) 'Transnationalism from below', Comparative Urban & Community Research, Vol. 6, New Jersey: State University New Jersey.

Smith, Robert (1997) 'Reflections on migration, the state and the construction, durability and newness of transnational life' in Pries, Ludger (ed) *Transnationale Migration*, Soziale Welt, Sonderband 12, Baden-Baden: Nomos Verlagsgesellschaft, pp. 197-220.

Smith, Robert C. (1998) 'Transnational Localities: Community, Technology and the Politics of Membership within the Context of Mexico and U.S. Migration', in Smith, Michael Peter & Guarnizo, Luis Eduardo (eds) *Transnationalism from below*, Comparative Urban & Community Research, Vol.6, New Yersey: Transaction Publishers.

Spittler, Gerd (2002) 'Globale Waren – locale Aneignung', in Hauser, Schäublin, Brigitta & Braukämper, Ulrich (eds) *Ethnologie der Globalisierung*, Berlin: Dietrich-Reimer Verlag, pp. 15-30.

Spitulnik, Debra (2002) 'Alternative Small Media and Communicative Spaces', in Hyden, Goran, Michael, Leslie & Folu F. (eds), New Brunwick: Transaction Publishers, pp. 177-190.

Stark, O. & Lucas, R.E.B. (1988) 'Migration, Remittances, and the family', Economic Development and Cultural Change, Vol. 36, n° 3, pp. 465-481.

Stegmaier, Peter (2008) 'Normative Praxis: konsitutions- und konstruktionsanalytische Grundlagen', in Raab, Jürgen, Pfadenhauser, Michaela, Stegmaier, Peter, Dreher, Jochen & Schnettler, Berndt (eds) *Phänomenologie und Soziologie. Theoretische Positionen, aktuelle Problemfelder und empirische Umsetzungen*, Wiesbaden: VS Verlage für Sozialwissenschaften, pp. 263-272.

Stephen, Lynn (2007) 'Transborder Lives. Indigenous Oaxacans in Mexico, California, and Oregon', Durham & London: Duke University Press.

Sturken, Marita & Cartwright, Lisa (2004) 'Practices of Looking. An introduction to Visual Culture', Oxford: University Press.

Tardits, Claude (1960) 'Contribution à l'étude des populations bamiléké de l'Ouest Cameroun', Paris: Berger-Levrault.

Tazanu, Mbeanwoah Primus (2012) 'Being available and reachable: New Media and Cameroonian transnational sociality', Bamenda, Cameroon: Langaa Research & Publishing Common Initiative Group.

--- (2012a) '"They behave as though they want to bring heaven down". Some narratives on the visibility of Cameroonian migrant youths in Cameroon urban space', in Hahn, Hans-Peter & Kastner, Kristin (eds) *Urban Life-Worlds in Motion. African Perspectives*, Bielefeld: Transcript.

Tetang, Tchinda Josué (2007) 'ICT in Education in Cameroon. Survey of ICT and Education in Africa: Cameroon Country Report', CIA World Fact Book, June 2007, online.

Thompson, Kevin C. (2001) 'Watching the Stormfront: white Nationalists and the building of Community in Cyberspace', in Morton, Helen (ed) Social Analysis, Issue 45(1), Computer-Mediated Communication in Australian Anthropology and Sociology, Adelaide: Departement of Anthropology, pp. 32-52.

Thorsen, Dorte (2007) 'Junior-Senior linkages: Youngster's perceptions of migration in rural Burkina Faso', in Hahn, Hans Peter & Klute, Georg (eds) *Cultures of migration. African perspectives*, Berlin: LIT Verlag.

Toennies, F. (1988) 'Community and society', New Brunswick: Transaction Books.

Tomlinson, J. (1999) 'Globalization and Culture', Cambridge: Polity Press.

--- (2007) 'The Culture of Speed: The Coming of Immediacy', London: Sage Publications, Inc.

Trager, Lillian (1998) 'Home-town linkages and local development in South-Western Nigeria. Whose Agenda? What impact?' in Murray, Last et al. (eds) Africa. Journal of the International African Institute, Vol. 68, n° 3, 1994, London: IAI, pp. 360-382.

Treibel, Annette (2008 (1990)) 'Migration in modernen Gesellschaften. Soziale Folgen von Einwanderung, Gastarbeit und Flucht', München: Juventa Verlag.

Tsatsou, Panayiota (2009) 'Reconceptualizing "Time" and "Space" in the era of electronic media and communications', Platform Journal of Media and Communication, Vol. 1 July 2009, pp. 11-32.

Tufte, Thomas (2002) 'Ethnic Minority Danes between Diaspora and Locality – social uses of mobile phones and internet', in Stald, Gitte & Tufte, Thomas (eds) *Global Encounters – media and cultural transformation*, Luton: University of Luton Press, pp. 235-262.

Tufte, Thomas & Stald, Gitte (2002) 'Global Encounters – media and cultural transformation', Luton: University of Luton Press.

Turkle, Sherry (1995) 'Life on the screen. Identity in the age of the internet', New York: Touchstone.

Turner, Graeme (2011) 'www.flowtv.org,' Queensland University, April 8th, 2011, online.

Turner, Victor (1969) 'The ritual process. Structure and anti-structure', New York: PAJ Publications.

Urry, James (2002) 'Mobility and Proximity', Sociology 36(2), pp. 255-74.

Uimonen, Paula (2001) 'Transnational.dynamics@development.net. Internet, Moderization and Globalization', Doctoral Dissertation, Dep. of Social Anthropology, University Stockholm. Stockholm: Elanders Gotab.

Vertovec, Steven (2004) 'Cheap calls : the social glue of migrant transnationalism', Global Networks: A Journal of Transnational Analysis,.Vol. 4, pp. 219–224.

Vickers, Amy (2001) 'Reality Text', Guardian Online section, 24 May, 5.

Walton-Roberts, Margaret (2004) 'Returning, remitting, reshaping. Non-Resident Indians and the transformation of society and space in Punjab, India', in Jackson, Peter, Crang, Philip & Dwyer, Claire (eds) *Transnational Spaces*, pp. 78-103.

Watson, James (1998) 'Media communication. An introduction to theory and process', New York: Palgrave.

Warnier, Jean-Pierre (1996) 'Rebellion, Defection and the Position of Male Cadets: a Neglected Category', in Fowler, Ian & Zeitlyn, David (eds) *African Crossroads. Intersections between History and Anthropology in Cameroon*, Oxford: Berghahn Books, pp. 115-124.

--- (1993) 'L'esprit d'entreprise au Cameroun', Paris: Karthala.

--- (1993a) 'The king as a container in the Cameroon Grassfields', in Heintze, Beatrix (ed) Paideuma, Mitteilungen zur Kulturkunde, Frankfurt: Frobenius Gesellschaft E.V., pp. 303-320.

--- (1985) 'Echanges, développement et hierarchies dans le Bamenda pré-colonial (Cameroun)', Studien zur Kulturkunde, Nr. 76, Wiesbaden: Franz Steiner Verlag.

Weber, Max (1968) 'Economy and Society', in Roth, E. & Wittich, C. (eds), Berkeley: Univ. of California Press.

--- (1948) 'From Max Weber: Essays in Sociology', by Gerth, H.., Wright C. (eds), London: Rouledge & Kegan.

Weintraub, Jeff (1997) 'Public/Private: the limitations of a grand dichotomy', The Responsive Community, 7, 2, spring 1997, pp. 13-24.

Wellman, Barry (1999) 'Networks in the Global Village. Life in Contemporary Communities', Colorado: Westview Press.

--- (1988) 'Structural analysis: from method and metaphor to theory and substance', in Wellman, Barry & Berkowitz, S. D. (eds) *Social structures: a network approach*, Cambridge: Cambridge University Press, pp. 19-61.

Wellman, Barry, Carrington, Peter J., & Hall, Alan (1988) 'Networks as personal communities', in Wellman, Barry & Berkowitz, S. D. (eds) *Social structures: a network approach*, Cambridge: Cambridge Univ. Press.

Wellman, Barry & Guilia, Milena (1999) 'The network basis of social support: a network is more than the sum of its ties', in Wellman, Barry (ed) *Networks in the Global Village. Life in Contemporary Communities*, Colorado: Westview Press, pp. 83-118.

Wellman, Barry & Haythornthwaite, Caroline (eds) (2002) 'The Internet in Everyday Life', Oxford: Blackwell Publishing.

Wellman, Barry & Leighton, Barry (1979) 'Networks, Neighbourhoods and Communities', Urban Affairs Quarterly, N° 14, pp. 363-90.

Wilding, Raelene (2006) '"Virtual" intimacies? Families communicating across transnational contexts', Global Networks 6, 2 (2006), pp. 125-142.

Wiles, Janine (2008) 'Sense of home in a transnational social space: New Zealanders in London', Global Networks 8, 1 (2008).

White, Mimi (2004) 'The attractions of Television. Reconsidering liveness', in Couldry, Nick & McCarthy, Anna (eds) *Media Space.*

Place, scale and culture in a media age, London & New York: Routledge, pp. 75-90.

Ytreberg, Espen (2009) 'Extended liveness and eventfulness in multi-platform reality formats', New Media Society 2009, 11:467, London: Sage publications.

Zhao, Shanyang (2003) 'Toward a taxonomy of copresence', in Presence: Teleooperators and virtual environments, Vol. 12, Issue 5, October 2003, online.

--- (2008) 'Identity construction on Facebook. Digital empowerment in anchored relationships', Journal Computers in Human Behaviour, Vol. 24, Issue 5, Sept. 2008.

Zifonun, Darius (2008) 'Widersprüchliches Wissen. Elemente einer soziologischen Theorie des Ambivalenzmanagements', in Raab, Jürgen, Pfadenhauser, Michaela, Stegmaier, Peter, Dreher, Jochen & Schnettler, Berndt (eds) *Phänomenologie und Soziologie. Theoretische Positionen, aktuelle Problemfelder und empirische Umsetzungen,* Wiesbaden: VS Verlag für Sozialwissenschaften, pp. 307-316.

Zlotnik, Hania (2004) 'International migration in Africa: An analysis based on estimates of the migrant stock', United Nations DESA / Population division, online.

Internet sources / State August 2011

https://www.cia.gov/library/publications/the-world-factbook/geos/cm.html
http://www.bamendauniversity.com and
http://en.wikipedia.org/wiki/Bamenda (source dated 7.12.2009)
http://www.internetworldstats.com/africa.htmcm
http://en.wikipedia.org/wiki/Demographics of Cameroon
http://en.wikipedia.org/wiki/Bamenda

ITU (International Telecommunication Union) 2009. Information Society Statistical Profiles Africa.

www.itu.int/ITU-D/ict/material/ISSP09-AFR final-en.pdf

http://www.itu.int/ITU-D/ict/material/Youth 2008.pdf

http://www.un.org/esa/population/migration/turin/Symposium Turin files/P09 Dumont&Lemaitre.pdf

http://www.dirsi.net/english/files/backgroundper cent20papers/070216--dunn.pdf

http://www.populationlabs.com/Cameroon Population.asp

http://www.nationsencyclopedia.com/Africa/Cameroon-MIGRATION.html

http://web.worldbank.org/WBSTIE/EXTERNAL/COUNTRIES/AF RICAEXT/CAMEROONESTN/0,,contentMDK:22340869-menuPK:50003484-pagePK:2865066-piPK:2865079-theSitePK:343813,00.html

htttp://web.worldbank.org

http://www.bc.edu/bc_org/avp/soe/cihe/inhea/profiles/Cameroon.h tm

http://flexcominstitute.com

http://www.paulscomputerinstitute.com

http://lauratebusinesscollege.com

http://www.iomdakar.org/profiles/content/migration-profiles-cameroon

http://agendia.jigsy.com/entries/economy/cameroon-2005-census-results-smack-of-diabolic-geo-political-planning

http://the-news-from-cameroon.com/article.php?category_id=1&article_id=1551

http://www.237online.com/2008062193/Actualites/Economie/camer oonian-diaspora-exhorts-govt-to-encourage-local-investment.html

http://www.un.org/esa/population/publications/wpp2000/annex-tables.pdf

http://www.itu.int/ITU-D/ict/statistics/Gender/index.html

http://www.acrwebsite.org/volumes/display.asp?id=11855

http://www.geohive.com/cntry/cameroon.aspx

http://siteresources.worldbank.org/INTAFRICA/Resources/CMR_Economic_update.Jan.26.11.pdf

http://www.aes.com/sonel

http://www.izf.net/upload/Documentation/Cartes/supercartes/camer
   oun.png

Appendix

Fieldwork 2009 (28.06. - 01.11.09)

Interviews 2009

I conducted 52 formal interviews with New Media users during my field stay in 2009. Since I have not quoted of all of the interviews, I will not list all the interviewees here. All the names are invented.

With most of the interviewees in this range I followed up a long-term relationship. Only a few of these interviews were recorded. A range of four of the interviews was added to this group in 2011. Nine of the interviewees I interviewed again – with other key points – min 2010/11. This series of interviews was dedicated to examine youth's internet and mobile phone uses in their daily lives.

Occasional Internet Users

Name: Andrea / Date of the interview: 03.10.11 / Indication O (Interview added 2011)

Person: Female. 24 years. She lived in her family's compound, with 3 younger sisters and brothers, her grandmother and parents. Three elder sisters and brothers were abroad. A sisters children also lived in the compound – she was in South Africa. A brother was in the US and another in Belgium. Andrea was doing a nursing formation. She was not often going to the cyber café, since they had internet access in their house.

Relation to the interviewer: Andrea was not that close, but I saw her regularly within a long period of time.

Situation of the interview: In her home, the grandmother and her sister's children were also partly present.

Name: Caroline / Date of the interview: 04.10.11 / Indication O (Interview added 2011)

Person: Female. 23 years old. She studied at BUST, near Bamenda, doing nursing. Her father died long ago, the mother just recently. She used to live in the SW Province. She had three brothers, one in Bamenda the others in Dschang. She was a born again, her church members were important to her, and she also worked for the church.

Relation to the interviewer: I had not seen Caroline all that often, but regularly. I know her since 2008. I also went to visit her at the students hostel where she lived.

Situation of the interview: In a restaurant in Bamenda, we were a bit under time pressure because she had to leave soon.

Name: Cynthia / Date of the interview: 14.08.09 / Indication O

Person: Female. 37 years old. She was working in a nursing department in a school in Batibo. She was only in Bamenda during holidays and sometimes weekends. She had a house in Bamenda and in Batibo. She was not married and childless, despite her age. She had eight siblings, all of them in Cameroon in different cities. A range of cousins were in the US. She had tried to go there several times but her visa requests were denied.

Relation to the interviewer: I did not know Cynthia closely. She was a friend of a friend of mine, and she wanted to talk to me, so I had decided to make an interview out of it.

Situation of the interview: In her house.

Name: Elene / Date of the interview: 02.10.11 / Indication O (Interview added 2011)

Person: Female. 23 years old. She was married with two children, four and six years old. She and her husband rented a house in Bamenda. She used to work as a hairdresser in Douala. Then they moved here because her husband had a job, now he was jobless again. At the time of the interview she went to afternoon classes to complete her A-levels. Before she used to work at a call box. Now she was taking care of a neighbour's shop and also sells credit.

Relation to the interviewer: When Elene still used to do callbox in town, we saw each other often. Later on I visited her a few times in her house.

Situation of the interview: In the shop where she was working, partly interrupted by clients.

Name: Ernest / Date of the interview: 11.08.09 / Indication O
(→ see interview 2010/11)

Person: Male. 25 years old. He lived with his mother and younger brothers in their family compound. He was so to say the father of the house. The father was back in the village, did not see the family often and was not responsible. He had a computer repair business in town, and doing all kinds of other businesses to keep himself up and greatly contributed to his siblings school fees and livelihood.

Relation to the researcher: It was a very close relationship, he was also a good friend of other friends of mine. I also knew his mother well. I had sometimes been invited in their house, and saw Ernest very often in town. We were also calling each other regularly when I was in Switzerland.

Situation of the interview: I have lead so many conversations with Ernest before. For the interview we met in a restaurant.

Name: Felix / Date of the interview: 30.07.09 / Indication O
(→ see interview 2010/11)

Person: Male. 24 years old. His family was in the SW Province, he was alone in Bamenda, rented a room. He was the eldest of nine siblings. All of them were schooling in the village living with his parents. At the time of the interview he still worked at a petrol station. He contributes to his siblings school fees. Later on he went to Kumba to sell shoes. During the time of the interview his sister migrated to Sweden.

Relation to the researcher: I met Felix often in town, visited him at the petrol station. He held close contact until he left Bamenda to Kumba.

Situation of the interview: in Dallas Cabaret, which was at day time empty and quiet.

Name: Hans / Date of the interview: 16.07.09 / Indication O

Person: Male. 34 years. He was staying close to his mother adjacent their compound. The father had left the family. He had five elder and younger siblings. Only one of them was still in the house. The others were in other cities in Cameroon. He claimed to work as an architect.

He had a cousin in the UK. He was obsessed to find „a white lady" in the internet.

Relation to the interviewer: I had a difficult story with Hans. Since he was following me for a while I thought I could at least also profit and do an interview with him. Later on he became more and more intrusive and in the end I had to cut contact.

Situation of the interview: In a restaurant.

Name: Immaculate / Date of the interview: 17.08.09 / Indication O

Person: Female. 23 years old. Her family was in Yaoundé, she grew up there. She lived with an aunt in Bamenda. Her mother was in Yaoundé with three younger siblings. Her father was absent. She could more or less sustain herself. She had A-levels. She was doing call box and also selling books and magazines. She took every opportunity to earn money by doing publicity, and so on. Her dream was to have her own marketing office.

Relation to the researcher: in 2009 I met Immaculate, I often spent time with her on her callbox. Later on she went to Yaoundé, where I visited her once. The contact faded out then.

Situation of the interview: On her call box, sometimes interrupted by customers.

Name: Kaspar / Date of the interview: 18.08.09 / Indication O

Person: Male. 21 years old. He lived with his family in Bamenda, parents and three younger and one elder sibling. He wanted to move out. He was working at Camtel, and could earn his livelihood, and contribute to his siblings' school fees.

Relation to the interviewer: I was not very close to Kaspar. Especially later on I did not see him again, he had also temporarily left Bamenda and did not work with Camtel again. I was an acquaintance of his boss.

Situation of the interview: In a restaurant.

Name: Miller / Date of the interview: 27.08.09 / Indication O

Person: Male. 32 years old. He stayed with his parents Upstation. His family was relatively well off. He had an elder brother in South Africa, one elder and two younger sisters were in Bamenda. His father was a director in a school, where Miller currently worked as teacher. He had a BA in economics, and wanted to further his education, and at best he wanted to study abroad to do a masters. He was not very internet versed.

Relation to the interviewer: I hardly knew Miller. I met him at a friend's wedding. He was following me up, and I decided to an interview with him. After that I had hardly seen him again.

Situation of the interview: In a restaurant.

Name: Nick / Date of the interview: 08.07.09/15.07.09 / Indication O

Person: Male. 22 years old. He stayed in Bamenda alone. His parents were in Kumba, also two younger sisters. An elder brother was in Douala. At the time of the interview he studied at a boarding school. Later on he was temporarily working in Douala, and then he went to a higher professional school at the outskirts of Bamenda. He came from a relatively well-off background, and his parents sponsored his education.

Relation to the interviewer: He was a friend of Moussa*, with whom I was very close. I was not so close with Nick though, but we had seen from time to time.

Situation of the interview: In a restaurant, in different sessions. It was not easy to interview him, he was somehow distracted and talking about many other things.

Name: Patience / Date of the interview: 10.07.09 / Indication O

Person: Female. 21 years old. She was Nick's* cousin. She stayed in her family's compound with an aunt and five younger sisters. The mother died. Together with her sisters and her aunt they cooked and sold food every day. They sustained themselves with it. She had A-levels and dreamt of studying abroad.

Relation to the interviewer: I was not close to Patience, I had only met her once before our interview, introduced by Nick*. After that I occasionally met her at the eating place.

Situation of the interview: In her compound.

Name: Richard / Date of the interview: 31.07.09 / Indication O

Person: Male. 20 years old. He stayed with his mother and eight younger sisters in Bamenda, the father was in the village. He had a room a bit separate in the compound. He would like to move out. He was repairing cars, doing an apprenticeship in town. Apart from that he worked as motor bike driver. His earned money he saved because he wanted to go abroad. Later on he was victim in a "migration scam" and lost a huge amount of money.

Relation to the interviewer: I visited him a few times in their compound, also in town at his workplace. Altogether we were not that close but saw each other regularly.

Situation of the interview: In a restaurant.

Name: Serge / Date of the interview: 30.07.09 / Indication O

Person: Male. 21 years old. He lived in his family's compound, with parents and two younger brothers. One elder sister was married and living in another city in Cameroon. At the time of the interview he had just finished A-levels and was jobless. He was trying to find a government job, since his mother worked as a teacher in a government school. He was hardly internet literate but currently tried to bolster his knowhow by looking for jobs in the internet.

Relation to the researcher: I was not really close to him, only met him a few times. I had the impression that he always wanted to profit from me. The relationship faded out with time.

Situation of the interview: In the cyber café where he used to browse.

Name: Valeria / Date of the Interview: 18.07.09 / Indication O
(→ see interview 2010/11)

Person: Female. 34 years old. She rented her own room in Bamenda. Her mother's compound was close by. She had four younger and elder sisters, all in Bamenda. Later on one of her elder sisters went abroad to the US. She was unmarried and childless. She had a BA level. At the time of the interview she was jobless, later on she had a temporary job as a teacher, only to be jobless again later on. She was sustaining herself with little temporary jobs, most of all with small contributions from her family. She was hardly internet literate.

Relation to the interviewer. Valeria was one of my closest friends. I saw her often, be it in town or at home, and when I was in Switzerland, I regularly called her. When in Bamenda, I taught her to use the internet.

Situation of the interview. I had so many conversations with Valeria. This interview had been conducted in a cyber café.

Regular Internet Users

Name: Anastasia / Date of the interview: 01.08.09 / Indication R (→ see interview 2010/11)

Person. Female. 26 years old. She was married with two children. Her father died some years back, her mother was in the village. She was the eldest of 5 sisters and brothers. One younger brother was in town, the others in other cities in Cameroon. She worked as a graphic designer in a printing workshop, where she was not well paid. Her husband had a job out of town, she only saw him at the weekend. Her two children were 3 and 6 years old. Because of them she could not go so often to the cyber café anymore as she used to do.

Relation to the interviewer. I knew Anastasia since 2003. That time she used to work in a cyber café. We were close to each other. I had often been invited to her home.

Situation of the interview. In her house.

Name: Barbara / Date of the interview: 12.09.11 / Indication R (Interview added 2011)

Person. Female. 23 years old. She stayed with her family in their compound, the parents, younger and elder sisters and brothers. Only one brother was out in Dschang. The mother was a teacher, and they also had a small restaurant in town, where also Anastasia worked daily. At the moment of the interview she was doing an internship with a foreign NGO in the health sector. She had a first degree in linguistics. Since then she had done internships and also fostered her formation in design. She was designing cloths and had written play scripts for films. Her dream was to further her formation in this field, possibly abroad.

Relation to the interviewer: I was not that close with Anastasia, but we saw each other regularly. Also she was always very busy, it was not easy to meet her. She was a close friend of Ernest*.

Situation of the interview: In her compound, because in the restaurant it would be too noisy. Partly other family members were present during the interview.

Name: Bertina / Date of the interview: 27.07.09 / Indication R

Person: Female. 28 years old. Her parents were in the SW Province, her three sisters were also with them. She lived in Bamenda with her grandmother of whom she also took care. She worked as a teacher in the outskirts of Bamenda. In the afternoons she either came to the cyber café, every second day, or she went to church, she was a full gospel. She was communicating with her sister, who was abroad, had applied for universities, and was planning to go abroad.

Relation to the interviewer: I contacted her in the cyber café, then I met her for the interview. She was at least at first a bit reluctant. After that we hardly saw each other again.

Situation of the interview: In her house.

Name: Clovis / Date of the interview: 24.07.09 / Indication R

Person: Male. 34 years old. He was married with two small children, stayed in Bamenda. The parents were in the village. He was having a paper and packaging business. He was often travelling in Cameroon. He went to the adjacent cyber café regularly, mainly for communication with friends abroad.

Relation to the interviewer: I met Clovis regularly since I frequented the adjacent cyber café. Then we always had a chat.

Situation of the interview: The interview was rather unplanned, during a power failure in the cyber café where he used to work.

Name: Christian / Date of the interview: 06.07.09 / Indication R (→ see interview 2010/11)

Person: Male. 25 years old. He had three younger siblings. He stayed with them and the parents in the family's compound in Bamenda. He had A-levels, is jobless and seriously intended to go abroad. This was also why he regularly visited the cyber café. Apart from that he was doing small jobs in order to earn some money. Later on, Christian had made several attempts to migrate, one visa was denied, then he travelled to Austria with a faked business visa and was sent back, and from end 2011 he was finally studying in the UK.

Relation to the interviewer: We were very close, had regularly seen each other and always stayed in contact by internet. I had also seen him in the UK.

Situation of the interview: In the cyber café where he used to browse.

Name: Emil / Date of the interview: 22.09.09 / Indication R

Person: Male. 23 years old. He had seven sisters and brothers. His mother was now in the village, and he did not know the whereabouts of his father, who had divorced from the mother long ago. He grew up with his mother and her new husband, who had not treated him well. His teacher in school had sponsored his further education. He was currently studying history in Yaoundé. Later on he had the problem that his benefactor stopped paying his study fees and he was struggling to continue his formation.

Relation to the interviewer: I was not that close to Emil, but saw him regularly. Also when he went to Yaoundé, I had seen him there a few times.

Situation of the interview: In a café near the student hostel where he stayed in Yaoundé.

Name: Herbert / Date of the interview: 20.07.09 / Indication R

Person: Male. 23 years old. He had three brothers and a sister, he was the second last. One elder brother was out in Yaoundé. The parents were here in Bamenda. He stayed in an aunts compound, had his own room there. He had not concluded his A-levels. At the time of the interview he worked in a nearby cyber café, later on he lost that job. I was suspecting him to scam.

Relation to the interviewer: I only met Herbert a few times. He wanted to know me, that is why I decided to interview him. Later on the contact faded out.

Situation of the interview: In a restaurant, partly in his home.

**Name: Ivo / Date of the interview: 03.07.09/16.07.09 /
Indication R** (→ see interview 2010/11)

Person: Male. 28 years old. He had grown up with his aunt, his parents died when he was very young. His two direct brothers were in Douala. The aunt's siblings were like his own siblings, they stayed in the compound. At the time of the interview he was doing an internship in a lawyers chamber in Bamenda. He had a degree in law. His dream was to go abroad to study. Later on in 2012 he went to Nigeria.

Relation to the interviewer: I was very close to Ivo, I knew him since 2003. I knew the family and friends and had often been a guest in their compound. We had always kept contact also through phone and email when I was in Switzerland.

Situation of the interview: In a restaurant.

Name: Julius / Date of the interview: 14.08.09 / Indication R

Person: Male. 35 years old. He had 5 brothers and sisters, three of them in Bamenda, the others in other cities in Cameroon. His mother lived in Bamenda, the father died. Other relatives were also in Bamenda. He rented his own room. He was worried that he was not yet married. He worked since a few years as a taxi driver. His dream was to have his own taxi. He went twice a week to the cyber café, mainly because he chatted with relatives abroad, and he was trying dating.

Relation to the interviewer: I got to know him when driving in his taxi. I had not seen him often. He got more and more intrusive and I had later on tried to avoid him.

Situation of the interview: In a restaurant.

Name: Lambert / Date of the interview: 25.08.09 / Indication R

Person: Male. 25 years old. He had three younger siblings, they were still schooling. The parents were also in town. He rented his own room. He had his own barbing studio in Bamenda. He could sustain himself through it, but could not – so far – contribute to his siblings school fees. He was going regularly to the cyber café to chat with two friends abroad and he was also browsing.

Relation to the interviewer: We were relatively close, we had seen each other regularly. I visited him at times in his barbing studio. Also we had at times seen each other in evenings when going out.

Situation of the interview: In a restaurant, in order not to be disturbed by customers.

Name: Matthias / Date of the interview: 03.08.09 / Indication R

Person: Male. 25 years old. He stayed with the mother in their family compound in Bamenda, during semester holidays. The father died. Only the youngest brother was still in. The others were in other cities in Cameroon. At the time of the interview he was studying in economics in Yaoundé. During semester break he went to the cyber café almost daily, because he had applied for universities abroad.

Relation to the interviewer: I was not too close to Matthias. But we had seen each other once in a while. He was a close friend of Manfred*.

Situation of the interview: In a restaurant.

Name: Miranda / Date of the interview: 13.07.09 / Indication R

Person: Female. 20 years old. She lived with her four siblings and parents in Bamenda. At the time of the interview, she just completed A-levels. She was currently trying to obtain a government job. Apart from that she worked in a friends' market place to earn a little money. She came regularly to the cyber café, because she had time and she liked chatting with a few friends abroad and in Cameroon.

Relation to the researcher: We were not close, we had only seen each other a few times. My friend in the cyber café encouraged her to have an interview with me.

Situation of the interview: In the back office of the cyber café.

Name: Olaf / Date of the interview: 30.07.09 / Indication R

Person: Male. 30 years old. He lived with his girlfriend in his mother's compound. The mother recently died. The father was alive, lived in Bamenda, but his relation to him was tensed. They were discussing about inheritance matters. Since he was coming from a polygamous home, he had many siblings. Apart from a few, most were in the US. He

513

came from a well off background. At the time of the interview he worked for Orange. Before he had a job in Douala. Later on he married and had a child. He started an import business with a brother in the US.

Relation to the interviewer: We have regularly seen each other. Also when in Switzerland we were in contact through email and Facebook.

Situation of the interview: In a restaurant.

Name: Patricia / Date of the interview: 05.10.11 / Indication R

(interview added 2011)

Person: Female. 27 years old. She was married and had four children, the eldest 14 years, the youngest twins were six years old. The husband worked in Dubai. At the time of the interview she had not seen him for 3 years, but he then came back for a while. Her parents were also in Bamenda, her siblings were in other cities in Cameroon, one brother was here. Some relatives were abroad. She was coming from a well off background. She worked as a mobile phone repairer in the hardware field. So far she was the only women in Bamenda in this business. She went regularly to the cyber café to chat and email with her relatives abroad.

Relation to the researcher: I only knew Patricia since 2010, but since then we had been in regular contact, I visited her at her job site, but also at home, her family in the village, and we sometimes went out together.

Situation of the interview: In her brother's - also a mobile phone repairer - office, because in her job site we were always interrupted.

Name: Precious / Date of the interview: 05.09.09 / Indication R

Person: Female. 26 years old. She lived in her parents compound. She had a daughter of six years who also lived with them. She is not (yet) married to the child's father, who was abroad for studies. She had three brothers and one sister, the brothers were all abroad. Later on also her sister married and went with her husband to China. She had a degree in linguistics. Currently she worked as a secretary at the council. She came from a well off background. Her dream was to study abroad. She came regularly to the cyber café, be it to chat with her siblings or boyfriend abroad, and also partly for work purposes, because they did not have internet in the office.

Relation to the researcher: I was not too close to her, I did not see her so often, also because she was very busy. But I had also stayed in contact with her when I was in Switzerland. She was a close friend of Ivo*.

Situation of the interview: In a restaurant. Her daughter was also present.

Name: Pride / Date of the interview: 09.09.09 / Indication R

Person: Male. 29 years old. He was the second youngest among four siblings. The sisters were in Douala, the younger brother studying in Yaoundé. His parents were in Bamenda. He rented his own room. He had A-levels, but called himself an architect. He had his own business office. He had relatives abroad, for whom he was constructing houses. He could sustain himself and also maintained his parents.

Relation to the interviewer: I was not very close to Pride. Later on our contact faded out.

Situation of the interview: In a restaurant.

Frequent Internet Users

Name: Claudia / Date of interview: 22.07.09 / Indication: F

Person: female, 23 years old. She stayed in her family's compound. She had just completed her first degree in linguistics. She was jobless at the moment. She had younger siblings, all of them were schooling, one younger sister was in UK. She was currently trying to go abroad.

Relation to the interviewer: I only knew her superficially, I had seen her once before the interview.

Situation of the interview: I met her at a cyber in old town. She was coming here regularly. Norbert, who works here in the cyber café, had asked her if she was ready to talk to me.

Name: Clement / Date of interview: 31.08.09 / Indication F

Person: male, 25 years old. His father died in 2002, the mother was in Limbe. He grew up there, he came to live with his sister in Bamenda, because he had difficulties in school, and the sister thought it would be better for him to join her. Today he stayed alone, rented a small flat. He had two other sisters and one brother who were younger than him, they were in Limbe and Douala.

Relation to the interviewer: He worked at a cyber café as a technician. We chatted often when I was going there to the cyber café to browse or to meet him or other people.

Situation of the interview: We met for an interview in the cyber café, when there were not many customers around, but there were breaks in between. The interview had been completed in different stages.

Name: Cletus / Date of the interview: 26.09.09 / Indication: F

Person: male. 26 years old. He was the second in his family, had four other sisters and brothers. His parents were in Bamenda, at the moment he lived in their compound, since he did not earn money. He wanted to move out and rent alone as soon as possible. He had studied Law in Buea and specialized in property rights, at the time of the interview he made a formation with the UN. He had been travelling abroad several times.

Relation to the interviewer: He was a friend of other friends of mine. He lived close by where I often spent my evenings. Since he was at times in Yaoundé, I had not seen him so often, but I was often in contact with him through the internet, when in Switzerland.

Situation of the interview: In the cyber café where he used to work.

Name: Clothilde / Date of the interview: 24.08.09 / Indication F (→ see interview 2010/11)

Person: female. 23 years old. She worked in a cyber at Foncha street. She was married, her husband was working in an NGO in the agriculture field, She had two kids, 3 and 1 year old. Her parents and family members were back in the village, one uncle was in Douala, before, he had been a few years abroad.

Relation to the interviewer: I met Clothilde often in the cyber café and also visited her at home several times. I am the "godmother" of her youngest child.

Situation of the interview: Partly in the cyber café, partly at her home.

Name: Dennis / Date of the interview: 17.08.09 / Indication F

Person: Male. 30 years old. He stayed with his mother, he was the father of the compound and head of the family, since his father recently died. His four younger siblings stayed in other cities in Cameroon. He was working as a graphic designer, and managed the property of his father. He was trying to go abroad since a longer time, but he had so far not succeeded. He had a range of friends and distant relatives abroad.

Relation to the interviewer: We were not very close, but good acquaintances, almost neighbours. Dennis was a good friend of Ernest*.

Situation of the interview: In the cyber close to his house, where he currently works.

Name: Divine / Date of the interview: 13.08.09 / Indication F (→ see interview 2010/11)

Person: Male. 25 years old. He was an orphan and had grown up in Bamenda with his uncle. He lived in his compound. The uncle's children were much older than him. His uncle had sponsored his school formation up to advanced level. He had a cousin in the US. He had his own NGO and was very engaged in this work, which was however not paying.

Relation to the interviewer: We have been in contact regularly. He tried to integrate me into his NGO work.

Situation of the interview: In his uncle's house. I also met him several times in the cyber café.

Name: Festus / Date of the interview: 19.10.09/23.10.09 / Indication F

Person: Male. 21 years old. He was a carpenter and had his workshop together with his brother. He had cut contact with the paternal side of his family due to tensions. His mother was still alive and in the village. One of his sisters had been in Ireland, but she died recently. From 2010 Festus went to Ireland to take care of his sister's child there.

Relation to the interviewer: we were not that close, Festus was a friend of a research colleague. Since he went abroad soon after I had known him, we only met a few times. We stayed in contact through internet.

Situation of the interview: The first interview with Festus I did together with Primus. Then I did a follow up alone. Both interview were lead in his workshop.

Name: Linda / Date of the interview: 08.07.09 / Indication F

Person: Female. 22 years old. She lived in her family's compound, her father recently died. She had ten siblings, she was about in the middle. The younger ones lived with her. Since recently she had a job at the Baptist mission, now she had done the GCE exam to become a teacher. She has some good friends who just recently went abroad, since she had no work, and had time, she came to the cyber almost daily to chat.

Relation to the researcher: Linda was just an acquaintance. I met her briefly before conducting the interview. After some time our contact faded out.

Situation of the interview: In the cyber café where she usually frequents.

Name: Manfred / Date of the interview: 14.07.09 / Indication F

Person: Male. 27 years old. He was from a relatively well-off family. He was the youngest of seven siblings, either working with the government or abroad. He rented his own room not far from his parents compound. He took care of himself, by pursuing several businesses, supported by his relatives abroad. I also suspected that he was scamming.

Relation to the interviewer: I met Manfred relatively often since he was always around in leisure places in my neighbourhood. Also I knew most of his friends.

Situation of the interview: The first interview was conducted in a cyber café - belonging to his brother - where he worked temporarily, another part had been conducted in a restaurant.

Name: Simon / Date of the interview: 04.07.09 / Indication F
(→ see interview 2010/11)

Person: Male. 24 years old. His parents lived in the village, he had younger siblings who also lived with them and schooled there. He had come to Bamenda on his own, had no relatives here. At the time of the interview he worked in a cyber café. When we got to know each other he worked in another cyber café, where I knew him through another friend. Later on they had not been in good terms with each other, and he left. The salary gave him a small but regular income. He could even support his family a bit.

Relation to the interviewer: Simon was close to me, we saw each other often, I visited him at his workplace. Also we were always in contact through internet when I was in Switzerland.

Situation of the interview: Partly in the cyber café, partly at his home.

Name: Titus / Date of the interview: 10.08.09 / Indication F

Person: Male. 24 years old. At the time of the interview he lived in his family's compound. Shortly after he moved together with his girlfriend, close to his family's compound, renting a room. He had an uncle, in whose business he was employed and earned a bit of money. He had two sisters and two brothers, the younger ones were still schooling, one elder sister was married in Bamenda and an elder brother lived in Douala. He had studied economics in Yaoundé. Beginning of 2012 he went to the US to study. Titus had the reputation of being involved in scamming and other illicit business practices.

Relation to the researcher: We saw from time to time, but not too often. But Titus was always a very interesting person to interview. He was also interested in the work I was doing.

Situation of the interview: Partly in the cyber café where he used to browse, partly in a restaurant.

Frequent internet users involved in scamming activities:

Name: Bryant / Date of the interview: 10.07.09 / Indication FS

Person: Male. 26 years old. He rented on his own, close to his parents compound. They lived there with two of his younger brothers. Another two sisters lived in other cities in Cameroon. He could maintain himself and also contributed to his younger sibling's school fees. His parents would not ask where the money came from. As far as I could evaluate, his activities were relatively "professional", e.g. hacking of visa cards, selling crude oil and real estates instead of puppies. His revenue was irregular, however at times considerable. During the time of the interview, he seriously tried to go abroad for studies, in order to stop his business, since he would be "too old for it" and needed to change his life.

Relation to the researcher: I got to know him as a friend's friend. He was open towards me from the beginning. We were not too close. At times we also chatted when I was in Switzerland.

Situation of the interview: In a restaurant

Name: Derrick / Date of the interview: 18.08.09 / Indication FS

Person: Male. 24 years old. At the time of the interview he still stayed with his aunt. Later on, he rented a room where he lived together with friends. Then they had internet in the house. His parents both died some time ago. His younger siblings also lived with the aunt. He said he tried to contribute to his sibling's school fees. From 2011 he started IT studies near Bamenda, and – as he said – stopped scamming.

Relation to the researcher: I met him relatively often, since he lived in my neighbourhood. Also I met in the cyber. I was very used to his friends, and was "hanging around" with them several times in their cyber cafés. Derrick was open towards me from the beginning and introduced me to many friends.

Situation of the interview: In several stages, in the cyber cafés.

Name: Elvis / Date of the interview: 31.07.09 / Indication FS

Person: Male. 20 years old. His mother lived in the US. He stayed with his sister, her husbands and children, and the grandfather. He also stayed partly with his friends. His family did not know what he was doing. His mother supported him regularly with money for his school, even though he was not schooling.

Relation to the interviewer: I did not have a close connection to Elvis. I even tried to avoid him later on, because I did not like his conduct towards me. I met him once in a while and he always tried to profit.

Situation of the interview: Partly in the cyber café, where he used to work, partly at his home.

Name: Jack / Date of the interview: 01.08.09 / Indication FS

Person: Male. 25 years old. He had older and younger siblings, who were schooling. His mother lived close to Bamenda, the father had left his family a few years ago. He had only few temporal jobs, earned his living mostly from scamming. He claimed to also contribute to his siblings school fees. During the time of the interview, he lived together with a group of friends – most of them also scammers – in a house they had collectively rented. Later on he moved to his elder brother to Yaoundé.

Relation to the interviewer: I was close to Jack. I knew many of his friends. Peter had introduced him to me. I also knew his elder brother in Yaoundé. We were also in regular contact through internet. I spent time with him and his co-residents and shared leisure activities. Jack was very open from the beginning.

Situation of the interview: Partly in cyber cafés, partly we met at Ayaba pool to talk, because it was a quiet place. At times we were alone, other times Peter or other friends were also present.

Name: Peter / Date of the interview: 29.07.09/30.09.09 / Indication FS

Person: Male. 27 years old. He had his family near Bamenda. The eldest brother was in the UK. Another elder brother was a business man in Douala. Two younger sisters studied in Buea. He stayed with his friends (Jack* a.o.) in the house they collectively rented. He had a small baby, but later on split up with his girlfriend. He said he felt responsible for the guys he was staying with. He was doing sports professionally (Volleyball) and was often mobile for sports events. He was not scamming himself but was the one who had played a considerable role to introduce me into scamming activities and many of his friends.

Relation to the interviewer: We had a close relation and met relatively often. However, later on he was owing me money and avoided then my contact. Also I heard that he went to stay permanently in Douala, so our relation stopped.

Situation of the interview: In several stages at Ayaba pool. Sometimes in presence of others, e.g. Jack*

Fieldwork 2010/11 (31.09.10.-09.01.11)

All together me and my field work assistant* have conducted 26 interviews in 2010/11, each of us did 13 interviews. Since I have not quoted from all of the interviews, I will not list all of the interviewees here. This series of interviews was dedicated to deepen our understanding about youth's social networks, notions of solidarity and support, and qualities of social ties, face-to-face and mediated.

Interviews 2010/11

The interviewees for this range of interviews I had already known previously and for a longer period of time. I will thus only add a situation update, compare to the list of interviewees in the previous section (fieldwork 2009). All of these interviews were recorded.

Name: Anastasia / Date of the interview: 13.11.10 / (→ see interview 2009)

Situation of the interview: In the printing shop where she used to work

Situation update: She had tried to obtain a government job and had failed her exams. She was planning to go to Gabon to work when her child was some months old: she was pregnant with her third child. Later on she lost the child, when it was one month old.

Name: Clothilde / Date of the interview: 26.09.10 / (→ see interview 2009)

Situation of the interview: In her house. Her children were also present.

Situation update: She did not work much in the cyber café any longer, because she had given birth to her third child just recently.

Name: Christian / Date of the interview: 01.11.10 / (→ see interview 2009)

Situation of the interview: In a restaurant.

Situation update: His forced return from Austria was just of recent. He was very depressed. Later on he went to the UK for studies.

Name: Divine / Date of the interview: 28.10.10 / (→ see interview 2009)

Situation of the interview: In his house.

Situation update: He had recently taken up a palm oil business, to finance his life and also to finance parts of his NGO work. Later on he was involved in an internship in another NGO.

Name: Ernest / Date of the interview: 18.11.10 / (→ see interview 2009)

Situation of the interview: In a restaurant.

Situation update: He had been involved in many different businesses recently, and built up his own job site.

Name: Felix / Date of the interview: 02.11.10 / (→ see interview 2009)

Situation of the interview: In his home

Situation update: He had just quit he business at the petrol station. Then he had built up a breakfast place. Later on he went to Kumba.

Name: Francis / Date of the interview: 09.11.10

Person: Male. 33 years old. He was the second of six direct sisters and brothers. One elder brother was in Belgium, the others were all in Cameroon. He stayed with his mother and a younger brother in their compound in Bamenda. The younger brother was mentally disabled. He stayed there because he felt responsible. His father was in Douala, had never really been responsible. He was coming from a polygamous home, he had steps, some of them were in the US. He was having A-levels. He was working in the marketing and event management field, together with a friend they had their business. It was running well in the sense that he could sustain himself and a bit his mother. The brother abroad also supported. He was often in the cyber café, to communicate with his steps, friends and brother abroad, but also to browse for job purposes.

Relation to the interviewer: I was very close to Francis. I knew him since 2003. We had many informal conversations and often spent time together, before having this interview. I knew his mum, siblings and friends.

Situation of the interview: In a restaurant, in order to have a quiet environment.

Name: Ivo / Date of the interview: 08.11.10 / (→ see interviews 2009)

Situation of the interview: In a restaurant.

Situation update: he was still doing the internship at the lawyers firm, having more responsibility now. He planned to further his studies in Nigeria, since his other plans to go to the US had not worked out.

Name: Lizette / Date of the interview: 06.11./13.11.10

Person: Female. 23 years old. She had a boyfriend with whom she stayed together. They had a three year old child. She was the third of six, one sister and four brothers. One elder sister was working in Bamenda and one elder brother finished studies and worked in Douala. Her parents were in Bamenda. Her relation with her father was a bit tensed. He had stopped paying for her education when she got pregnant. So she could never finish A-levels. She had a callbox, which was continuously becoming a bigger business, and the baby father had a small shop just adjacent. They could manage to sustain themselves. She hardly went to the internet, at the time of the interview she was doing a crash course to learn how to handle the computer.

Relation to the interviewer: I was very close to Lizette. She was almost my neighbour and her callbox was at the junction where I stayed in Bamenda. So I saw her daily and spent time with her. When I was in Switzerland we were calling each other regularly.

Situation of the interview: At her callbox, in two sessions, in between interrupted by clients.

Name: Moussa / Date of the interview: 13.11.10

Person: Male. 21 years. He was coming from a (polygamous) Muslim family (Hausa). He had seven direct sisters and brothers. The parents and his mother and co-wives were in the village, also some younger siblings. Two of his elder brothers were herdsman and stayed with their own families in Moussa's father's compound. Some step brothers were in Nigeria, another brother in Douala. Only Moussa and an elder step brother stayed in Bamenda. They were the only with higher school education. At the time of the interview he had finished college and was schooling in a higher professional school and also rented there. His father was financing his education. He dreamt of going abroad for further studies. He was regularly going to the cyber café to communicate and browse.

Relation to the researcher: I was very close to Moussa. I met him often in Bamenda, visited him in his school and we met in town. I also spent a few days with his family in the village. When in Switzerland I was always in contact with him through chat and Facebook.

Situation of the interview: In a restaurant.

Name: Simon / Date of the interview: 06.10.10 / (→ see interview 2009)

Situation of the interview: In his home

Situation update: Meanwhile he worked as a graphic designer at Maryland Printers. He was regularly paid, but he more and more dreamt of going abroad for studies.

Name: Valeria / Date of the interview: 26.09.10 / (→ see interview 2009)

Situation of the interview: In her home

Situation update: She was jobless again after having briefly worked in a school in a village. She was very insecure about her future. Also she was increasingly disturbed by health problems.

Interviews Field Ass. 2010/11

These interviews were conducted by my field assistant Eric Chefor (real name). I was listening to these interviews again and we discussed them together. I did not know any of his interviewees.

Eric Chefor – the field assistant

Person: Male. Age 26 years. At the time when we collaborated, he studied at the technical college in the outskirt of Bamenda. Before he had completed a BA in economics. His family was in Bamenda. He rented a room at the college. He had a wide range of interests, in a political and social field. He had several times written articles for the local youth magazines, and he was currently working over a critique of the introduction of the 5-FCFA-coins in Cameroon. These coins had never been circulated since their introduction, due to their non-value. People rather used them to produce jewellery and as decorative items.

He went up to the finance ministry to tell his story. He had been conducting interviews for Divine Fuh, when he conducted his fieldwork in Bamenda. I got to know Eric by coincidence, and not through Divine.

Relation to the researcher: We have often met to exchange about the research, in particular during the period he was conducting interviews for me.

Name: Franklin / Date of the interview: 29.12.10 / Indication EC

Person: Male. 24 years old. His father was in the US. He stayed with his mother. The number of sisters and brothers was not stated. He also had step siblings. He had a first degree and worked as a secretary.

Relation to the interviewer: Not very close. The interview style was very formal.

Situation of the interview: In a restaurant.

Name: George / Date of the interview: 08.11.10 / Indication EC

Person: Male. 21 years old. The father was absent, the mother died when he was young. He grew up with his aunt. He did not mention any direct sibling. He was a student. Apart from that he was doing small businesses to partly sustain himself.

Relation to the interviewer: George was Eric's close friend.

Situation of the interview: In a restaurant.

Name: Gladys / Date of the interview: 03.12.10 / Indication EC

Person: Female. 26 years old. She was staying in her parent's compound in Bamenda. She had siblings, and was the eldest. She was working as a journalist, could sustain herself and supported the family.

Relation to the interviewer: Not so close

Situation of the interview: In a restaurant

Name: Isidore / Date of the interview: 29.12.10 / Indication EC

Person: Male. 32 years old. He apparently lived on his own, and was not yet married. His family was in Bamenda, but also other cities in

Cameroon, and in UK and Belgium. He was working in a marketing enterprise and had his own businesses. He had made his MA in the UK. He said that he was not supporting other family members, that would not be his responsibility.

Relation to the interviewer: A distant friend
Situation of the interview: In a restaurant.

Name: Kenneth / Date of the interview: 18.12.10 / Indication EC

Person: Male. The age was not indicated (end twenty?). Rented on his own. His family was in Bamenda, he was not indicating the number of siblings. He worked as a teacher, sustaining himself and the mother.

Relation to the interviewer: A distant friend
Situation of the interview: In a restaurant.

Name: Kingsley / Date of the interview: 21.12.10 / Indication EC

Person: Male. 25 years. He stayed with his family in Bamenda when on semester break. He had two younger sisters. He had a BA, but was still studying for a MA.

Relation to the interviewer: A former class mate
Situation of the interview: In a restaurant

Name: Minette / Date of the interview: 25.11.10 / Indication EC

Person: Female. 32 years old. She was the eldest of her siblings, the number was not indicated. Her parents lived in Bamenda. She had a BA and worked as a journalist. She sustained herself and also contributed for her retired parents.

Relation to the interviewer: A close friend
Situation of the interview: In a restaurant.

Name: Pride / Date of the interview: 26.11.10 / Indication EC

Person: Male. 29 years. He was the first of three children. The father died. He stayed in the family compound. Some cousins were also living there. He was working as a film maker

Relation to the interviewer: An acquaintance

Situation of the interview: In a restaurant.

Name: Terence / Date of the interview: 20.11.10 / Indication EC

Person: Male. The age was not indicated (mid twenty?). He was coming from a polygamous home, they were altogether 23 children. He was one of the youngest. His parents had died, some step mothers were alive. They lived in Bamenda. He also lived in the family compound when on semester break.

Relation to the interviewer: A colleague from university

Situation of the interview: In a restaurant.

Interviews Diaspora

These interviews were conducted between 2007 and 2009, altogether 17 in number. I have followed up several of these interviewees over a long period of time. Since I have not quoted of all of the interviews, I will not list all the interviewees here.

Name: Aboubakar / Date of the interview: 16.10.2007

Person: Male. 25 years old. He came from the North of Cameroon. He was a Francophone and Fulbe. He was married to a Swiss. He had his parents and 12 siblings in Cameroon. He was the only of his family who was abroad. In Cameroon he had worked in an import/export business. In Switzerland he was currently doing an internship in a home for people with disabilities. For a while he was also teaching Fulfulde to students in the Institute for Social Anthropology in Basel.

Relation to the interviewer: I was not close to him. I had obtained his contact through a Swiss friend. After the interview we had not seen each other again.

Situation of the interview: In a restaurant.

Name: Alexandre / Date of the interview: 29.3./17.4./23.5./26.5./26.6./12.09.2007

Person: Male. Mid-fifties. He was in Switzerland since 12 years. He was a francophone. He was married to a Swiss, came to Switzerland through her. Now he was divorced. He had a job at Novartis. At the time of the interview he was jobless. He was a very active member in an African church in Basel, and in an ethnic association of the Bassa in Switzerland. He had active relations to Cameroon, doing business, set up a farm, also.

Relation to the interviewer: Alexandre was very open, and we met several times. He also invited me to his church and to a wedding. Over a period of time we met regularly. So the interview was conducted in several sessions. I had his contact through a Swiss friend.

Situation of the interview: In restaurants, or meeting at events.

Name: Alice / Date of the interview: 15.05.2009

Person: Female. 35 years. She was Anglophone, from Bamenda. She was in Switzerland since 6 years. She was married to a Swiss and they had a child, and she expected her second child soon. She also brought her daughter from Cameroon, who was 14 years old. When she came to Switzerland she had applied for asylum, and then got to know her husband. In Cameroon she had an MA in linguistics, now she was working in an old people's home. Later on she involved herself with NGO work in Cameroon.

Relation to the interviewer: After our interview we did not see each other for two years, then we got into regular contact again. She lived in Bern.

Situation of the interview: I went to visit her in her home in Bern, where I also got to know her family.

Name: Blaise / Date of the interview: 10.11./08.12.2007/06.03./25.05.2008

Person: Male. 29 years old. He was a francophone, from Bafoussam. His family was in Cameroon. He was in Switzerland since 2 years. He had applied for asylum, and lived in fear that it could be denied. Through some friends from other African countries here, he got into difficulties with the police, as he said. He had an accident in Cameroon, and he still suffered from the consequences.

Relation to the interviewer: We were not close. I knew him through a Swiss friend and Africans from Ivory Coast. After some time I cut contact because he got a little bit intrusive.

Situation of the interview: In a restaurant, meeting at events.

Name: Daniel / Date of the interview: 18.02.2007

Person: Male. 22 years. He was in Germany since one year, where he was studying. He was a francophone, from Douala. In Cameroon he was working as a teacher. Then he had decided to further his studies abroad. He was having a small part time job which gave him a bit of money. He tried to support his family, since they expected a lot of him, as he said.

Relation to the interviewer: I got to know Daniel (together with Jean*) through a German friend. When I spent a few days in Stuttgart, I decided to meet his Cameroonian friends and do an interview. After that we have not seen each other again, but had although stayed in contact through email for a while. Also I had followed up the interview in an email interview.

Situation of the interview: At my friends home in Stuttgart.

Name: Godlove / Date of the interview: 19.05.2009/23.02.2010

Person: Male. 27 years old. He was in Switzerland since 2 years. He had applied for asylum. He was the youngest of 10 siblings. His mother died long ago, the father was absent. So he grew up with his elder sister in Limbe. He could not support his family, he just tried to send something from time to time. He tried to keep himself busy by using any opportunity for job programmes, also. In 2012 he was still in Switzerland in an unsettled state, even though they denied his application for asylum, the sojourn was prolonged due to the elections in Cameroon in late 2011.

Relation to the interviewer: I was very close to Godlove. We had met several times after our interviews. I had been visiting him in Biel. I got to know him through Paul*, at a meeting of the Swiss Anglophone Association which was held in Biel.

Situation of the interview: In a restaurant.

Name: Jean / Interview 18.02.2007

Person: Male. 28 years old. He was a francophone and came from Douala. He was coming from a well off background. He was in Germany since 8 years, came here as a student. Now he was working. He did not need to support his family, they were fine, as he said. His parents and his siblings were in Cameroon.

Relation to the interviewer: I got to know Jean (together with Daniel*) through a German friend. After that we had not seen each other again, but had although stayed in contact through email for a while. Also I had followed up the interview in an email interview.

Situation of the interview: At my friends home in Stuttgart.

Name: Louis / Date of the interview: 30.03./03.04./18.05.2009

Person: Male. Beginning 40ties. He was in Switzerland since about 10 years, and was married to a Swiss. He was doing a range of jobs and formations. Also he had a DVD shop in Basel. He had a MA in economics. He was in contact with many Cameroonians and other African students in Basel, sometimes he hosted them, and helped them to arrange themselves with the new environment. Also he was active in local Basel politics. He was in active relations with people in Cameroon, his family and beyond. His mother was still alive, he supported her. His father had died. His siblings were also in Cameroon.

Relation to the interviewer: We met several times. Louis was very helpful, indirectly through him I got access to other Anglophones and the Anglophone Association. I had been invited by him and his wife to his home and vice-versa.

Situation of the interview: In a restaurant, and in his home.

Name: Marguerite / Date of the interview: 12.09.2007

Person: Female. 50 years old. She was francophone and came from Douala. She was the eldest of 14 children. She was married to a Swiss, and was in Switzerland since 8 years, but in Europe since 20 years. She had left behind two children with her sister in Cameroon, when she went abroad. Later on the children joined her in France. She had got the French citizenship now. She had also a child of 6 years with her Swiss husband. She did not work any longer. She took care after their house and garden, and her child. She was active in a Cameroonian women's group association, but left due to tensions in the group.

Relation to the interviewer: We were not close. After the interview we had not met again. I got to know her through Alexandre.

Situation of the interview: In a restaurant.

Name: Maurice / Date of the interview: 06.06.2007

Person: Male. 28 years. He was a Francophone. He was in Switzerland since 7 years, before that he was for a short while in France, where he was married to a French. They had a son. Now he was divorced and had married again, a Cameroonian. They want to have children. His father and one sibling lived in Cameroon, the mother in France and all other siblings (another five) in different European countries. He had not much time and money to often go back to Cameroon.

Relation to the interviewer: We were not close. I met him at Alexandre's wedding invitation.

Situation of the interview: In a restaurant.

Name: Maximilliam / Date of the interview: 07.06.2007

Person: Male. 38 years old. He was a Francophone, and had been in Europe for 20 years, he had for many years lived in Germany, and since two years in Switzerland. He was married to a German, but lived separated now. He had his life here in Europe, he said. His parents had both died and of his siblings only one was still in Cameroon. Some other siblings were as well abroad. He tried to regularly go back to Cameroon, but at times he had no time for it.

Relation to the interviewer: We were not close, we briefly met before the interview, after that we had not seen each other again.

Situation of the interview: In a restaurant. He also brought a colleague with him, since they wanted to go out.

Name: Paul / Date of the interview: 22.03.200/23.03.2010

Person: Male. Beginning 40's. He was in Switzerland for almost 10 years. He was married to a Swiss. At the time of our first meeting he did an internship with the UN in Geneva. Usually he lived in Biel. He had gone through different formations and jobs since he was here. His family was in Bamenda. He was very much involved with diverse businesses, his NGO in Cameroon, and as the president of the Swiss Anglophone Association in Biel. He was travelling to Cameroon as often as he could. Later on he divorced from his wife, and spent more time in Cameroon than in Switzerland. He possibly wanted to go back entirely.

Relation to the interviewer: I met Paul twice in Switzerland, in Geneva and Biel. Since he was so mobile, I saw him several times in Cameroon. I got his contact from Louis*.

Situation of the interview: In his home.

Name: René / Date of the interview: 01.07./06.11.2007/28.02.2008

Person: Male. 23 years old. At the time of our first meeting he was in Switzerland since 7 months. He had applied for asylum. First he was in Interlaken and then in Thun. His family was from Bamenda, he had four siblings. None of them was abroad. He tried to support them a bit, but it was difficult. He had A-levels and has done odd jobs in Cameroon. That he came to Switzerland, was a coincidence, because he met somebody who brought him here. Later on René married a Cameroonian who he got to know in Switzerland. His application for asylum got denied and he had to leave to Cameroon. He then tried to get a visa again for reunion with his wife who stayed in Switzerland (status C), which got denied. Meanwhile he won an American green card and went to the US, where he lived from 2011.

Relation to the interviewer: I was very close to René. I obtained his contact through his cousin, whom I got to know in 2003 in Bamenda and who informed me that his cousin went to Switzerland, whereof I contacted René. After having met several times in Switzerland in Basel, Thun and Bern, we also met several times when I was in Cameroon and he was also back. I knew his family, parents and siblings. Also when he went to the US we stayed in regular contact through internet and also calls.

Situation of the interview: In restaurants, in my own and his home.

www.ingramcontent.com/pod-product-compliance
Lightning Source LLC
LaVergne TN
LVHW042123070326
832902LV00036B/560